Changing the Rules

Changing the Rules

Technological Change, International Competition, and Regulation in Communications

Robert W. Crandall and Kenneth Flamm
Editors

THE BROOKINGS INSTITUTION / Washington, D.C.

1775 Massachusetts Avenue, N.W., Washington, D.C. 20036

Library of Congress Cataloging-in-Publication Data

Changing the rules: technological change, international competition, and regulation in communications / Robert W. Crandall and Kenneth Flamm, editors.
p. cm.
Includes index.
ISBN 0-8157-1596-X (alk. paper). ISBN 0-8157-1595-1 (pbk.: alk. paper)
1. Telephone. 2. Telecommunication. 3. Telecommunication policy. 4. Telephone—United States. 5. Telecommunication—United States. 6. Telecommunication policy—United States. 7. Telecommunication systems. I. Crandall, Robert W. II. Flamm, Kenneth, 1951–
HE8731.C47 1988 88-28793
384'.041—dc19 CIP

9 8 7 6 5 4 3 2 1

The paper used in this publication meets the minimum requirements of the American National Standard for Information Sciences—Permanence of Paper for Printed Library Materials, ANSI Z39.48-1984.

Set in Linotron Times Roman
Composition by Harper Graphics
Waldorf, Maryland
Printing by Arcata Graphics
Kingsport, Tennessee

THE BROOKINGS INSTITUTION

The Brookings Institution is an independent organization devoted to nonpartisan research, education, and publication in economics, government, foreign policy, and the social sciences generally. Its principal purposes are to aid in the development of sound public policies and to promote public understanding of issues of national importance.

The Institution was founded on December 8, 1927, to merge the activities of the Institute for Government Research, founded in 1916, the Institute of Economics, founded in 1922, and the Robert Brookings Graduate School of Economics and Government, founded in 1924.

The Board of Trustees is responsible for the general administration of the Institution, while the immediate direction of the policies, program, and staff is vested in the President, assisted by an advisory committee of the officers and staff. The by-laws of the Institution state: "It is the function of the Trustees to make possible the conduct of scientific research, and publication, under the most favorable conditions, and to safeguard the independence of the research staff in the pursuit of their studies and in the publication of the results of such studies. It is not a part of their function to determine, control, or influence the conduct of particular investigations or the conclusions reached."

The President bears final responsibility for the decision to publish a manuscript as a Brookings book. In reaching his judgment on the competence, accuracy, and objectivity of each study, the President is advised by the director of the appropriate research program and weighs the views of a panel of expert outside readers who report to him in confidence on the quality of the work. Publication of a work signifies that it is deemed a competent treatment worthy of public consideration but does not imply endorsement of conclusions or recommendations.

The Institution maintains its position of neutrality on issues of public policy in order to safeguard the intellectual freedom of the staff. Hence interpretations or conclusions in Brookings publications should be understood to be solely those of the authors and should not be attributed to the Institution, to its trustees, officers, or other staff members, or to the organizations that support its research.

Foreword

IN 1984 a federal antitrust suit against the American Telephone and Telegraph Company was settled when AT&T consented to a breakup that allowed it to enter businesses other than regulated communications. This settlement was predicated on two beliefs: first, that competition should discipline market participants in telecommunications whenever possible, and second, that AT&T should be free to pursue the potentially convergent technologies of computers and telecommunications.

Three years after that decision, the Brookings Institution organized a conference to examine the changes brought about by procompetitive policies in the U.S. telecommunications industry, the degree to which those changes were spreading to other developed economies, and the convergence in technology between computers and communications. The conference was conceived and sponsored in part by the Nynex Corporation. Participants came from business, government, and universities in the United States, Europe, and Japan. This volume contains revised versions of the papers presented at that conference.

Robert W. Crandall and Kenneth Flamm are senior fellows in the Economic Studies and the Foreign Policy Studies programs, respectively, at the Brookings Institution. They wish to thank John A. Arcate of the Nynex Service Company for his support and assistance in organizing the conference. They also wish to thank Carol Delaney of Brookings for managing the conference; Sarah Bales, Jonathan Galst, Daniel A. Lindley III, and Elizabeth Schneirov for research assistance; Jeanette Morrison, Barbara de Boinville, James McEuen, and James R. Schneider for editing the manuscript; Victor M. Alfaro, Patricia Nelson, and Almaz S. Zellecke for verifying its factual content; Susan L. Woollen for editorial assistance; Florence Robinson for preparing the index; David Rossetti and Ann M. Ziegler for helping with conference logistics and other secretarial support; Thomas T. Somuah and Janet E. Smith for secretarial assistance; and Kathryn A. Ho for administrative support.

The Brookings Institution is grateful to the Nynex Service Company for its financial support of the conference and this volume. Additional support was provided by the John D. and Catherine T. MacArthur Foundation.

The views expressed here are those of the authors and editors and should not be attributed to the organizations whose assistance is acknowledged or to the officers, trustees, or other staff members of the Brookings Institution.

BRUCE K. MAC LAURY
President

January 1989
Washington, D.C.

Contents

Tables

Figures

Overview

Robert W. Crandall and Kenneth Flamm

THE PAST TWO decades have seen dramatic change in both the computer industry and the telecommunications sector of the U.S. economy. Technological change in computer equipment has continued at a dizzying pace, with price declines averaging 20 percent a year. Though its technical progress has not been quite so dazzling, communications equipment has also borrowed heavily from the same electronics revolution that has propelled computers.

There are other similarities between the communications and computer industries. Both spend a large share of revenues on research and development and deserve their description as high-technology industries. For long periods each sector had historically been dominated by one giant U.S. company—International Business Machines for computers and American Telephone and Telegraph for communications. And both IBM and AT&T spent most of the 1970s defending themselves from landmark antitrust actions by the U.S. government.

In the past decade the hegemony of IBM and AT&T has withered as new competitors have appeared to exploit new technologies. Indeed, it was new technologies developed immediately after World War II—microelectronics, computers, microwave communications—that provided the opportunities for the first wave of high-technology firms. Many of the recent newcomers began to chip away at the assumptions of natural monopoly that underlay the government's protection of telephone carriers from competition, while others began to dispel IBM's aura of invincibility.

Communications as a Protected Monopoly

Traditionally, the communications business has been a regulated monopoly. In most countries, a state company controls local and long-distance services and buys its equipment from favored national suppliers. In the United States, AT&T is a private corporation that before 1984 included twenty-two local Bell operating companies, the only national long-distance service (AT&T Long Lines), and its own equipment supplier (Western Electric). The pre-1984 Bell operating companies had about 80 percent of the nation's local telephone subscribers; a variety of independent companies controlled the rest.

Until the 1970s there was essentially no competition for local or long-distance telephone carriers. Local companies enjoyed state public utility grants of monopoly franchises; AT&T was a de facto monopolist of long-distance services, protected by the U.S. Federal Communications Commission from competitive entry. In the 1950s, however, a new technology—microwave communications—transformed the prospects for competition in long-distance services. By the mid-1960s, entrepreneurs began to consider offering such competition, and one, Microwave Communications, Inc. (MCI), even filed a request with the FCC for permission to supply a limited form of long-distance service. In 1969, after six years of administrative proceedings, the FCC consented, thereby opening the door to competition, if only a crack. Over the next decade the door swung open much wider. Drawing on new digital technologies, entrants sought to offer customers new data transmission service, more sophisticated terminal equipment, and a variety of services—such as packet switching—that AT&T was either unable or unwilling to offer.

By the mid-1970s, AT&T could no longer argue convincingly that it was a "natural monopoly"—a firm that could offer all modern communications at the lowest cost. It had a large, prestigious research organization, Bell Laboratories, but also had an enormous amount of obsolete customer equipment on its books and an aging analog technology that was being threatened by those who would offer the new digital technology so crucial for the transmission of data. AT&T was forced to defend its monopoly by arguing that the offering of new technology by aggressive entrants would force AT&T to raise local rates as it lost business and retired obsolete equipment. In the end the new entrepreneurs and new technology won.

Once the FCC let competitors into the long-distance and the terminal equipment markets, AT&T attempted to fight back in the marketplace. This aggressive defense attracted the concern of the U.S. Justice Department, which brought a major antitrust suit against AT&T in 1974. After more than seven years, this suit was settled by a consent decree that required AT&T to divest itself of all its operating companies. In return, AT&T was freed from the restrictions of a 1956 decree that had kept it from entering any business other than regulated communications.

In 1984 AT&T completed the divestiture of its operating companies. The divested Bell operating companies (BOCs) are managed through seven new regional holding companies (RHCs), sometimes called regional Bell operating companies (RBOCs).[1] AT&T is now free to compete in the computer and other high-technology electronics markets through its two main subsidiaries: AT&T Communications, which offers long-distance (including some intrastate) voice, data, and video transmission in the United States and internationally, and AT&T Technologies, Western Electric's successor. Because the Bell operating companies are now free of AT&T, they may purchase their capital equipment from any competitive source. As a result, IBM, other computer companies, and other communications equipment manufacturers may now compete to sell the equipment used by the firms that supply local telephone service to 80 percent of Americans. In previous decades AT&T bought most of its equipment from its own Western Electric subsidiary.

In addition, the new high-tech manufacturers may sell their equipment to the new competitive long-distance carriers or even offer long-distance service themselves. Thus a slight crack in the monopoly facade of AT&T caused by technological advances has now evolved into full-scale competition for communications equipment and long-distance services. Only local service remains a protected, franchised monopoly (in most states).[2]

Technological Convergence?

The recent attempts by AT&T and IBM to invade one another's markets have been remarkably unsuccessful. Despite the obvious similarities in the basic technologies of the two industries, large computer firms like

1. Nynex Corporation (New York and New England), Bell Atlantic Corporation, BellSouth Corporation, American Information Technologies (Ameritech) Corporation, Southwestern Bell Corporation, US West, Inc., and Pacific Telesis Group.

2. The modified final judgment reconfigured the Bell System, redefining basic exchange areas and making them much larger. Before divestiture, an *exchange* was an area with a single uniform set of charges for telephone service, and calls between points in an exchange area were *local* calls. The judgment defined an exchange to be more or less equivalent to a U.S. government statistical unit, the Standard Metropolitan Statistical Area (SMSA). The Bell System territory was divided into about 160 of these exchanges, called *local access and transport areas* (LATAs), ranging from large metropolitan areas to entire states.

Under the judgment, the BOCs provide regulated telecommunications services *within* LATAs (intra-LATA), and AT&T and other interexchange carriers provide services *between* LATAs (inter-LATA). Thus inter-LATA calls may be within a single state as long as they are between terminals in different LATAs. The new division does not match the predivestiture differences between AT&T Long Lines and BOC operations, between intrastate and interstate jurisdictions, or between toll and local services. For further details, see R. F. Rey, ed., *Engineering and Operations in the Bell System,* 2d ed. (Murray Hill, N.J.: AT&T Bell Laboratories, 1983), chap. 1.

Apple, Digital Equipment Corporation (DEC), or Unisys are not major players in either the communications equipment or telecommunications services market, nor are communications companies like Northern Telecom, GTE, CIT-Alcatel, or Plessey important competitors in the computer industry. Even IBM has achieved a significant presence in communications only through its acquisition of Rolm, a large private branch exchange (PBX) manufacturer.

Only small computer network specialists—like Bolt, Beranek, and Newman, Inc. (BBN), 3COM Corporation, or Telenet Communications Corporation—have successfully managed to straddle the two technologies, and even their success has been rather modest. It appears that specialization and product differentiation are as important as technological convergence in explaining the structure of these two industries. Technological giants, exploiting economies of scale and scope, are forced to coexist with smaller, entrepreneurial firms that specialize in product niches that the larger companies have ignored or been slow to exploit. Historical examples from the computer industry include timesharing, minicomputers and later microcomputers, computer networks, and artificial intelligence applications.

The lack of convergence thus far in the computer and communications markets and the increasing importance of competition in stimulating technical change in these industries are not unique to the United States. This volume contains papers that address not only technological convergence between computers and communications in the United States, but also the attempts by other countries to exploit potential convergence. The volume also details the changes in the structure of U.S. telephone markets, the changes forced by U.S. regulatory policies and rapid technical progress on telecommunications policy in other countries, and the consequences of the increasing internationalization of equipment markets. In addition, we focus on some of the problems created by rapid technological advances and market competition in setting technical standards and protecting national security.

Innovation, Regulation, and Competition

The volume begins with a paper by Kenneth Flamm, who first presents a primer on information technology, tracing out the technical links between computer and communications systems. His paper provides several measures of the rate of technological progress in the various elements that make up these systems.

Flamm finds that, contrary to popular belief, technical change has been much slower in telecommunications equipment than in the computer industry, and that among telecommunications equipment, transmission technology—rather than switching technology—has improved the most rapidly. Surprisingly, Flamm's price indexes for switching equipment show a markedly greater decline after January 1, 1984, when AT&T's ownership of operating companies expired. All this suggests that market structure has been a chief determinant of technical innovation.

Flamm concludes that both recent history and economic logic argue for a more rapid pace of innovation in telecommunications in the future, but with rather different implications for the architecture of the network than has been suggested.

His analysis raises questions about the extent to which economies of scope based on spillovers between computer and communications technologies can be used as a rationale for easing entry by telephone operating companies into unregulated computer markets. And he notes that changes in the composition of research and development investment in the new, more competitive communications industry, historically a vitally important source of significant basic research within U.S. industry, may have unexpected consequences for America's technology base.

The two countries with the most aggressive policies of promoting competition in telephone services are Japan and the United States. Leonard Waverman of the University of Toronto looks at the recent changes in the U.S. long-distance, or interexchange, market, focusing especially on two questions: first, whether the breakup of AT&T has sacrificed economies of multiservice operations (economies of scope), and second, whether competition in long-distance markets is feasible given the economies of scale in these markets.

Waverman concludes that the econometric evidence is of little value in discovering economies of scope in telecommunications. He notes that these economies are certainly present in one respect because it is clearly more efficient to originate and terminate both local and long-distance calls over the same local subscriber loops. But this fact fails to prove that it is more efficient for a single firm to own both the long-distance circuits and the local circuits. As long as the local circuits are priced reasonably, there may be no benefit from having one firm offer both local and long-distance service.

The evidence on economies of scale in long-distance services is equally unhelpful, according to Waverman, but the recent development of fiber optics suggests the existence of very large-scale economies in constructing such fiber networks. Just as competition is becoming intense in the long-distance market, with AT&T's share falling below 80 percent, technology may indeed be creating the conditions for a natural monopoly. Fortunately, Waverman points out, aggressive investment in fiber optics by many new carriers may create enough excess capacity to allow long-distance services to be "contestable" in a most unusual manner. In the near future, new entrants will be able to lease transmission capacity at very low rates, thereby constraining any potential monopoly power of AT&T.

The next paper, by Robert Crandall, examines the proper role of the "bottleneck" local-service Bell operating companies after the 1984 divestiture. Crandall argues that the theory behind the divestiture has a certain plausibility with one exception: there is only minor evidence of deliberate cross subsidization of competitive services by the regulated Bell operating companies before divestiture. In fact, regulators were pursuing the opposite course: deliberately shifting costs from regulated local services to the more competitive long-distance services in the early 1970s, just as entry into long distance was becoming a real possibility.

At this juncture, the antitrust suit and the divestiture are history. An important issue that remains, however, is the proper role of the divested Bell operating companies. Under the antitrust decree, these companies are barred from equipment manufacture, long-distance services (outside their local franchise areas), and ill-defined "information services." Despite a considerable increase in competition in the equipment and the long-distance markets since divestiture, the court has refused to allow the operating companies to reenter these markets because they control access to most local customers. Only when bypass—the use of non-telephone-company circuits to reach local customers—becomes a reality is the court likely to relax these restrictions.

Now that there are seven regional Bell holding companies, some argue that the abuse of their bottleneck power is much less likely. Regulators will have more yardsticks with which to compare the Bell operating companies. Hence Crandall suggests that allowing them to enter equipment manufacture and information services may be a risk worth taking to gain additional competitors in these markets. Ironically, allowing them to reenter interstate long distance from local service may well enhance the economic welfare of users as long as regulators persist in subsidizing local service out of long-distance revenues.

The recent Japanese experience is remarkably similar to that of the United States in the 1970s. Professors Tsuruhiko Nambu, Kazuyuki

Suzuki, and Tetsushi Honda explain that Japan has recently privatized its telephone company—Nippon Telegraph and Telephone (NTT)—and admitted entry into certain long-distance and local services. After demonstrating that the economics of NTT operations are quite similar to those of AT&T before divestiture, the authors look at the policy dilemma facing the Japanese government ministry regulating this industry.

If free competition on dense long-distance and local routes is permitted, prices may be driven so low that NTT will not be able to continue offering low-priced services to small subscribers in high-cost areas. They suggest that one solution may be to impose access charges on the new competitors for originating and terminating calls over NTT circuits. Unfortunately, these access charges will then encourage the private competitors and large users to seek technologies for bypassing the NTT circuits. To many U.S. analysts, this pattern seems strikingly familiar.

Problems Raised by Liberalization

Apart from regulatory quandaries, the decisions to liberalize telecommunications markets raise several other problems. For instance, technical standards are no longer simply a matter for arbitration among postal, telegraph, and telephone administrations (PTTs), but a serious competitive issue among competing vendors of services and equipment.

Stanley M. Besen and Garth Saloner review the emerging theoretical and empirical literature on technical standards. In telecommunications, competing international coalitions of rivalrous firms are jointly meeting to set standards like the integrated services digital network (ISDN) and the U.S. open network architecture (ONA) for telephone services. Standards are increasingly a dimension of the competitive strategies of firms in high-technology industries, a weapon to be used on one's opponents in the same way as price or quality—to gain a commercial advantage. Furthermore, even if everyone gains from standardization, there is no guarantee that the "right," socially optimal standard will be chosen. Besen and Saloner conclude that the different paths toward standardization in telecommunications being taken in the United States and Europe are an intriguing test of whether regulated competitors or competing regulators are likely to come closer to optimal results.

Ashton B. Carter examines yet another interaction between technical change and the telecommunications structure, the constraints imposed on the system by the requirements of U.S. national security. Carter sees the issue in terms of two factors: military needs shaping the organization and structure of commercial networks, and the national security implications of the relative competitive position of American firms in the underlying technology base vis-à-vis their foreign competitors.

Though the first issue, in particular, was raised by the U.S. Department of Defense when divestiture was taking place, Carter sees little in the way of a threat that could not be resolved by an appropriate department initiative. The real issue, suggests Carter, is whether such programs should be imposed administratively, as a hidden cost of doing business for commercial firms, or whether they should be bought and paid for like any other explicit Defense program. The technological dependence issue, argues Carter, is as much an economic one as a narrow security threat and should be explicitly argued in those terms, not transformed into a backdoor industrial policy by the Defense Department.

Pressures for Change in Other Markets

Experiments in deregulation and more liberal entry are transforming not only domestic telecommunications markets, but the rules of the game for international competition. Eli Noam of Columbia University argues that the telephone industry in most countries is driven by ''political télématique''—the tight government control of a public telephone monopoly and its national equipment supplier(s). This monopoly traditionally has been used to cross-subsidize inefficient postal operations and has been resistant to entry attempts by domestic entrepreneurs and foreign communications services providers. Noam argues that these cozy national monopolies are coming under competitive attack as businesses seek more efficient communications paths using the technology made possible by the electronics revolution.

In the United States, Japan, and the United Kingdom, entry by new service providers has been permitted, but in most other industrialized countries, the PTT retains its tight-fisted control over the telephone network. Noam argues that disintegration of this political control of telephony is inevitable because the demand for communications by world business will lead to a migration of enterprises toward those countries with the most open and dynamic telephone industries. Moreover, equipment suppliers in protected markets will fall further and further behind as they are frozen out of the open markets because their own politicians do not reciprocate by allowing telephone equipment imports.

Ashoka Mody of Bell Laboratories (now with the World Bank) provides a different international perspective on electronics and telecommunications equipment supply. He focuses on the rise of certain newly industrializing countries (NICs) as exporters of electronics and telephone equipment. His paper analyzes the conditions for success in developing these high-technology sectors, focusing on the need to accumulate expertise sequentially in certain production processes and to exploit economies of scale while at the same time maintaining a healthy dose of domestic competition.

Among the NICs, Korea has been the most successful in developing its electronics industries. Mody argues that this success is the result of large Korean conglomerates focusing on semiconductor production early in the development of their electronics capabilities despite government opposition to that strategy and despite little trade protection. These large conglomerates then developed production capability in relatively simple consumer electronics products before attacking more sophisticated products. Other NICs, such as Brazil, failed to manufacture components on a large scale domestically, relying heavily on imports and production by transnational corporations instead.

At present, Korea, Taiwan, and Singapore appear to be the NICs with the greatest potential for developing strong telecommunications equipment industries. It is no coincidence that these countries are also the most dynamic in developing their own internal telephone service sectors. As a result, these countries will have large internal markets for telephone equipment. Whether they will be able to penetrate the markets in developed countries for sophisticated telephone equipment is a question Mody cannot answer at this juncture.

Nestor Terleckyj searches for international linkages as part of his analysis of the determinants of productivity growth in the computer and telecommunications industries. It is particularly intriguing that he finds none in communications, in distinct contrast to the situation in the market for computers. After meticulously constructing and presenting historical data on research and development, input use, production, and productivity growth in these two sectors, Terleckyj estimates a dynamic econometric model intended to track investment, productivity, prices, and output. He finds some similarities but many differences in the patterns of growth in the two industries, and little in the historical record that suggests strong links between the two sectors.

Conclusion

Nine distinct perspectives on an industry in flux raise as many questions as they answer. What is clear is that technology and market structure, mediated by the political and regulatory systems, have interacted in myriad ways. Recent changes in regulatory systems have left us in a more complicated, more interesting, and uncertain new world. The linkages between communications markets and other related markets may now develop in ways previously foreclosed by regulatory boundaries. Deregulation and the growing push for international standards mean those same newly liberalized markets will increasingly be integrated into a single international marketplace.

Still, as technological barriers fall, and national markets become easier to enter from within, new sorts of political obstacles may arise to block access from outside national boundaries. In part, the importance of the technology is at fault: every nation wants its companies to be major players on the international scene. Discrimination in favor of local firms through manipulation of standards, procurement, and data-flow regulations is rapidly replacing the more obvious control mechanisms of the national monopolies of the past.

Two forces are working against such technological protectionism. First, companies are finding it advantageous to reach out across national boundaries in their search for business partners and allies. Such multinational alliances undercut explicitly nationalist policies with increasing frequency. Perhaps more important, it is clearly in the interest of users of information, around the world, to have unimpeded access to the global network. In the final analysis, gains from using the new technology to gather and process information are likely to outweigh the small fraction of the benefit from technical advance that is captured as profit by producers of hardware and software. We can only hope that reason prevails, that bargains are struck, and that the international network of tomorrow extends beyond the boundaries of the national systems of today.

Innovation, Regulation, and Competition

Technological Advance and Costs: Computers versus Communications

Kenneth Flamm

EXAMINING THE particulars of technological change in communications sheds light on three significant issues. First, many observers within industry contend that a "convergence" of computers and communications systems—resulting in technological spillovers and economies of scope—is reshaping the economics of both industries. The belief was ratified by American Telephone and Telegraph's rapid entrance into the market for general-purpose computers after the Bell System's divestiture in 1983 and by International Business Machine's phased entry into the communications marketplace.[1] The efforts of the Bell regional holding companies to sell computer products and services reflect the same thinking. If convergence is indeed taking place, one would expect to find that the rates and directions of technological change in the two industries are similar. If there is little resemblance, then attempts to alter regulatory policy to permit regional holding companies to enter computer markets will require stronger justification than the advantages to be gained by presumed spillovers from their investments in communications research and development.

Second, if rates of innovations in computers and communications do not seem all that much alike, it is important to ask why. In terms of technological intensity, measured by R&D investment as a percentage of sales, communications equipment (at 9 percent) is second only to computers (at 12 percent) among America's commercial high-technology industries.[2] Understanding the forces generating this investment and the

Without implicating them in my errors, I would like to thank Ray E. Albers, Frederick T. Andrews, John Arcate, Alan Chynoweth, Rosanne Cole, Robert Costrell, John Healy, Edward Irland, Charles Jackson, Robert Lloyd, F. M. Scherer, Jack Triplett, and Clifford M. Winston for their many useful comments.

1. For more details about the divestiture and the new companies, see the "Overview" at the beginning of this volume.

2. The figures are for broad groupings of American industries for 1980 and are taken from the National Science Foundation, *Research and Development in Industry, 1982* (Government

effects of the investment is important in charting the future of the American economy. One key historical difference between computers and communications has been the pervasive effect of regulation on the communications industry and the minimal role it has played in computers. The interaction between regulatory policy and technological change is important not only in interpreting history and theory, but also in guiding decisions on public policy as the United States considers what impact divestiture will have on R&D activity within the communications equipment industry, a critically important source of basic research within American industry.

Finally, a variant on the "convergence" thesis was used in an influential analysis by Peter Huber, who interpreted AT&T's decision to divest as a reaction to basic, ongoing change in the structure of the U.S. communications network.[3] Huber noted that the design of the network hinges on economic choices between two basic building blocks: transmission links (trunks) and switches between trunks.[4] He argued that rapid advances in the technology of digital switches, which he identified with computers, will change the present hierarchical, centralized network into a decentralized, "geodesic" network. In his view, AT&T's willingness to divest was driven by its anticipation of these changes and its desire to be a relatively unregulated participant in constructing this radically new network. Huber's interpretation of the present and his vision of the future depend, of course, on assumptions about the pace of technological change for the various network elements, assumptions that can be tested against the recent historical record. If the evidence contradicts his assumptions, analysts may have to alter this vision of the future of the network.

Elements of Computer and Communications Systems

Since the computer industry shares with communications systems many elements in a common technology, there is a widespread inclination to anticipate a convergence of the two industries. Because this assumption has shaped public policies regulating them, it is useful to understand the ways in which communications systems and computers mirror one another.

Printing Office, 1984), p. 23. Only the aircraft and missile industry, with significant support from the Defense Department, spends a greater share (14 percent) of its sales on R&D. See also note 45.

3. See Peter W. Huber, *The Geodesic Network: 1987 Report on Competition in the Telephone Industry* (Washington, D.C.: U.S. Department of Justice, Antitrust Division, January 1987), chap. 1.

4. More and longer lines (trunks) mean that fewer switches are necessary and vice versa.

A communications system transmits information from a source to a recipient. It has three major components. At the ends of the system are *terminal equipment* (telephones, graphics displays, fax machines, and so forth). A *transmission system* sends the information from source to destination through a communications channel, or circuit. Some of the transmission systems are analog, that is, they send information as a continuous modulated waveform; others are digital and transmit voice signals that have been converted into streams of binary digits (bits). Unless every source or recipient of information sent through the system has a direct path to every other user (an extraordinarily costly though extremely reliable alternative), *switches* must route traffic from a local distribution system, through the main arteries or trunks that link large regions, then back into a local distribution system to its final destination. The communications network is made up of both the transmission system and the switches installed at junctions (nodes) between channels.

Since the 1940s it has been known that all information (voice or data) can be encoded as a sequence of binary digits (that is, taking on one of two values, 0 or 1). A pure communications system merely transmits the bits from source to destination without altering the bit stream. An information processing system, or computer, however, operates on the bits and alters the bit stream. The information-handling capacity of the communications channel is measured by its *bandwidth*, the number of bits that can pass through it per second. The capacity of a computing system is similarly measured by bandwidth, in this case the rate at which bits can be processed.

In many respects a computer resembles a highly compressed communications system. A stream of bits is taken in through an input device, transmitted to a processor, and sent back out to its destination, an output device. The *input-output devices* connect the central processing unit to the external world and are, in effect, analogous to the terminal equipment of a communications system. These devices include terminals (graphics displays, teletypes, consoles), printers, card readers and punches, and various kinds of analog-digital interfaces that convert electrical signals into binary data. In place of the communications system's transmission lines and switches, the computer has a *central processing unit*, which accepts the incoming bit stream, acts on it according to some stored instructions (a program), then sends it back out. The transmission system is, of course, highly compressed in a computer, as electrical pulses travel over interconnected paths between components. These pulses are switched between transmission paths according to its program. The computer also has a high-speed storage device, or primary *memory*, that holds instructions and other useful information, and generally a hierarchy of secondary memory devices that provide a larger and more permanent, if slower, store for instructions and data. The crucial difference between the computer and a communications system, however, is that a central processing unit, unlike a conventional communications switch, is more than a traffic cop; it alters the bit stream as it passes through.

Technological Progress in Computers

In recent years various studies have estimated annual changes in computer prices, and though the estimates differ, the studies agree on the trends.[5] When adjusted by the GNP price deflator, real (inflation-adjusted) prices for mainframe computers of a given speed and capacity decreased more than 20 percent a year in the 1970s, even though technological progress seems to have been somewhat slower than in earlier decades. In the 1960s the decreases exceeded 25 percent a year. Since the late 1950s the trend in annual decline may have been as as high as 28 percent.[6] Although these figures refer only to mainframe computer systems, there is a widespread belief within the industry that for supercomputers and for minicomputers and microcomputers the decreases may have been even greater.[7]

One approach to determining the price of information processing capacity is to measure bits of information processed per unit of time, or bandwidth. The bandwidth of a computer system can then be thought of as a function of the effective bandwidths of the components and subsystems within a computer. Within this framework, technological change has reduced the price of a computing system by reducing the price per unit of output of its subsystems and components.

The results of this approach (described in appendix A at the end of this volume) are generally consistent with those reported in studies of computer

5. For a comprehensive survey of such studies, see Jack E. Triplett, "Price and Technological Change in a Capital Good: A Survey of Research on Computers" (U.S. Department of Commerce, Bureau of Economic Analysis, 1986).

6. See Kenneth S. Flamm, *Targeting the Computer: Government Support and International Competition* (Brookings, 1987), pp. 27, 222.

7. Sketchy evidence suggests that prices for minicomputers may have fallen more rapidly still. One study shows a nominal price decrease in excess of 60 percent a year from 1960 to about 1971 for small minicomputers, and a 63 percent annual decline from 1970 to 1974 for special-purpose minis. See C. B. Newport and Jan Ryzlak, "Communication Processors," *Proceedings of the IEEE*, vol. 60 (November 1972), pp. 1321–32; and Lawrence G. Roberts, "Data by the Packet," *IEEE Spectrum*, vol. 11 (February 1974), pp. 46–51.

price using other methodologies. The methodology also has one great advantage—economists' well-developed body of production theory can be brought to bear. In particular, improvements in price performance for a system can be broken down into the contributions of each of its components and subsystems, which enables one to examine technological change in such related products as general-purpose computer systems and digital switches that have benefited from technological advances in certain common components.

Figure 1 shows the price indexes for both component and system bandwidth on a logarithmic scale. The sharpest decreases, in the mid-1960s, were in the cost of central processor bandwidth. This pace slowed considerably in the late 1960s and has generally lagged behind the rate of overall progress in later years. (An exceptional year, 1978, saw widespread introduction of a new line of large-scale IBM processors.) Internal memory costs began to drop in the early 1960s, plunged in the late 1960s and early 1970s as the ferrite core memory was replaced with integrated circuit

Figure 1. *Price Indexes for Computer Components, 1957–78*

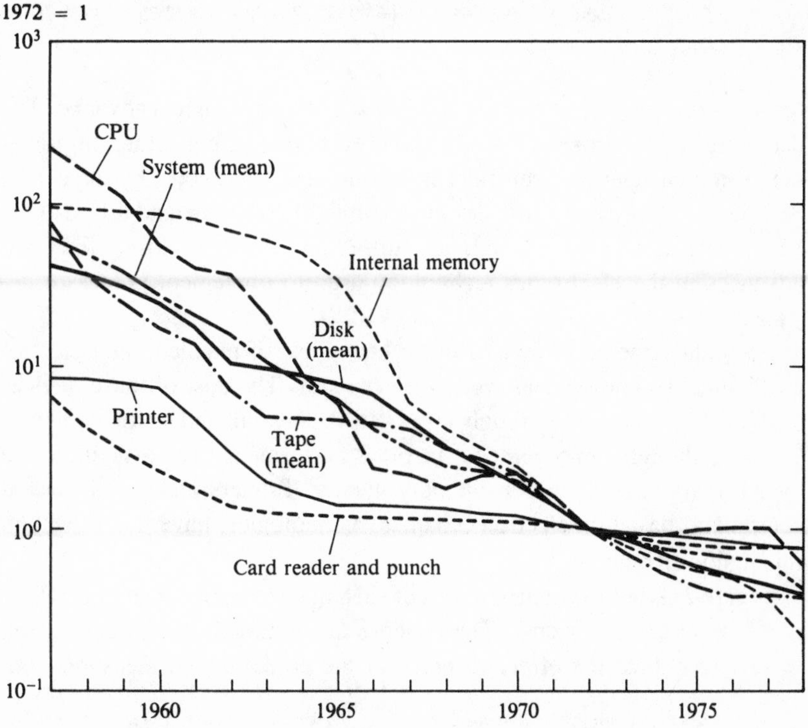

1972 = 1

Source: Author's calculations; see appendix A.

Figure 2. *Improvements in Computer Cost, by Component, 1957–78*

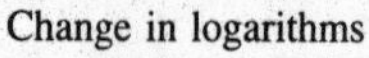

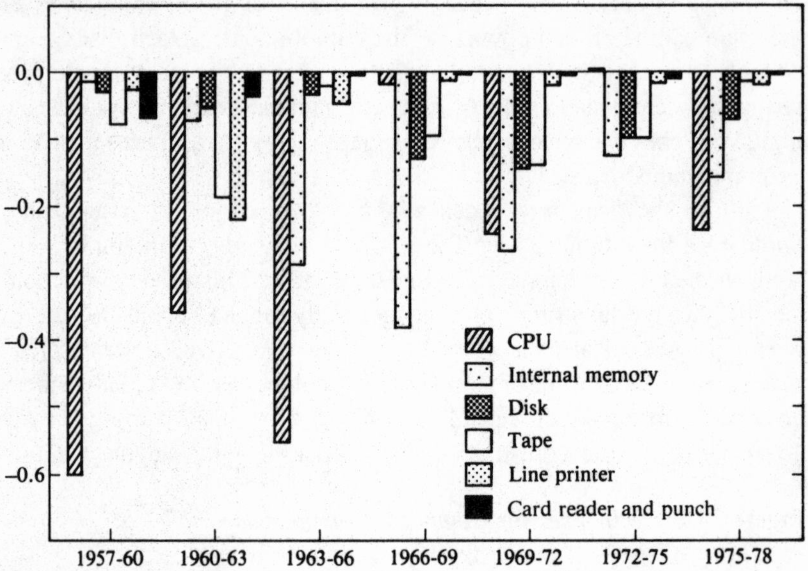

Source: Author's calculations; see appendix A.

technology, and generally have set the pace of technological advance. The declining cost of processor bandwidth drew both on architectural improvements and cheaper electronics costs through the late 1960s; by the end of that period, however, most design features of today's standard business computer had been introduced, and further progress was closely linked to advances in the electronics technology used to implement processor designs.[8]

Secondary memory devices make heavy use of magnetic and electromechanical technology, as well as electronics. The cost of these devices in the 1960s and 1970s roughly tracked the overall index for computer systems, though improvements in price performance for disk drives in recent years have been considerably greater. Printers and punched card equipment, based on electromechanical technologies, have improved at a much slower rate.

Figure 2 shows the contributions of each major component to the overall decrease in computer cost. The changes are summed over three-year intervals to reduce the effect of noise in the crude annual measures. Im-

8. See Kenneth Flamm, *Creating the Computer: Government, Industry, and High Technology* (Brookings, 1988), chap. 2, for a concise summary of these improvements.

provements in central processing units contributed most to price decreases through the mid-1960s but were then surpassed by the contribution of improvements to internal memory. Line printer improvements were most significant in the early 1960s, disk drives in the late 1960s and early 1970s. Tape drives improved most in the early 1960s and early 1970s. These results generally agree with assessments made at the time.[9]

In short, improvements in central processors have represented the single largest, and earliest, contribution to advances in computer system performance. The most important source in recent years has probably been innovation in memory (a fact that augurs significant repercussions on the computer industry from the unprecedented price increases for memory chips registered in 1988). Advance in electronic components is reflected in the large contributions of both central processor and memory improvements.[10] Advances in magnetics technology are third because of their contribution to the performance of secondary storage. Improvements to electromechanical components such as printers and punched card equipment have been much less rapid and less important.

Technological Progress in Communications Systems

Historically, digital representation of analog voice signals was first incorporated into transmission systems and only later applied to switching systems. A brief review of the development and diffusion of new technologies is necessary to understand cost decreases in these functions.

Innovations in Transmission

The transmission technologies in the modern Bell network can be traced to the early 1920s, when the first carrier transmission systems, so-called type A, B, and C carriers, were installed. Voice signals require a 3-kilohertz band of frequencies for good-quality transmission. Carrier systems were developed to transform voice frequencies into signals carried on a wider band at higher frequencies, then to convert them back to voice frequencies for distribution at their destinations. Carrier signals can also be transmitted over the same open wire, twisted pairs in cables, and coaxial

9. See, for example, the remarks of J. Presper Eckert, then vice president of Univac, cited in "Improvements in Hardware Performance," *Datamation*, vol. 12 (January 1966), p. 34.

10. One simple analysis suggests that of a roughly five orders of magnitude improvement in CPU performance, three orders are due to improved electronics and two to better processor design. R. W. Hockney and C. R. Jesshope, *Parallel Computers: Architecture, Programming, and Algorithms* (Bristol, England: Adam Hilger, 1981), p. 3.

cables used for voice frequency, and new circuits can be mined from existing ones. Thus carrier systems were piggybacked on existing voice frequency circuits.

From the 1940s on, major investments were made to add transmission capacity using these piggyback systems. In the 1950s transmission systems using microwave radio (the TD systems, followed by TH radio in the 1960s, and AR6A in the late 1970s) joined carrier systems as the source of expansion in long-distance circuit capacity. Carrier transmission was also developed for short-haul trunks using open wire (type O) and twisted pairs (type N). Microwave radio came to short-haul trunks in the 1960s (the TL/TM systems).[11]

Digital transmission systems, which convert analog voice signals to a digital bit stream, were first built for military use during the Second World War.[12] Later military systems multiplexed numerous such channels onto a single, time-shared line. Advances in electronics first made digital transmission commercially viable in the early 1960s for short hauls (T1 carrier on twisted pairs in cable) and in the mid-1970s for long distance (T4 carrier on coaxial cable).

Government users installed the first practical fiber-optic cable system for digital data transmission in the mid-1970s.[13] The first Bell System commercial fiber-optic route (the FT3 system) went into service in 1979. Radical decreases in fiber cost have since extended the technology's reach to transcontinental trunks.

Figure 3 shows the diffusion of these successive waves of technology in the Bell System. The expansion of network capacity was dominated by carrier on cable in the 1940s and 1950s and microwave radio from the late 1950s to the late 1970s. After the mid-1960s, digital systems became significant. Since 1980 a decrease of about 70 percent annually (an order of magnitude every two years) in the cost of high-capacity optical fiber has meant that digital transmission systems using fiber optics have come

11. These systems are described in detail in E. F. O'Neill, ed., *A History of Engineering and Science in the Bell System*, vol. 7: *Transmission Technology, 1925–1975* (Murray Hill, N.J.: AT&T Bell Laboratories, 1985), pp. 318–21.

12. The first digital transmission system (SIGSALY) digitized and scrambled wartime transatlantic conversations between Winston Churchill and Franklin Roosevelt. The first modern computers were secret systems built to decrypt intercepted German communications. The parallel with the development of digital computers was physical as well as theoretical. The bulky circuitry required by SIGSALY matched that of early computers. It was installed in the basement of Selfridge's department store in London and connected to the Cabinet war rooms by underground cable.

13. The system was installed in 1976 by the National Aeronautics and Space Administration at the Kennedy Space Center; see "NASA's Optical Cable Withstands Florida Elements," *Government Computer News* (November 20, 1987), p. 90.

Figure 3. *Transmission Plant Mileage, by Type, 1925–80*

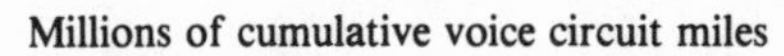

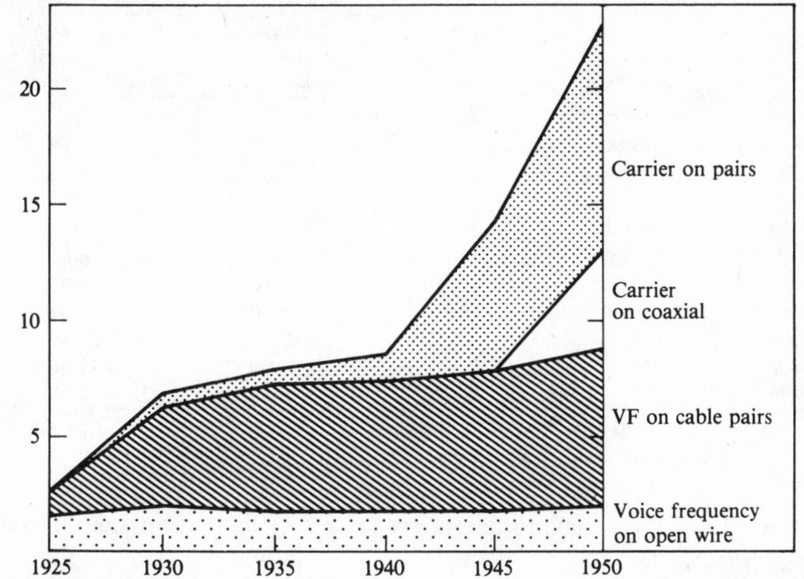

Millions of cumulative voice circuit miles

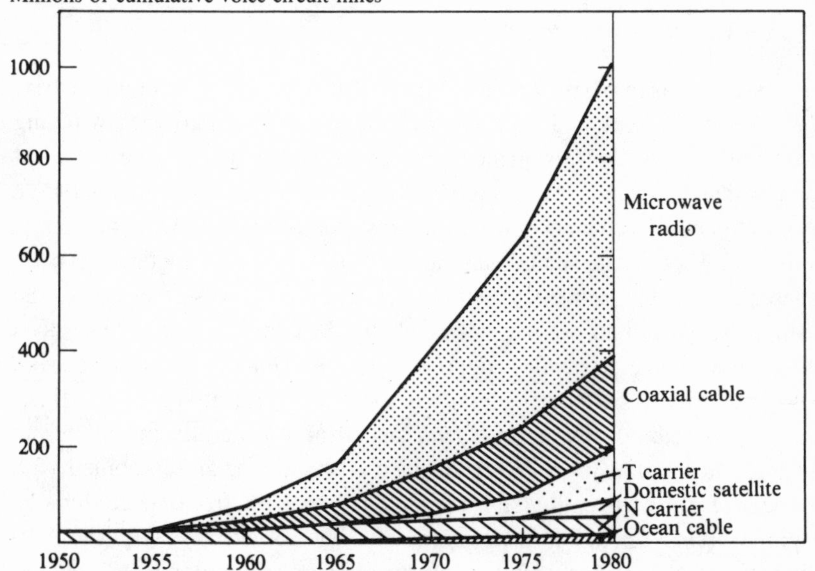

Source: E. F. O'Neill, ed., *A History of Engineering and Science in the Bell System*, vol. 7: *Transmission Technology, 1925–1975* (Murray Hill, N.J.: AT&T Bell Laboratories, 1985), pp. 784–85.

Table 1. *U.S. Market for Selected Communications Transmission Equipment, by Type of System, 1978–88*
Millions of dollars unless noted otherwise

Year	*Microwave equipment*	*Satellite earth stations*	*Fiber-optic equipment*	*Fiber-optic as percent of all investment*
1978	177.5	122.2	18.5	5.81
1979	199.6	142.0	38.7	10.18
1980	228.5	176.8	92.3	18.55
1981	403.1	137.5	171.2	24.05
1982	441.3	171.9	344.5	35.97
1983	442.0	n.a.	n.a.	n.a.
1984	470.0	473.0	879.0	48.24
1985	526.0	596.0	750.0	40.06
1986	576.0	900.0	883.0	37.43
1987	588.0	902.0	1,101.0	44.42
1988	670.0	905.0	1,104.0	41.21

Source: *Electronics* (January issues, 1979–88).
n.a. Not available.

to represent a substantial share of spending on new transmission investments (table 1).[14]

Innovations in Switching Systems

A switch has two basic components, a controller and a switching matrix. Information is sent and received over lines feeding into the switching matrix; the controller sets up the connections. The connections may consist of physically discrete paths (*space-division switching*) through separate communications channels, or time slots (*time-division switching*) on a single physical path shared among multiple channels. In time-division switching, voice frequency signals are sampled at discrete times, the samples assigned to a time slot, and the data then reassembled into a continuous voice frequency signal at the other end of the transmission system. Modern switches generally embody a hybrid of space- and time-division switching. The term *digital switch* now generally means one in which voice signals or data, encoded in a bit stream, are combined with some element of time-division switching. The shift from space-division to time-division coincided approximately, but not exactly, with the transition from electromechanical to electronic switch components.

14. See Robert J. Sanferrare, "Terrestrial Lightwave Systems," *AT&T Technical Journal*, vol. 66 (January–February 1987), p. 101.

The Bell System had introduced three broad classes of electromechanical switches by the late 1930s.[15] The least sophisticated, the so-called step-by-step switches, were based on relatively inflexible controller hardware that translated customer-dialed digits into a fixed path through the network. If the path was unavailable, a busy signal resulted. In 1921 the Bell System installed its first panel switches. The switches, rotors, and relays in the controller could set up a particular talking path independent of the dialing sequence. This ability made the system more flexible and increased network efficiency. In 1938 Bell introduced the first crossbar switch, in which the precise talking path used was centrally determined by more flexible controller hardware. This innovation eventually meant that by reconfiguring this *common control*, call setup and routing procedures could be periodically changed in an existing switch.

The next major innovation was to replace the fixed, hardware logic that controlled the crossbar switching matrix with a high-speed, electronic digital computer. Call setup procedures could be changed cheaply and almost instantly by replacing the software stored in the computer's memory, resulting in unprecedented flexibility and efficiency. The first commercial electronic switching system featuring a digital computer controller with stored program control was a switchboard (public branch exchange, or PBX), the 101-ESS, put into service in 1963. The first full-scale commercial central office switch with stored program control, the 1-ESS, was installed in 1965.

Although the 1-ESS was cheap and fast, it still depended on space-division switching. (The switching network for the 101-ESS had actually contained some time-division switching, largely because special signals and switching functions required for a central office were not necessary in PBX applications.) However, while the 1-ESS was being developed, Bell Laboratories agreed to develop a full-scale electronic system with a time-division switching network for the Army Signal Corps.[16] Though the design never went into production, the knowledge gained in working on it was later applied to designing Bell's first central office time-division switch, the 4-ESS.

Time-division switching entered the commercial mainstream in 1970 in a French switching system, the E10, which, however, did not incor-

15. See G. E. Schindler, Jr., ed., *A History of Engineering and Science in the Bell System*, vol. 3: *Switching Technology, 1925–1975* (Murray Hill, N.J.: AT&T Bell Laboratories, 1982), chaps. 3, 4.

16. Ibid., pp. 262–66.

porate stored program control.[17] The first commercial system that contained both an electronic switching matrix and stored program control appears to have been the IBM 2570, announced in 1969 and marketed in Europe, but this small exchange used space-division switching.[18] The first time-division switch that also included stored program control was Western Electric's 4-ESS, placed into service in 1976.

One of the major advantages of a digital switch is that it eliminates the need for costly interfaces and analog-digital conversion hardware when connected to a digital transmission system. By 1976 the use of digital transmission systems, notably T1, was reasonably widespread. Potential savings in interfacing these systems with the 4-ESS, as well as the inherent advantages of the switch's small size, low power consumption, and reliability compared with electromechanical network elements, argued for introduction of the new technology.

These technical developments did not always require installation of new equipment. After 1969, for example, existing installations of the no. 4A crossbar switch were retrofitted with stored program controllers.[19] And shortly after the 4-ESS went into service in 1976, its more powerful processor was used to upgrade the capacity of the older 1-ESS system.[20]

The diffusion of these new switching systems within the Bell System proceeded slowly. As figure 4 shows, it was not until the early 1960s that the volume of lines handled by the older step-by-step and panel technologies was surpassed by crossbar systems, and not until the mid-1970s that the 1-ESS made significant inroads into crossbars' volume. Table 2 shows the distribution of Bell toll-switching capacity, by type of system, in 1967, 1976, and 1983.

The slow pace of introduction is striking to anyone familiar with the computer industry. The no. 5 crossbar switch, based on relays, was introduced in 1948 and manufactured until 1976. The first no. 4A crossbar went into production in 1950; the last one was installed in 1976. Both systems were upgraded over this period, but electronic systems to replace them, although often developed as far as the prototype stage, were never sent into production (the 4-ESS replaced them in 1976). The first core

17. See Amos E. Joel, Jr., "Digital Switching—How It Developed," *IEEE Transactions on Communications*, vol. 27 (July 1979), p. 954.

18. David R. Jarema and Edward H. Sussenguth, "IBM Data Communications: A Quarter Century of Evolution and Progress," *IBM Journal of Research and Development*, vol. 25 (September 1981), p. 403; and "IBM Phone Calls Europe," *Business Week* (April 19, 1969), p. 39.

19. Schindler, Jr., ed., *Switching Technology*, p. 321.

20. The greater processor capacity also made it possible to add a higher-capacity switching network. See ibid., pp. 291–95.

Figure 4. *Distribution of Automatic Switching Systems Serving Bell System Lines, 1930–73*

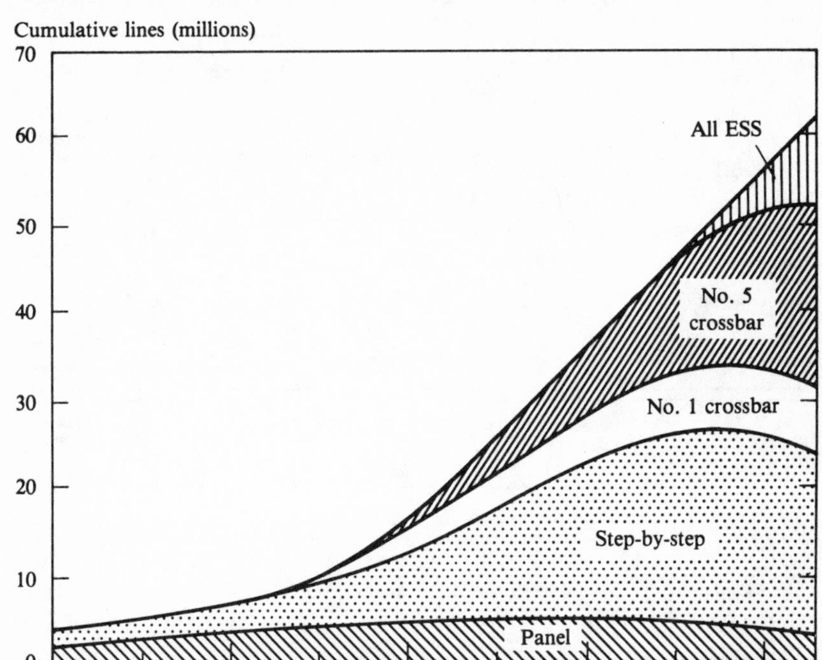

Source: Bell Telephone Laboratories, *Engineering and Operations, 1977* (Murray Hill, N.J., 1977), p. 455.

memories were introduced into electronic switches around 1971, just as the computer industry was completing a transition from core to semiconductor memory. A group to study the application of integrated circuits to switching was not even formed until the late 1960s.[21]

There were similar lags in introducing software innovations. High-level programming languages were not used to develop software for electronic switches until the mid-1970s. Standardization of processor architectures and machine instructions also came a full decade later (the mid-1970s) to digital switches than it did to digital computers.[22]

21. AT&T sometimes also moved slowly to integrate the new computer technology into its routine business tasks. Special-purpose relay computers for accounting applications were built in the late 1950s, as the rest of the world was making the transition from vacuum tube to transistor computers. Paper tape was not replaced by magnetic tape for billing information until after 1974. See ibid., pp. 142, 157, 169, 202, 282, 379.

22. See D. L. Carney and others, ''Architectural Overview,'' *AT&T Technical Journal*, vol. 64 (July–August 1985), pp. 1348–50; and R. F. Rey, ed., *Engineering and Operations in the Bell System*, 2d ed. (Murray Hill, N.J.: AT&T Bell Laboratories, 1983), p. 433. By the early

Table 2. *Switches Installed in the Bell System, January 1967, 1976, 1983*

	Toll switches					
	Number in place[a]			*Capacity*[a]		
Type	*1967*	*1976*	*1983*	*Erlangs 1967*	*Millions of CCS/BH*[b] *1976*	*Million terminations 1983*
Step	500 (41)	361 (32)	86 (12)	330 (4)	2.7 (10)	1.50 (27)
Crossbar tandem	237 (19)	210 (19)	126 (18)	1,700 (18)	6.1 (24)	
5 crossbar	407 (33)	341 (31)	194 (27)	1,100 (12)	2.5 (10)	
4/4A crossbar	73 (6)	177 (16)	62 (9)	6,200 (66)	14.4 (56)	
1/1A-ESS	...	23 (2)	166 (23)	...	0.2 (1)	1.25 (21)
4-ESS	...	...	84 (12)	...	...	3.00 (52)

	Local switches			
	Number in place[a]		*Capacity*[a] *(millions of lines)*	
Type	*1976*	*1983*	*1976*	*1983*
Step	1,669 (17)	800 (8)	18.3 (27)	6.1 (7)
CDO (Step)	4,388 (44)	3,700 (38)	4.7 (7)	5.1 (6)
Panel	55 (1)	...	0.9 (1)	...
No. 1 crossbar	310 (3)	180 (2)	6.7 (10)	4.1 (5)
No. 5 crossbar	2,700 (27)	2,120 (22)	26.5 (39)	23.5 (27)
1/1A-ESS	637 (6)	1,750 (18)	9.8 (15)	42.1 (49)
2/2B-ESS	185 (2)	700 (7)	0.8 (1)	4.4 (5)
3-ESS and 104-RSS	...	420 (4)	...	0.6 (1)
Time-division	...	110 (1)	...	0.2 (2)

Sources: A. E. Ritchie and L. S. Tuomenoksa, "System Objectives and Organization," *Bell System Technical Journal*, vol. 56 (September 1977), p. 1018; Bell Telephone Laboratories, *Engineering and Operations in the Bell System, 1977* (Murray Hill, N.J., 1977), p. 285; and R. F. Rey, ed., *Engineering and Operations in the Bell System*, 2d ed. (Murray Hill, N.J.: AT&T Bell Laboratories, 1983), pp. 461, 463.

a. Numbers in parentheses are percentages.

b. CCS/BH = Hundreds of call seconds per busy hour (peak load switched).

1960s, a high-level language (JOVIAL) was being used for real-time military command and control applications, but an experiment using JOVIAL to program the 1-ESS was unsuccessful. The experimental BLISS language was developed for use with switching systems but never put into production use. (Telephone interview with Victor Vysottsky, October 12, 1987.) High-level languages by nature produce less-efficient code than a well-tuned assembly-language program and therefore make less-efficient use of processor cycles. As processing power became relatively cheaper, however, it became more cost-effective to substitute brute processing cycles for programmer effort.

The 5-ESS, introduced in 1981, was the first Bell System switch to use a single processor and software architecture for installations of varying sizes. Previous Bell switches had used incompatible processors and assembly-language programming, necessitating development of all required software from scratch whenever a new switch was introduced.

In short, technologies that were actively pursued and developed in the Bell Laboratories were actually introduced into the operating network at an exceedingly slow and conservative pace. The contrast with the computer industry is especially striking: Bell Lab researchers were active and important players in computer systems research, yet actual implementation in the telephone system of many of the advances they participated in lagged behind the computer industry a full decade.

Measuring the Effects of Technological Advance on Costs

Casual observations are suggestive, but a serious attempt to measure the pace of technical change in communications equipment requires close scrutiny of a variety of often fragmentary and always messy sources of data. While it is often impossible to measure cost reduction and performance improvement with great precision, the data that can be assembled do uniformly point to one conclusion: overall, quality-adjusted communications hardware costs dropped much less rapidly than those of computers. And within communications systems, it was transmission cost—not costs for switching, which are usually identified with computers—that fell the most. These conclusions suggest rather different directions for the evolution of the structure of the network than have been indicated, for example, by the Huber Report, a point to which I shall later return.

System Costs

First, consider the overall effect of new technologies on the cost of transmission and switching. Costs have clearly decreased, but progress appears to have been considerably less rapid than in computer systems.

Between 1940 and 1950, the period of most rapid change, average book costs per circuit mile for communications channels in the AT&T Long Lines Department (those facilities separately allocated to the toll network) fell by two-thirds, from about $150 to $50 in current dollars (figure 5). These book capital costs include both toll switches and transmission systems, but suggest that the steepest declines in economic costs occurred when rapidly expanding carrier systems were superimposed on the existing cable infrastructure and new circuit miles mined from old lines.

Hardware costs (line plus terminations) for data communications on a 300-mile private line show virtually no change through the mid-1960s, then plunge by a full factor of ten in 1965–66 (figure 6). After virtually

Figure 5. *AT&T Long Lines Average Capital Cost per Circuit Mile, 1930–80*

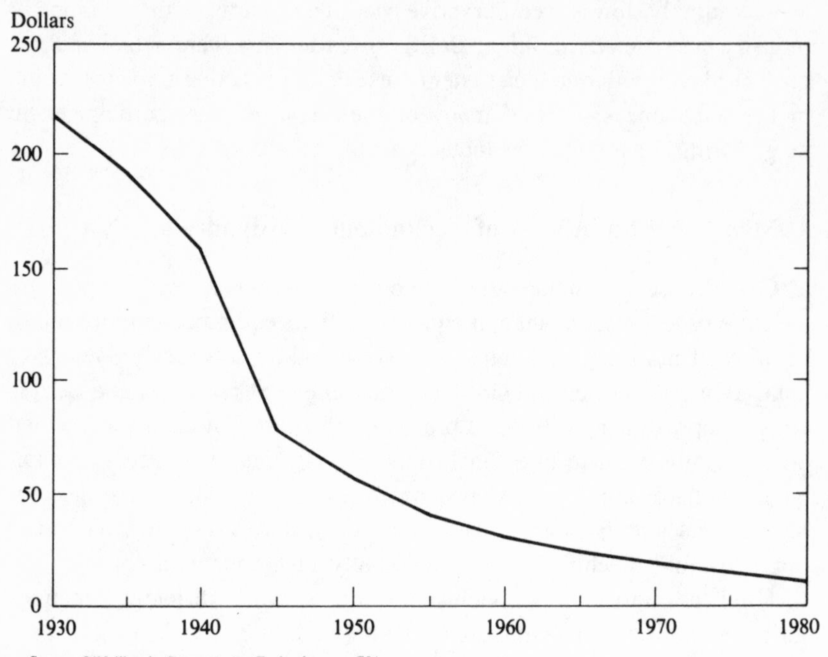

Source: O'Neill, ed., *Transmission Technology*, p. 781.

no change through the early 1970s, they decreased again by a factor of ten in 1973–74. From 1955 to 1975, costs fell about 20 percent a year. However, most of the decrease came as higher data transfer rates over voice-quality circuits were used in successively faster generations of modems. At a constant transfer rate of 300 bits a second, costs on a 300-mile line fell from $94 to $62 a month between 1962 and 1975, a decrease of a little over 3 percent a year in current dollars. So most of the savings came from improvements in the technology used to convert analog signals to digital bit streams at low error rates, not in improvements to the transmission system equipment and switches.

Finally, the U.S. Defense Department's Advanced Research Projects Agency in the early 1970s showed data communications costs falling by two-thirds from 1960 through 1967, about 15 percent annually.[23] Again, most of the decrease was associated with higher data transmission rates, not lower line costs.

23. Roberts, "Data by the Packet," p. 48.

Figure 6. *Data Communications Costs on a 300-Mile Private Line, 1955–79*[a]

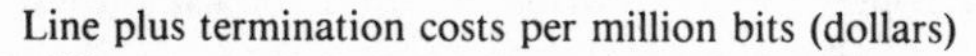

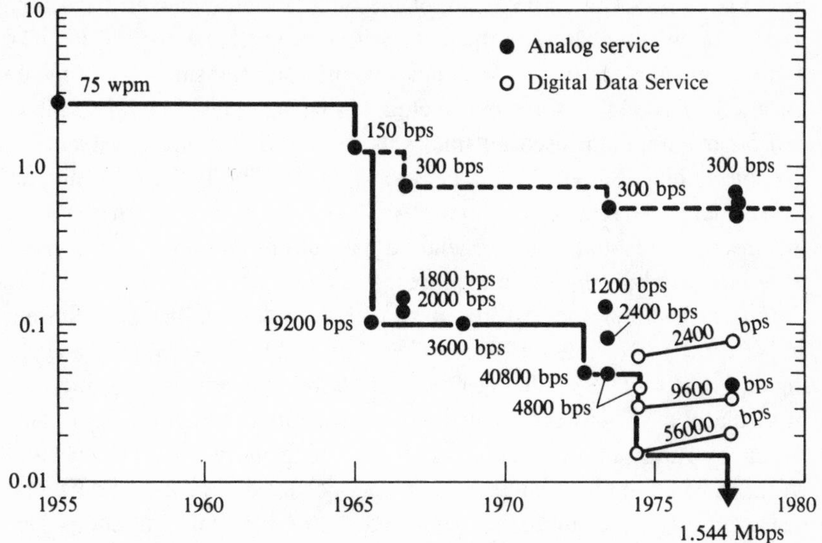

Source: Montgomery Phister, Jr., *Data Processing Technology and Economics*, 2d ed. (Bedford, Mass.: Digital Press, and Santa Monica Publishing Co., 1979), p. 549.
Units: wpm = words per minute; bps = bits per second; Mbps = million bits per second.

In short, data communications costs seem to have decreased, but not as much as computing costs, which probably fell more than 20 percent annually. After factoring out the effect of higher modem speeds, which reflect improvements largely independent of transmission and switching costs,[24] a rather small decrease, perhaps 3 to 5 percent a year, in the cost of communicating over the network was produced in recent decades. Yet a major decrease in capital costs for the network does seem to have occurred in the 1940s and 1950s, before computers had come into widespread use. Could improvements in transmission rather than switching have set the pace for technological advance at the outset of the postwar era? To consider the question, one must examine the evidence available, imperfect though it is.

Matched-Model Price Indexes

The most straightforward approach to examining cost changes in switching and transmission is to study the so-called telephone plant indexes

24. But not entirely, since higher rates of transmission require less-noisy, higher-quality lines.

(TPIs) produced by AT&T for planning purposes and for use in rate cases. These price indexes used a bundle of typical purchases in each of AT&T's major capital accounts to weight price changes for capital goods installed in the network. The components of the bundle were changed from time to time to reflect changes in the composition of investment spending. The TPI for inside plant (which includes switches and transmission hardware installed in central offices but excludes towers, poles, conduits, cable, and other equipment used outside) was, before divestiture, the basis for the communications equipment deflator published by the U.S. Commerce Department's Bureau of Economic Analysis. It was used in the national income accounts and was the standard measure of communications hardware cost used in almost all analyses.

Figure 7 shows movements in annual averages for the Bell System TPIs from 1946 to January 1983.[25] Costs for both transmission and switching equipment were stable until the mid-1960s, when the rate of increase of virtually all prices accelerated. The acceleration coincided with some increase in the rate of inflation, but something more may have been at work. In fact, a major change in regulation occurred in 1965 as the Federal Communications Commission switched from informal continuous surveillance of the Bell System to a more stringent system of public investigations and hearings.[26] One may reasonably suspect that the change in regulatory procedures may have been linked to a structural change in price-setting procedures within AT&T at the time.[27]

To the extent that these series are to be believed, they indicate that through 1977, nominal transmission equipment prices generally rose more slowly than those for switching equipment.[28] Electronic switches are clearly an exceptional case, with prices rising at a much slower rate than a composite switch index through the early 1970s, then matching changes in it as electronic equipment came to dominate expenditures for switches. Since the mid-1970s, switching costs have risen more slowly than costs for transmission equipment. Within the category of transmission equipment, circuit equipment prices have risen the least since World War II.

25. The actual indexes, the components of which are classifiable as either switching or transmission equipment, are available upon request from the author.

26. See William G. Shepherd, "The Competitive Margin in Communications," in William M. Capron, ed., *Technological Change in Regulated Industries* (Brookings, 1971), pp. 99–100.

27. Nestor Terleckyj, in his paper in this volume, also finds indications of structural change in the mid-1960s.

28. Another set of Bell equipment price indexes, reported in a 1969 McKinsey & Company consulting study, suggests a similar trend through the mid-1960s. See *A Study of Western Electric's Performance* (New York: AT&T, 1969), p. 80.

Figure 7. *The Bell System Telephone Plant Indexes, 1946 to January 1983*

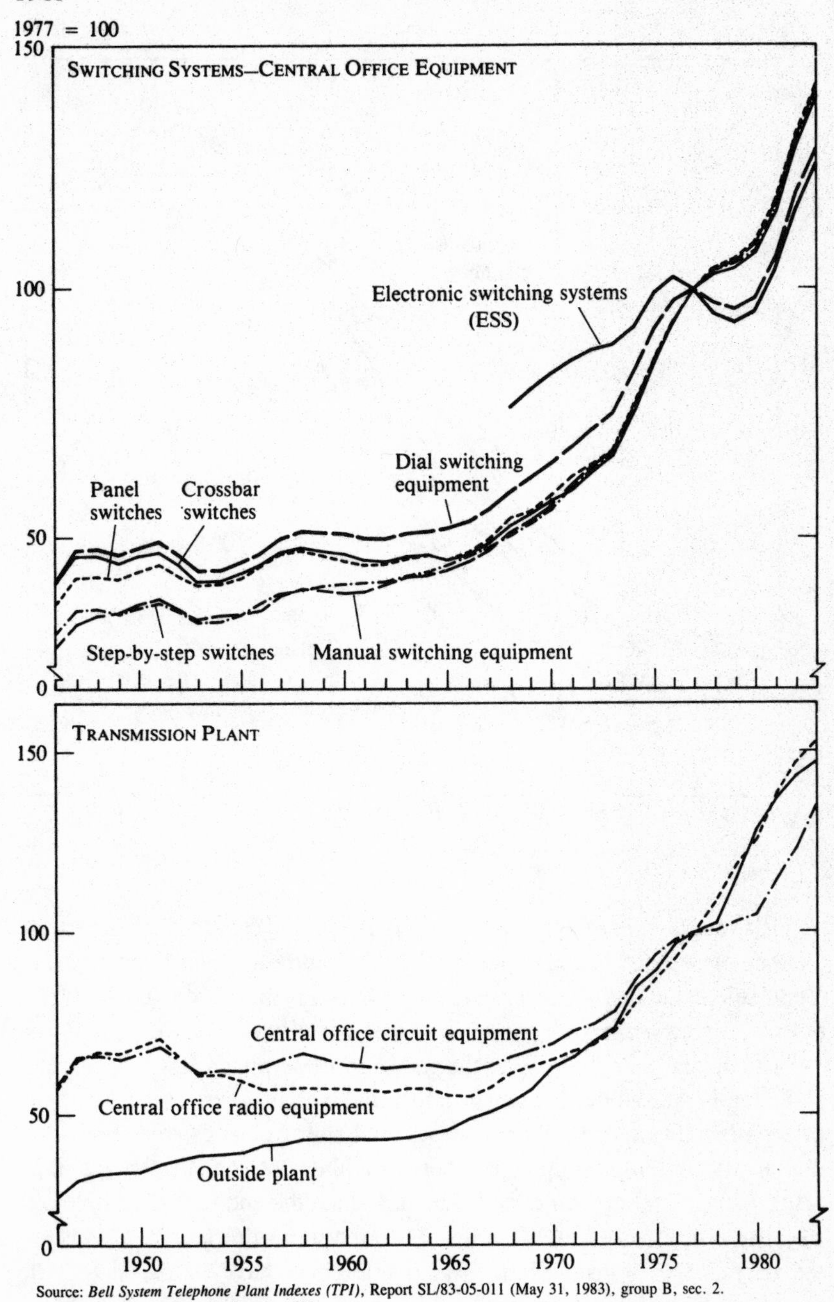

Source: *Bell System Telephone Plant Indexes (TPI)*, Report SL/83-05-011 (May 31, 1983), group B, sec. 2.
a. Annual averages except for January 1983.

Figure 8. *Comparison of Telephone Plant Indexes for AT&T and New York Telephone, 1977 to January 1986*[a]

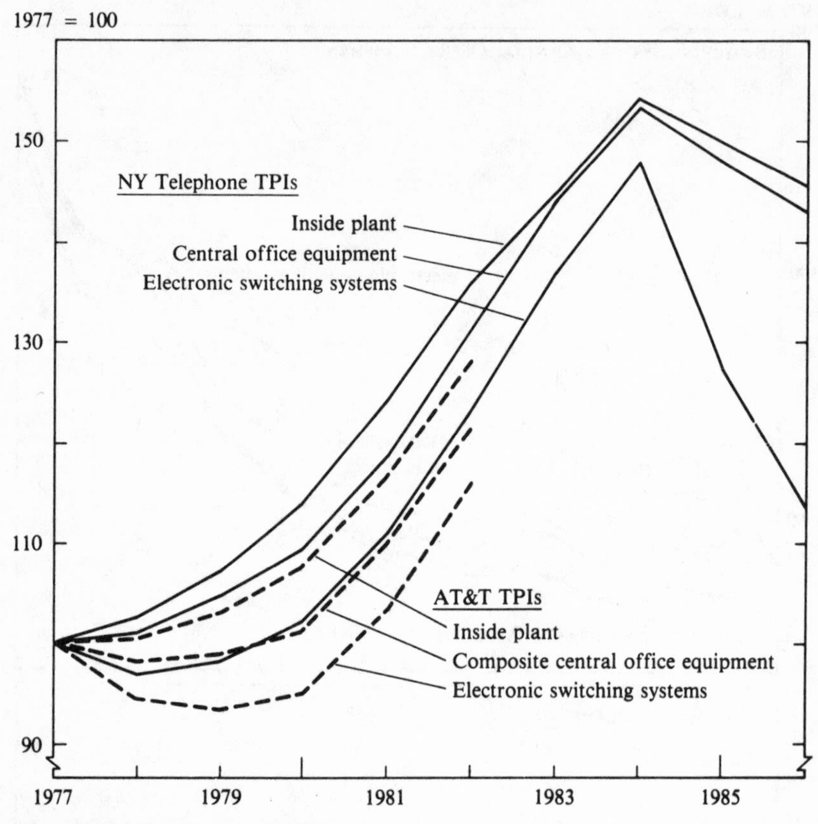

Sources: Same as figure 7; and unpublished Nynex data.
a. Annual averages except for January 1986.

TPIs for New York Telephone closely match indexes for the entire Bell System over their shared range, 1977–82 (figure 8). The trend in New York Telephone indexes after divestiture began is most provocative, however, showing a sharp decrease in prices for electronic switches. From 1984 to January 1986, in fact, the index of these prices fell 25 percent.

Recently introduced matched-model indexes of telephone equipment prices maintained by the U.S. Bureau of Labor Statistics show costs for digital transmission equipment other than fiber optics rising only a little faster than digital central office switches since the end of 1985 (table 3). Because costs for fiber-optic transmission systems dropped much faster, the decrease in transmission costs overall may have outpaced that in switching costs in recent years.

Table 3. *Producer Price Indexes for Communications Equipment, January 1985–88*
1985 = 100

Item	*1986*	*1987*	*1988*
Digital transmission equipment[a]	n.a.	n.a.	113.1
Digital central office switches	100.1	105.9	107.3
Analog PBX	n.a.	98.9[b]	n.a.
Digital PBX	100.0	90.5	96.2
Under 400 lines	100.0	n.a.	93.0
Over 400 lines	100.0	n.a.	99.5
Other switching equipment	100.8	136.8	136.6

Source: U.S. Bureau of Labor Statistics, *Producer Price Indexes*, January issues; for 1986, p. 118; for 1987, pp. 123-24; and for 1988, pp. 121-22.
n.a. Not available.
a. Excludes fiber optics.
b. As of December 1986 (January figure unavailable).

The major deficiency of these series is their treatment of technical change. When a new model is introduced, it is linked to the existing price index—in effect, only price changes following its introduction have any impact on the composite index. Any one-time, discontinuous drop in cost associated with its introduction is ignored. If the cost of older models immediately dropped to match that of the new equipment of identical quality, or if new equipment were introduced when it just matched the functionality of older vintage technology at roughly the same cost, this would not be a problem. However, in practice, prices for older equipment often continue to reflect the higher production costs of older technologies. Purchases of the older equipment gradually disappear from the purchase bundle rather than dropping in price to keep pace with lower cost per unit of capacity of the new products. In computers, for example, linked price indexes of matched models fall much less steeply than quality-corrected prices based on the characteristics of new models.[29]

In short, linked price indexes of matching models may capture intra-generational price decreases, but they probably miss the abrupt intergenerational decreases in prices that occur as new types of products are introduced.

Figure 9 contrasts Bell Communications Research (Bellcore) estimates, discussed later, of the cost of a 15,000-trunk central office from 1965 to 1983 with the composite TPI for electronic switching systems that begins

29. See David W. Cartwright, Gerald F. Donohoe, and Robert P. Parker, "Improved Deflation of Computers in the Gross National Product of the United States," Working Paper 4 (Washington, D.C.: U.S. Department of Commerce, Bureau of Economic Analysis, 1985); and Rosanne Cole and others, "Quality-Adjusted Price Indexes for Computer Processors and Selected Peripheral Equipment," *Survey of Current Business*, vol. 66 (January 1986), pp. 41–50.

Figure 9. *Comparison of Estimates of Switch Prices by Bell Communications Research with the Electronic Switching System (ESS) Component of Telephone Plant Indexes, 1965–83*

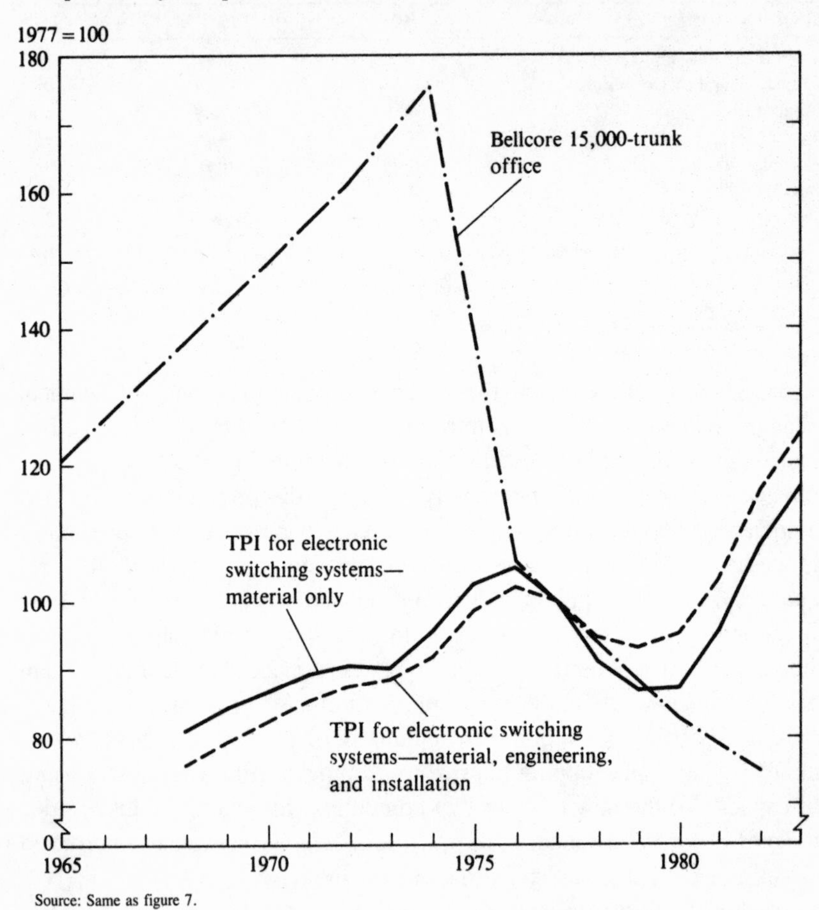

Source: Same as figure 7.

in 1968. A version of the TPI that reflects only material costs is plotted along with the index including specification and installation costs, but this does not affect the conclusions. The Bellcore index uses the price of a no. 4A crossbar switch before 1976 and a 4-ESS switch afterward. An abrupt decrease in switching costs coincided with the introduction of the 4-ESS, followed by gradual declines that match up closely with the TPI through 1979. The rise in the TPI in the early 1980s probably reflects increasing costs for smaller electronic switches included along with the 4-ESS in that index.

Figure 10. *Linked Indexes of Computer and Switch Prices, 1967–83*

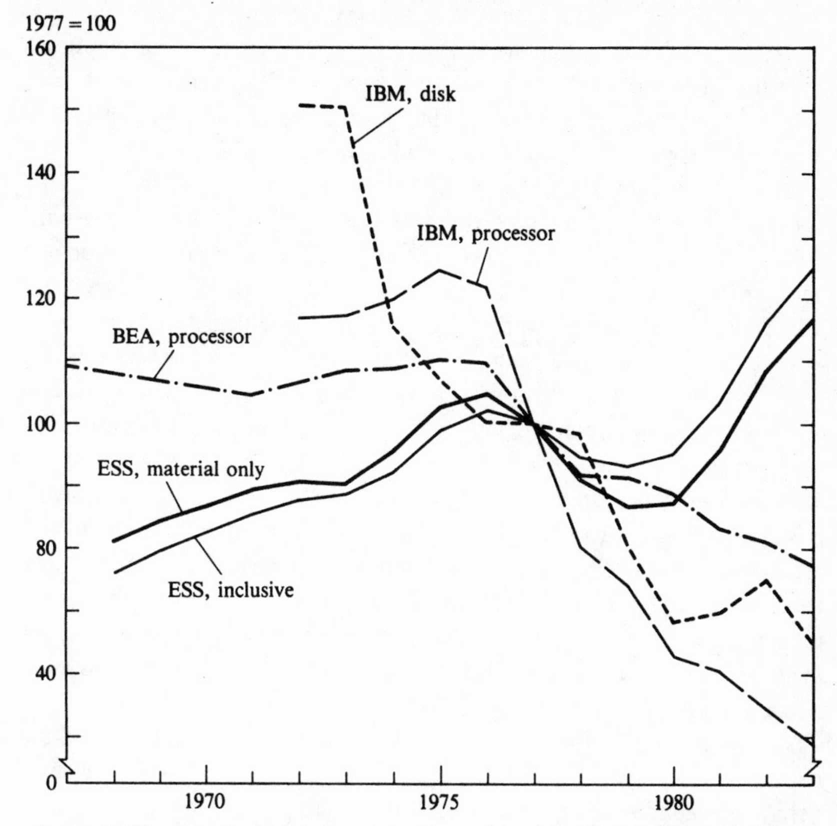

Sources: For TPIs, same as figure 7. For computer prices, Rosanne Cole and others, "Quality-Adjusted Price Indexes for Computer Processors and Equipment," *Survey of Current Business*, vol. 66 (January 1986), pp. 41–50.

Despite the limitations of a linked matching-model index such as TPI in measuring price performance of a good undergoing significant technological change, some comparison with other such indexes may be useful. Figure 10 presents two matched-model indexes for computer processor prices, one produced by the Bureau of Economic Analysis, the other by IBM; a matched-model index of computer disk-drive prices; and the TPI (materials only) for electronic switching systems. Even linked indexes of matched models for both sets of prices show computer processor and disk-drive prices decreasing much more rapidly than those of electronic switches.

Incremental Capital Costs

Since matched-model price indexes consistently underestimate cost reductions resulting from technological advance, other methods are needed.

One way of measuring quality-adjusted costs directly is to calculate an incremental cost for new switching and transmission capacity added to the telecommunications network. If one can come up with a simple, direct measure of total switching or transmission capacity installed in the network, the incremental costs associated with expanding that capacity at any given moment can easily be calculated.

THE ECONOMICS OF TRANSMISSION SYSTEMS. When analyzing economic returns from using distinct transmission technologies, one must consider two forms of economies of scale: economies of cross section and economies of haul. Unit costs per circuit mile of transmission capacity will generally decline as both the cross section (the density of circuits on a route) and the distance spanned by the route (the haul) are increased.

Economies of cross section on a route of fixed length stem from two sources. First, although high-capacity transmission systems have traditionally required large minimum investments per route mile compared with low-capacity systems, costs per circuit on these high-capacity systems have been much lower than on low-capacity systems (which, however, require investments per mile of a much smaller absolute size). Second, switches can concentrate low-volume traffic on cheaper, high-volume channels and share circuits among large numbers of users. The cheaper switches become relative to trunks, the greater the number of layers of concentration and sharing of circuits that becomes economical.

Because requests for telephone service at any instant can exceed the capacity of the system to serve them, however, assumptions about user behavior when denied access to the system are also required. Telephone traffic engineering practice customarily fixes this probability of being blocked at some very low level, then calculates the number of circuits required to service some offered traffic load.[30] With a variety of assumptions about call arrivals and blocked calls, a nonlinear relation between traffic carried per trunk and total number of circuits on a route is obtained. Large circuit groups can carry much more traffic per circuit than very small groups (figure 11); hence economies of cross section in providing communications capacity would result even if hardware cost per circuit did not decline with increases in system cross section.

30. User load to a network is defined as call arrivals per second times average call duration in seconds. The unit of load so calculated is known as an *erlang*, in honor of A. K. Erlang, one of the inventors of traffic theory. In the Bell System, load was equivalently measured in hundreds of call-seconds per hour (CCS, with 36 CCS equal to 1 erlang). See Sushil G. Munshi, ''Random Nature of Service Demands,'' in John C. McDonald, ed., *Fundamentals of Digital Switching* (Plenum Press, 1983), pp. 41–68; and M. T. Hills, *Telecommunications Switching Principles* (MIT Press, 1979), chap. 4.

Figure 11. *Average Trunk Occupancy at Different Blocking Percentages*

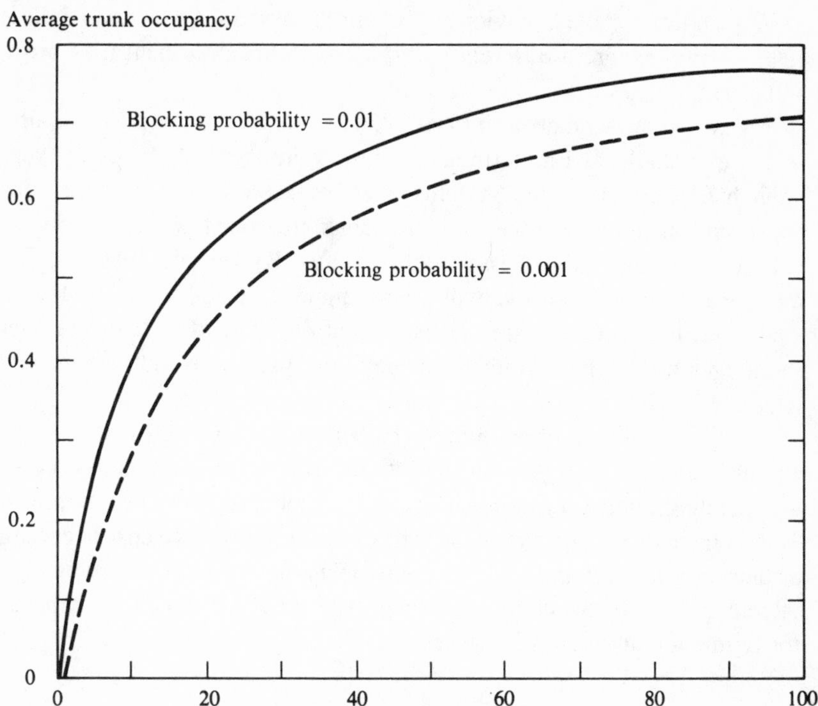

Source: A. E. Joel, Jr., "Circuit-Switching Fundamentals," in John C. McDonald, ed., *Fundamentals of Digital Switching* (Plenum Press, 1983), p. 9.

After a reasonable number of trunks have been added to a circuit group, however, each additional circuit increases traffic per circuit very little—indeed traffic per trunk remains nearly constant. This fact has been used to argue that, with microwave technology, adding trunks to circuit groups has little economic impact once moderate-size circuit groups are achieved.[31] But the hardware-driven economies of cross section that are realized when higher-capacity transmission equipment is used remain.

This discussion of economies of cross section has so far ignored the costs of terminating the transmission system and interfacing it to switching

31. Waverman developed a numerical example for a particular microwave system, in which marginal cost per CCS (1/36 erlang) falls more steeply, then becomes flatter faster than marginal cost per circuit. He argued that this reduced the importance of economies of scale. See Leonard Waverman, "The Regulation of Intercity Telecommunications," in Almarin Phillips, ed., *Promoting Competition in Regulated Markets* (Brookings, 1975), pp. 201–39.

or other equipment. For long hauls, termination costs are generally small relative to total cost and thus relatively unimportant in explaining technology choices or the behavior of costs in the system. Short and medium hauls, however, are a different story, and economies of haul may prove important.

Figure 12 shows that in the late 1970s voice frequency twisted pairs, which essentially had no termination cost, were the best choice for very short hauls, digital carrier systems (metallic cable at the time) best for short- and medium-distance runs, and analog carrier systems (carrier on cable, microwave radio) best for longer hauls. By the mid-1980s, digital on optical fiber had effectively displaced digital on metallic cable. Despite somewhat higher termination costs, digital on optic fiber cost less over the longer runs, where voice frequency wire pairs were not economical (figure 13).

The potential use of seemingly higher-cost technologies at a given circuit capacity is therefore not as irrational as it might seem. Costs shown in appendix figure A-1 exclude terminal equipment and must be interpreted as incremental cost per circuit per unit distance, or average cost over long distances. For short and medium hauls, technologies with relatively high incremental costs per circuit mile may well be attractive if less expense for terminal equipment is required.

Figure 12. *Transmission System Relative Costs, Late 1970s*

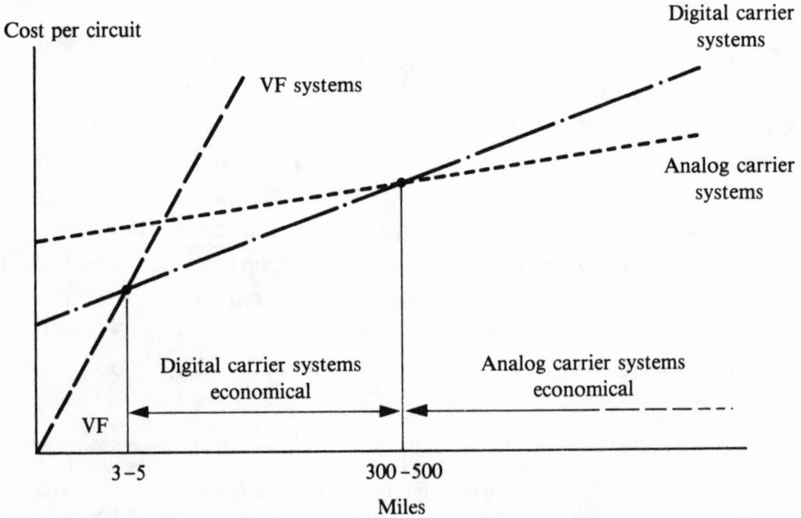

Cost per circuit

Digital carrier systems

VF systems

Analog carrier systems

Digital carrier systems economical

Analog carrier systems economical

VF

3–5

300–500

Miles

Source: O'Neill, ed., *Transmission Technology*, p. 781.

Figure 13. *Installed First Cost per POTS Circuit, Mid-1980s*[a]

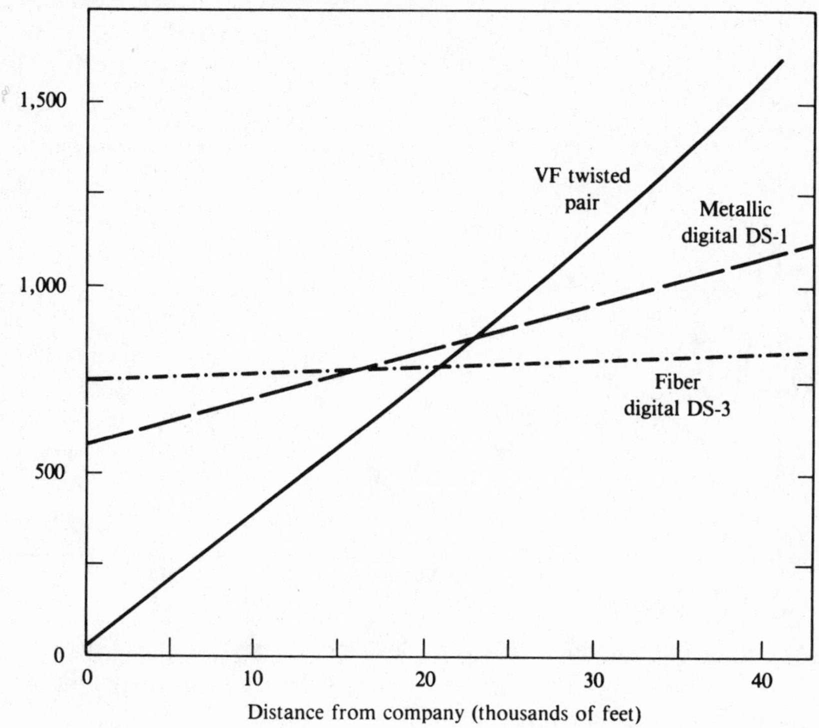

Source: Bell Communications Research.
a. POTS, or plain old telephone service, is "no-frills" voice service.

To summarize, then, accurate measurement of decreases in transmission costs must take into account the total capacity of the equipment and resulting economies of cross section as well as the physical distance the system is intended to traverse and possible economies of haul. This complicates any attempt to measure prices adjusted for quality or performance.

MEASURING INCREMENTAL COSTS. Figure 14 charts an attempt to calculate incremental capacity costs (change in capital cost per change in capacity) using AT&T data. Switching activity is assumed proportional to total local calls plus ten times the number of long-distance messages.[32]

32. The rationale for the factor of 10 applied to long-distance messages is that the duration of long-distance calls is about 2.5 times that of a local call, and that a long-distance call goes through perhaps 4 times as many switches, 2 local switches at each end and 2 toll switches; a local call requires only a single switch. Therefore the total erlangs of installed switching capacity required for a long-distance message would be about 10 times that of a local call.

The 2.5 figure comes from noting that the average duration of a local call was about 3 minutes in the Bell System in the 1970s, compared with 7.5 to 8.5 minutes for a long-distance

Figure 14. *Incremental Capacity Cost Measures, 1951–81*

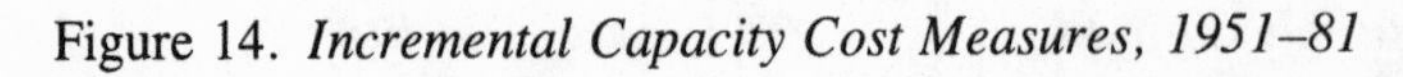

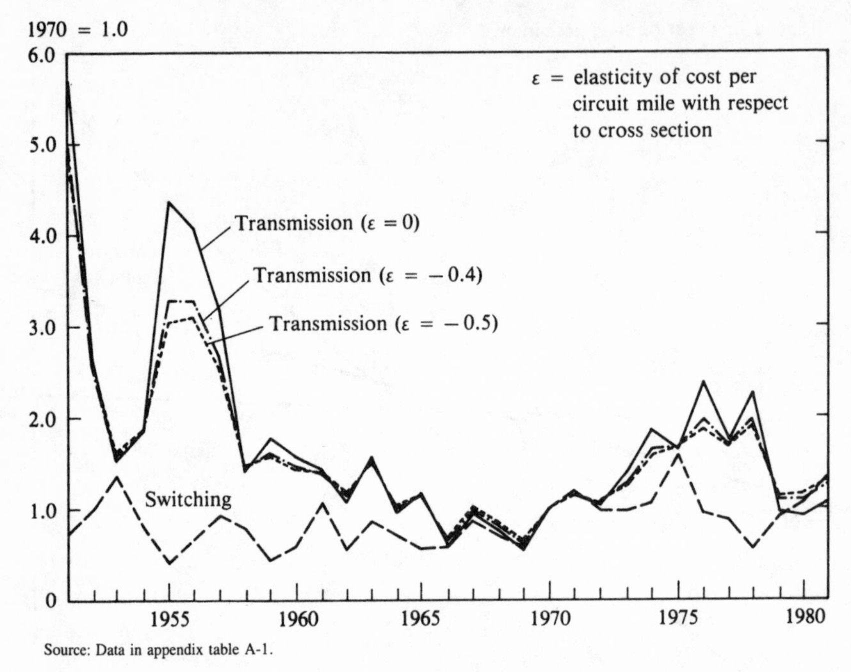

Source: Data in appendix table A-1.

Measuring transmission capacity is less straightforward, since average cross section—and cost—may vary greatly for any given addition of circuit miles to the network.

By capitalizing on the observed historical stability in the relationship between costs per circuit mile and the cross section of a transmission link, it is possible to control for the effects of variation in cross section on incremental transmission cost. (The particulars are given in appendix A.) One can then calculate a constant cross-section cost index, α, and attribute declines in α to the effects of technological change.

The calibration of index α requires assumptions about the elasticity of the cost per circuit mile with respect to cross section. Values in the -0.4 to -0.5 range are suggested by the data shown in appendix A, while a value of zero would be appropriate if there were no economies of cross section. Figure 14 shows the results of this analysis: for transmission, α is plotted; for switching, cost per increment in switching capacity.

Granting that the measures of switching and transmission cost are appropriate in some crude sense, there are other practical problems in in-

call. See Schindler, Jr., ed., *Switching Technology*, p. 295; and AT&T, *Bell System Statistical Manual, 1950–1981* (June 1982), p. 805.

terpreting figure 14. For one thing, there is generally a lag between the time an investment is made and its fruition as operational capacity. This may be expected to give the ratio of investment to capacity a certain jagged look from year to year as last year's spurt in investment is realized in this year's increase in capacity. A second problem is that the switching measure is based on actual traffic, not capacity. Since new investment spending tends to be lumpy and anticipates future growth in demand, capacity typically jumps ahead of current demand, slows while demand catches up, then jumps forward again. This, too, will give incremental cost ratios based on current traffic a certain jagged appearance. Third, changes in route length may contribute to change in the constant cross-section transmission cost index. A spurt in construction of higher-cost, short-haul routes would tend to move the index up; a shift to longer hauls might move it down.

Finally, new transmission capacity based on carrier systems has typically mined the current physical plant to create new circuits. As circuit capacity has expanded, cheap increments based on mining existing physical structures have generally been followed by more expensive, "greenfield" construction. For a given generation of technological advances, one might expect marginal increments to transmission capacity to get progressively more expensive, only to drop again with the next wave of technological progress. This, too, will give these ratios a volatile, up-and-down movement.

Nonetheless, the trend in these ratios broadly tracks the trend in transmission and switching plant costs. Figure 14 shows a dramatic decline in incremental transmission costs in the 1950s (about 9 percent a year, from a three-year average centered on 1952 to a three-year average centered on 1965). The pace of advance slowed after the early 1960s, achieving less than a 1 percent annual decrease from 1965 to 1970. Costs then turned around, increasing more than 11 percent annually from 1970 to 1976, only to drop more than 5 percent annually from 1977 to 1981 (the advent of fiber optics may well have figured in this decrease).

Reductions in average cross section have coincided with the spikes in incremental cost, particularly in the early 1950s and mid-1970s (see appendix A). Since installation of lower-capacity links probably accompanied a shift to shorter route lengths, the spikes may in part reflect bursts of investment in shorter-haul transmission facilities.

Incremental switching costs show an average annual decline of 2 percent in the 1950s, and increase at 8 percent annually from the mid-1960s through the mid-1970s. A sharp decrease from 1975 to 1978, coinciding with the introduction of the 4-ESS, was followed by an increase through the early 1980s, as was the case with the TPIs.[33]

Thus the incremental cost measures show transmission costs dropping faster, or increasing more slowly, than switch prices through the postwar period, except for the mid-1970s, when switches set the pace for technological advance. The 1950s seem to have seen the sharpest decreases, with less rapid decreases in the 1960s. Virtually none of the decreases in incremental costs measured here show up in the TPIs before 1970. After 1970 incremental costs increase markedly, but not as much as the TPIs indicate. Overall then, incremental costs track a much faster rate of technological progress than the TPIs suggest.

Further Evidence on Switching Costs

Other data seem to confirm a gradual increase in switch costs from the early 1970s through the early 1980s.

THE BELLCORE ESTIMATES. Specialists at Bellcore have estimated the cost of a representative central office switch in offices of varying sizes, ranging from a small exchange with 2,500 lines to a toll switching center with 60,000 trunks. These series are shown in figure 15.

Estimates interpolated from these numbers on the basis of the TPIs have been omitted in figure 15, since, as has been observed, the TPIs are untrustworthy measures of performance-adjusted price. And, although Bellcore specialists apparently took pains to compare similar configurations over the years, abrupt changes in cost from one model to a new and not entirely identical model did occur, and features present in newer switches were sometimes unavailable in older models. In other cases, price differentials reflect only the different configurations of trunk and line interfaces added to the same basic switch. The prices from 1965 to 1974 for the largest switches, for example, pertain to the no. 4 crossbar with different sets of add-ons, while switch prices for these offices since 1975 are based on the 4-ESS. The Bellcore estimates are presented in appendix B at the end of this volume.

Prices per connection for the large, no. 4–based switching centers were significantly lower than those for smaller switches used by local offices. The shift to a digital switching matrix in big systems in 1976 dramatically lowered the costs of the largest systems, while prices for the smaller

33. Compound annual growth rates (three-year centered averages) were −2.0 for 1952–65, 8.0 for 1965–70, 3.6 for 1970–76, −0.6 for 1976–80, and 3.4 for 1976–81.

Figure 15. *Estimates of Switch Prices, 1965–84*

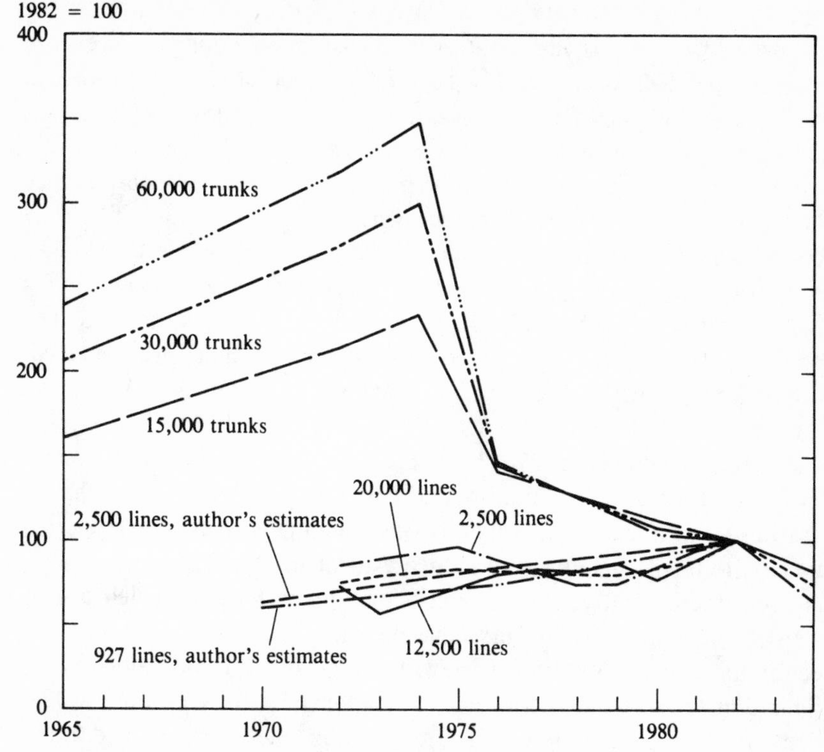

Sources: Author's estimates and Bell Communications Research data in appendix table B-1.

switches continued to increase gradually. Digital switching was not generally implemented in these smaller switches until the early 1980s.

The last two price series shown in figure 15 are independent estimates I have constructed for small switches (2,500 and 927 lines, respectively), configured in a manner as close as possible to the Bellcore 2,500-line setup. These estimates, consistent with Bellcore's small-switch price series, were arrived at through the statistical techniques described next.

AN ECONOMETRIC ANALYSIS, 1970–82. Ideally, one would like to estimate a quality-adjusted switch-price time series econometrically, based on actual transaction prices for switches and supplementary information on their characteristics and configuration when sold.[34] Unfortunately, in what was essentially a captive market until recently, and a very small

34. From the viewpoint of performance, the crucial characteristics of a switching system are the number of incoming calls that can be processed (a limitation dependent on the power of the processor or controller in the switch), the traffic load (in erlangs) that can be handled by the

market at that, with a limited quantity of machines produced every year, there is no published data on switch costs. Not even catalog prices are available, because installed switches are custom-configured; and even if transaction prices were to be discovered in some archive, it is unclear how one would interpret what was often a transfer price between two divisions of a large, regulated monopoly.

The closest thing to a competitive market in switches probably existed in the rural heartlands of America, where many small central offices continue to be operated by smaller, independent telephone companies. Much of the equipment purchased by these companies was financed with assistance from the Rural Electrification Administration of the U.S. Department of Agriculture.[35] Selected information on switch prices and characteristics is collected by the REA to monitor these loans. A request under the Freedom of Information Act was used to obtain these figures, and data were assembled for 1970, 1976, and 1982. Switch prices were treated as being generated by a multi-output cost function, with switch characteristics—lines, trunks, and options—as the outputs. A loglinear approximation to this cost function was then estimated.

From these data, estimated prices for a given type of switch can be constructed to measure quality-adjusted changes in prices. I chose two configurations: a 2,500-line office with a 12:1 line-to-trunk ratio—a configuration for which Bellcore has also produced estimates—and a 927-line office with 76 trunks, which used the mean values for lines and trunks in the sample of contracts. For 1970, prices reflect a switch with common control and automatic number identification (ANI). For 1976 the switch

---

switching matrix, and the number of lines or trunks that can be physically interfaced to the switching network. In practice, switching systems are engineered so that the number of lines or trunks that can be connected is rarely a binding constraint on the system; for given processor power, maximum call attempts and traffic load are linked by a relation involving the average duration of a call, which varies with customer mix. Since the advent of stored program control, an additional characteristic of a switch has been the range of so-called features (speed calling, call waiting, call forwarding, and so on) that it supports. Since these are written into the control software, processor power may be a good proxy for the potential range of features that can be added to a switch of given size.

For constant offered traffic load per customer line or trunk, and constant customer mix, cost per line is a measure of the cost of switch functionality. This, perhaps, is why cost per line is commonly used as a summary measure of the cost of switching functionality. See Bell Communications Research, *LSSGR Issue 1*, December 1984, sec. 17, "Traffic Environment"; and Groupe des Ingénieurs du Secteur Commutation du CNET (GRINSEC), *North-Holland Studies in Telecommunication*, vol. 2: *Electronic Switching* (Amsterdam: Elsevier Science Publishers, 1983), pp. 287–90.

35. The loan operations of the REA are described in Barry P. Bosworth, Andrew S. Carron, and Elisabeth H. Rhyne, *The Economics of Federal Credit Programs* (Brookings, 1987), pp. 121–26.

includes processor control, an alternative that has all the features of a common control and is cheaper. For 1982, prices are available for switches with processor control and 35 percent digital trunking. The hypothetical office is built by Stromberg-Carlson on a negotiated contract and installed in Minnesota. The estimates and standard errors are shown in table 4. Full details are described in appendix A.

Figure 16 compares my estimates with those of Bellcore for the 2,500-line office with 35 percent digital trunking. Price changes in matched models are shown by solid lines; dotted lines show improved technology that has reduced cost while maintaining functionality. The estimated price for 1982 is virtually identical to the Bellcore estimate for a similar office.

THE TRANSITION TO DEREGULATION, 1982–85. The most significant question to ask in order to understand what drove technological change in communications equipment is, What impact did divestiture have on the cost of public switching equipment? To analyze this issue, I obtained procurement data covering 1982–85 on magnetic tape from the REA. Because these data were made available in a different and more detailed form than the data found in published reports, I used a somewhat different model to estimate quality-adjusted price changes (see appendix A). I used two second-order approximations (translog and quadratic) and two first-order approximations (loglinear and linear) to a multi-output cost function.

Table 5 reports the most relevant results. The price levels estimated at sample mean characteristics values using the four forms are close. Rates of change show greater variation, but the pattern remains relatively stable. Switch prices declined 1.5 to 7.5 percent in 1982–83, depending on functional form, and 13 to 15 percent in 1983–84. The decline in 1984–85 was a relatively steep 17 to 18 percent for the logarithmic forms and

Table 4. *Estimated Small Switch Prices, Selected Years, 1970–82*[a]
Thousands of current dollars

Switch	*Year*	*2,500-line office*[b]	*927-line office*[c]
Common control, ANI[d]	1970	477 (60.5)	213 (34.3)
Common control, ANI[d]	1976	725 (93.0)	309 (29.1)
Processor control	1976	634 (74.7)	263 (40.5)
Processor control	1982	799 (65.3)	374 (28.7)
Processor control, 35% digital trunks	1982	764 (65.3)	357 (29.2)

Source: Author's calculations.
a. Numbers in parentheses are approximate, heteroskedasticity-consistent standard errors.
b. Hypothetical office with a 12:1 line-to-trunk ratio.
c. Hypothetical office with 76 trunks; lines and trunks correspond to average in sample.
d. Automatic number identification.

Figure 16. *Comparison of Estimated Switch Prices with Bell Communications Research Data for a Hypothetical 2,500-Line Office, 1970–85*

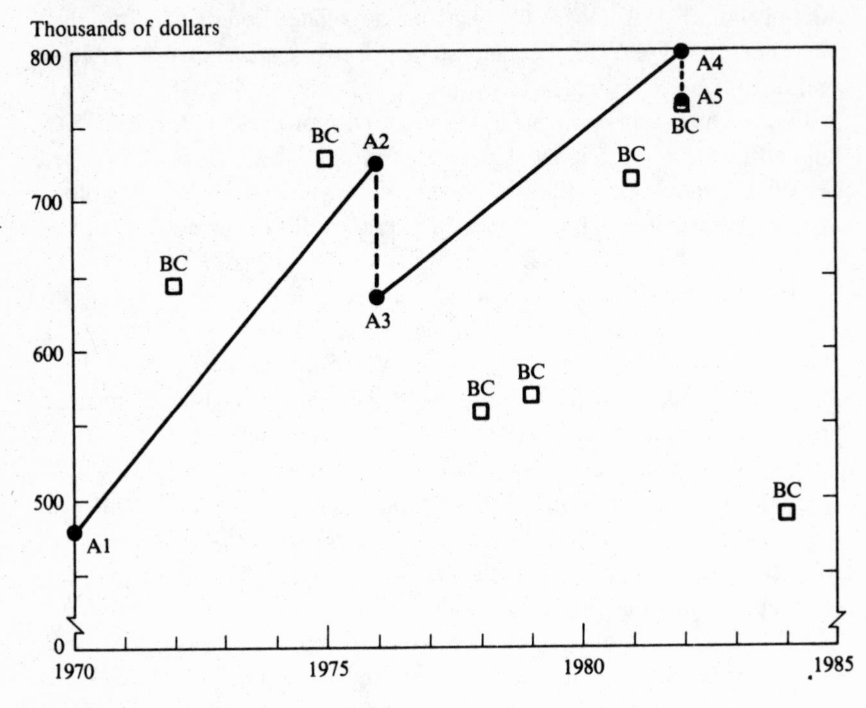

Legend

A1 Common control, automatic number identification, author's estimate

A2 Common control, automatic number identification, author's estimate

A3 Processor control, author's estimate

A4 Processor control, author's estimate

A5 Processor control, 35% digital trunks, author's estimate

BC Bellcore estimates

Sources: For A1–A5, author's econometric estimates for a hypothetical 2,500-line office with 35 percent digital trunking and a 12:1 line-to-trunk ratio; see text discussion. For BCs, Bell Communications Research historical price estimate for 2,500-line switch is described in appendix table B–1.

9 to 10 percent for the linear and quadratic approximations. Geometric rates of decline over the entire period were more tightly clustered at 9 to 12 percent.

Most interestingly, switch prices began a steep decline after divestiture in 1983. Depending again on the functional form used, the decrease either tapered off in 1984–85 or accelerated even further.

A similar analysis was undertaken for contract costs for remote-switching terminals. RSTs, essentially slave switches that operate under

Table 5. *Estimated Price Levels and Rates of Change of Small Central Office Switches, 1982–85*[a]

Year	Functional form			
	Quadratic	*Translog*	*Linear*	*Loglinear*
		Switch price (thousands of dollars)		
1982	458 (18.2)	451 (24.6)	476 (24.4)	455 (24.4)
1983	451 (16.2)	433 (23.2)	440 (17.1)	445 (19.8)
1984	382 (19.0)	371 (23.3)	384 (20.8)	378 (20.1)
1985	347 (19.9)	314 (19.1)	347 (18.5)	311 (18.0)
		Percent rate of change		
1982–83	−1.57 (3.15)	−3.96 (3.29)	−7.52 (3.78)	−2.11 (4.31)
1983–84	−15.36 (3.05)	−14.19 (3.41)	−12.88 (3.41)	−15.14 (3.36)
1984–85	−9.05 (4.66)	−15.45 (4.80)	−9.60 (3.97)	−17.62 (4.66)
		Compound growth rate		
1982–85	−8.84 (1.53)	−11.35 (1.42)	−10.03 (1.41)	−11.88 (1.69)

Source: Author's calculations.
a. Numbers in parentheses are approximate, heteroskedasticity-consistent standard errors.

control of a larger central office switch, are often used in sparsely populated but geographically extensive rural areas. Since they have only one measured characteristic, the number of lines under their control, approximating a cost function for them is much simpler than it is for standard switches. Table 6 shows estimated RST prices at the sample mean number of lines in 1982–85 for the same four functional forms as in table 5. Apart from generally slower decreases in cost, the results are similar to those for standard switches. In 1982–83 RST prices rose 4 to 5 percent. Immediately after divestiture they fell 7.5 to 9 percent, but either moderated or showed price increases in 1984–85. The compound annual rate of

Table 6. *Estimated Price Levels and Rates of Change of Remote-Switching Terminal (RST) Prices, 1982–85*[a]

Year	Functional form			
	Quadratic	*Translog*	*Linear*	*Loglinear*
		RST Price		
1982	137 (7.56)	133 (8.75)	128 (7.56)	151 (11.65)
1983	143 (7.32)	140 (8.35)	133 (7.42)	138 (9.76)
1984	130 (5.23)	127 (6.30)	122 (5.41)	128 (7.50)
1985	132 (6.34)	120 (6.30)	125 (6.52)	120 (8.32)
		Percent rate of change		
1982–83	4.02 (3.84)	5.24 (4.69)	4.36 (4.13)	−8.33 (4.73)
1983–84	−8.92 (3.61)	−9.21 (4.14)	−8.18 (3.94)	−7.51 (4.90)
1984–85	1.45 (3.93)	−5.90 (4.00)	2.56 (4.32)	−6.18 (4.81)
		Compound growth rate		
1982–85	−1.31 (1.37)	−3.48 (1.65)	−0.62 (1.49)	−7.34 (1.91)

Source: Author's calculations.
a. Numbers in parentheses are approximate, heteroskedasticity-consistent standard errors.

decrease was 1 to 3.5 percent for the preferred second-order functional approximations.

INTERPRETATION. The close fit between these price estimates and Bellcore estimates for 1976–82 indicates that my methodology offers reliable estimates of quality-adjusted prices for small switches. However, a comparison of results after 1982 suggests that estimated prices show some sensitivity to the functional form used.

By 1982 at least, Bell System purchases of small switches appear to have been made at market prices, as measured by what customers of the Rural Electrification Administration were paying for comparable equipment from non-Bell suppliers. These estimates, along with the fragmentary data presented earlier, suggest that prices for small switches rose at moderate rates through the 1970s and early 1980s.[36] After 1982, however, technological progress in small switches seems to have accelerated. Prices probably declined modestly in 1982–83; they plunged sharply in 1983–84, and then continued to fall in 1984–85, probably because of the increased competition accompanying divestiture.

The behavior of prices for remote-switching terminals merits further exploration. As was true for standard switches, a relatively steep price decrease in 1983–84 followed price increases in 1982–83. Estimated changes in 1984–85 vary from a small increase to a 6 percent decrease, depending on the functional form specified.

Why did RST prices seem to fall much less than central office switch prices when presumably they draw on the same technology and are sold in the same markets? There are at least two possible reasons. For one thing, RSTs will usually work only with a proprietary switch design produced by the same manufacturer and specifically built to accept a link to a particular RST. The RST market, then, is captive once a manufacturer for central office switches is chosen, and there may be some insulation from competitive pressures on price. Also, RSTs are controlled by more complex software running on the central office switch. The cost of producing software is generally believed to have fallen very slowly, and the additional software cost may have been bundled into the price of the RST, thus reducing the decrease when compared with the stripped central office switch.

Finally, all these estimates suggest rates of decrease far smaller than those experienced by computer systems. From 1982 to 1985 these cal-

36. The Bellcore data described earlier suggest that large switches may have behaved quite differently, with the introduction of a digital switching matrix in the 4-ESS sharply reducing costs after the mid-1970s, followed by a resumption of the upward trend a few years later.

culations give a 9 to 12 percent compound annual rate of decrease in prices for small switches at the sample mean configuration, far less than the 20 percent or so measured for computers. However, the evidence also suggests that the trend after 1983 has been toward steeper declines, so the possibility remains that switch and computer price decreases will be more alike in the future.

CAVEATS AND LIMITATIONS. There is a potential problem, which appears in several guises, in the estimates on switching costs. Not all technical characteristics of switches relevant to producers and their customers may be reflected in the limited measurements of my data set. Two factors are particularly important in assessing the price decreases of the program-controlled electronic switches: the reductions in the costs of maintaining and repairing them and the new kinds of features and services they make possible.

It is widely held that electronic switching systems reduce routine maintenance and operating costs. Estimates from within the Bell Telephone Laboratories and GTE suggest that the transition from electromechanical to stored program analog switches reduced annual maintenance and operating costs by 6 to 7 percent of the purchase cost of a switch, while the changeover from stored program analog to stored program digital switches dropped annual maintenance and operating costs by 2 percent of the purchase cost.[37] Based on a 10 percent discount rate, and a service life of ten or twenty years, the present value of such cost savings expressed as a percent of first cost is

	Discounted lifetime cost savings as percentage of first cost	
Annual cost savings as percentage of first cost	*10-year service life*	*20-year service life*
1	6	8.5
2	12	17
6	37	51
7	43	60

37. The GTE estimates come from Terence Robinson, "Depreciation Reserve Assessment," in Albert L. Danielsen and David R. Kamerschen, eds., *Telecommunications in the Post-Divestiture Era* (Lexington, Mass.: D.C. Heath, 1986), table 19-1. They pertain to operating costs as a fraction of purchase price for digital (15 percent), analog (16.9 percent), and electromechanical switches (23.3 percent) in 1984, when the article was written. Bell Laboratories' estimates refer to differences in maintenance costs in 1978 as a fraction of purchase price, and have been calculated by me from data published in Ronald A. Skoog, ed., *The Design and Cost Characteristics of Telecommunications Networks* (Murray Hill, N.J.: AT&T Bell Laboratories, 1980), table 10-1. I have assumed all switches are used at full capacity, with an average 18 CCS

By far the largest savings occurred after 1965, when stored program control replaced electromechanical systems; maintenance costs were reduced by an amount equal to 40 to 60 percent of initial cost. The changeover from an analog switching matrix to a digital system in 1976 resulted in savings of 10 to 20 percent of initial cost. The savings come from cheaper and much more reliable components and easier and cheaper diagnosis and repair of faults. These savings, however, occurred only when the new technologies were introduced. Once the changeover had been completed, lower maintenance costs were probably not a major influence on switching costs.

The cost savings on operations and maintenance were probably as important a factor for transmission systems as for switching systems. Annual operating costs for transmission systems can be estimated separately for the electronic and nonelectronic portions of the system; operating costs on the electronic portion are generally lower.[38] Precise calculations will vary with the exact technology examined, the length of haul, and the cross section, but significant reductions were definitely associated with use of microwave radio. Microwave radio had much lower operating costs than carrier on coaxial for the nonelectronic part, had roughly similar costs for the electronic part, and effectively substituted electronic for nonelectronic components in its cost structure.

A more complicated issue is the effect of new and better features and services on prices for digital switches. More and more of the cost has depended on improved software for programmable switches, and much of the added software has been related to new features. It is helpful to separate new features into outside services (including Centrex and custom calling services such as call forwarding and speed calling) that are offered to customers, and inside services (such as the ability to use the Common Channel Interoffice Signaling System to set up calls, route calls to operators, or dynamically reroute traffic) that enable more efficient management and use of the network by service providers. The cost of inside services is never directly seen on a customer's bill, but the cost of outside services is.

The features that allow better use of network capital presumably should be reflected in greater numbers of calls transmitted and switched for any given investment in network. Thus they ought to have shown up in the

---

load per line. Under these assumptions, annual maintenance costs go from 13.9 percent of purchase price for a no. 4 crossbar switch (electromechanical), to 6.6 percent for a 1-ESS (stored program control analog), to 5.0 percent with a 4-ESS (stored program control digital).

38. Skoog, ed., *Design and Cost Characteristics*, tables 6-1, 6-2, 6-6, 6-7.

productivity calculations in appendix A, which estimated incremental capital cost for each increase in transmission or switching capacity. Better switching software will show up as more traffic switched over the installed circuit miles of transmission plant. Furthermore, many additional improvements went into maintenance and testing software, and the effects showed up in maintenance cost, which has already been assessed.[39]

Also, some of the most important of these features, such as local automatic message accounting (LAMA), ANI, firebar capability, and digital and specialized trunk capability, were controlled for in the econometric approach outlined in appendix A. Thus the returns to this aspect of additional software investment should have been captured in at least some of the above analysis. Costs of the newer custom calling services were largely captured in the switch characteristics used in the econometric analysis of the REA data and were removed when a stripped switch cost was calculated.

That revenues from these outside services appear to be a small fraction of service revenues for local operating companies raises an interesting question: do all characteristics of a product necessarily have significant value to the customer?[40] A gold-plated computer might be pretty, but few users would regard its additional cost as worthwhile. In a free market, of course, few would be sold, and the question of case color would be irrelevant in analyzing improvements in computer price performance. (If any were sold, then clearly one must control for gold-plating in an analysis of computer prices.) In an equipment industry that is captive to a regulated customer, however, there is a possibility of gold-plating to increase the cost of the capital good, particularly when a fixed, regulated rate of return determines the total return on an investment in a new plant.

In short, unmeasured dimensions of technical improvements in switching systems have probably caused gains in price performance to be underestimated. However, the same problems also afflict such measures in computer hardware. Maintenance and operating costs have dropped with successive generations of components and diagnostic technology in com-

39. In the 1-ESS switch, for example, maintenance and plant applications accounted for about 45 percent of the code developed. See Schindler, Jr., ed., *Switching Technology*, p. 285.

40. In 1977 about 1.5 million lines out of a total of 75.1 million used these services in the Bell System. See ibid., p. 412; and U.S. Telephone Association, *Telephone Statistics, 1987* (Washington, D.C.: USTA, 1987), p. 2. I could find no published data on the fraction of revenues accounted for by these services, but sources in the industry suggest that the percentage is small. If demand for the services is relatively inelastic, of course, it is possible that the social return to providing them is high, so that even though revenues are low, substantial investments in making the services available might be socially worthwhile—they are then being subsidized by users of basic service.

puters. Increasingly complex operating systems often come bundled with computers, and software development costs have soared.[41] New and different features and applications have also been an important factor in greater computer use. All the caveats applicable to switches are also applicable to computers, and if the objective is a comparison between the two, the measures constructed here are reasonably reliable.

Summary

Few data are available that can be used to measure quality-adjusted prices for communications equipment directly for the years considered here. A handful of equipment models and technologies was developed for internal use within a regulated monopoly, then upgraded and improved in ways difficult to quantify. Nevertheless, one can arrive at some cautious conclusions based on the broad outline of the historical record.

First, like other indexes based on matched models, the Bell System TPI methodology does not deal effectively with goods undergoing rapid technological change. Its portrayal of prices is not consistent with the fragmentary evidence on the rate and direction of technological change in transmission and switching equipment. At best it seems to capture only intragenerational improvement in price performance.

Second, available evidence indicates much slower technological progress than was the case with computers.

Third, transmission equipment seems to have become cheaper faster than switches, especially in the late 1940s and 1950s. In spite of their resemblance to computers, switches have decreased in cost rather slowly, with the sharpest decline in 1976–79 when the Bell System introduced a large toll switch with both stored program control and a time-division electronic switching matrix. Because the older space-division switches used electromechanical switching networks, the decrease in prices may have reflected a slower rate of technological progress in electromechanical technologies, rather than the steep decrease in electronics costs that has driven computer prices. Also, Western Electric's widespread adoption of stored program control in its product line did not take place until the mid- to late 1970s.

41. For example, Phister calculates that about 46 percent of the development cost for a $100,000 processor in 1974 was software costs. See Montgomery Phister, Jr., *Data Processing Technology and Economics*, 2d ed. (Bedford, Mass.: Digital Press, and Santa Monica Publishing Co., 1979), p. 514.

Fourth, although the pace of technological progress in communications equipment has picked up since divestiture in 1983, the price performance of transmission technologies seems to have improved as rapidly as that of digital switches. There is little indication that switches are currently becoming cheaper relative to costs for transmission.

Finally, because the TPIs seem to have underestimated technological advance, distortions in economic analyses based on TPIs may have occurred. Since TPI-based price indexes are often used to estimate capital investment in communications services, this investment is probably underestimated as well. The observation that telephone services are a constant and shining exception to an otherwise dismal picture of productivity growth in the American economy would therefore seem to be based on overestimates of productivity improvement in communications services and underestimates of technical improvements in communications equipment.

Divergence of Computer and Communications Equipment Costs

To explain why technological advances in computers have outpaced those in communications hardware is a speculative task. The rate of diffusion of new transmission and switching technologies does not appear to have been slower in the United States than in Europe and Japan. In fact, the diffusion of digital transmission systems in the Bell network set the pace for the world.[42] Similarly, in electronic central office switches, Bell again led with the introduction of the 1-ESS in 1965. By 1976 the Bell network had hooked far more of its lines and trunks to such switches than had communications companies in other countries.[43] Only in adopting digital switching, pioneered in France, did Bell lag behind its domestic and international competitors.[44]

So, an explanation of the relatively slow rate of technological progress must go beyond factors specific to the United States. Furthermore, it was not for lack of investment in developing the relevant technology. Among civilian industries, U.S. communications equipment producers invest a share of their sales in R&D second only to computer manufacturers. And

42. See M. Robert Aaron, "Digital Communications—The Silent (R)evolution?" *IEEE Communications*, vol. 17 (January 1979), pp. 22–25.

43. See Amos E. Joel, Jr., ed., *Electronic Switching: Central Office Systems of the World* (New York: IEEE Press, 1976), app. B.

44. Ibid.; and Telephony/Market Research, *U.S. Carrier Spending in 1986: Wooing Business Users* (Chicago: Telephony Publishing Corp., 1986), fig. 3.

a high proportion of that R&D expenditure—three to four times as much as that of computer manufacturers—goes toward basic research. These firms (presumably Western Electric and the Bell Labs accounted for most of this basic research) accounted for 17 percent of all basic research performed by U.S. industry in 1979, compared with 9.5 percent of all R&D and less than 3 percent of all sales by R&D-performing firms.[45]

One explanation for the slower rate of progress is the effects of regulation and market structure. Since foreign communications firms have also usually been regulated monopolies, such an explanation is consistent with their experience as well. Two causes for the dampening effect of regulation on the introduction of technological innovations seem relevant: the Arrow effect and the effects of regulatory oversight of rates of return.[46]

In 1962 Kenneth Arrow showed that, all else being equal, an unchallenged monopoly has less incentive to innovate than a firm that, by virtue of a technical innovation, can acquire some monopoly power in a previously competitive market.[47] The telephone industry was, before divestiture, a reasonable approximation of just such an unchallenged monopoly.[48] Hence the Arrow effect might account for the slower introduction of innovations than was the case in the more competitive computer industry, even if the innovations were developed from the same basic technology.

Regulatory policy may offer an alternative explanation for sluggish innovation. Because telephone companies' rates were set to reduce a fixed rate of return on investment, the companies were implicitly encouraged to adopt excessively capital-intensive technologies. They also had no incentive to limit costs, such as those for research and development or capital investment. When a new technology was being developed, there would have been a tendency to select the most expensive choices in equipment design. Excessively durable, complex, and sophisticated machinery would be built, and one would expect a tendency to "gold-plate" equipment put into use. It is hard to say what effect this might have on the incentive to

45. National Science Foundation, *Research and Development in Industry, 1983* (GPO, 1985), pp. 34, 43. The absolute amount of basic research performed ($195 million) was matched only by industrial chemicals ($197 million) on a sales base that was more than 50 percent greater than in communications.

46. Both arguments are explored in somewhat different terms in Shepherd, "Competitive Margin in Communications."

47. See Kenneth J. Arrow, "Economic Welfare and the Allocation of Resources for Invention," in National Bureau of Economic Research, Special Conference Series, *The Rate and Direction of Inventive Activity: Economic and Social Factors* (Princeton University Press, 1962), pp. 609–25.

48. From time to time, of course, the Bell System did respond to competitive threats—for example, microwave transmission systems built by competitors in the 1950s and 1960s. But the regulatory framework made such challenges the exception, not the rule.

introduce particular innovations; that would depend on whether a new technology was more or less capital-intensive than its predecessors.

The indirect effects of regulation may also have been more predictable than the direct bias. Regulated firms have an incentive to push for lower statutory depreciation rates on their capital equipment, since such artificially low rates raise the net book value of investment, on which a return is calculated, above the reproduction cost of capital assets. This will raise the real return to stockholders. Rapid innovation increases the economic depreciation rate that ought to be charged against capital.[49] Thus if regulatory bodies allow relatively slow depreciation rates, the book value of existing, undepreciated equipment taken out of service will exceed its economic value, and rate-setting procedures using returns on book investment will reduce returns actually received when older equipment is replaced.

Suppose, for example, a new technology involves replacing old machinery with a new machine costing the same and producing an equivalent level of output. In terms of capital per unit of output, then, there will be no incentive to pick a more capital-intensive technology to increase absolute profits. The economic advantage of the new technology may be that it requires less use of other inputs than the old method does to produce equivalent output, but there will be no particular incentive to adopt or not adopt it if a regulated, fixed rate of return applied to the capital stock determines earnings.

Assume further that both old and new machines require identical annual depreciation and maintenance investments and have another, smaller depreciation rate that regulatory authorities use to compute allowed return on investment. Replacing the old machine with the new technology, because it writes down the book value of the rate base for all future years, reduces the present discounted value of future profits (see appendix C at the end of this volume). A profit-maximizing public utility would then have an incentive to stick with the older technology, despite its economic inferiority.

Historically, rates of capital equipment depreciation in the telephone industry have been very low. In the computer industry, new products have typically had a life expectancy of six or seven years. By contrast, in the

49. By increasing the user cost of capital, investment is deferred until later periods, when capital goods are more productive. For a useful discussion of optimal depreciation rates with technological progress, see William J. Baumol, John C. Panzar, and Robert D. Willig, *Contestable Markets and the Theory of Industry Structure* (Harcourt Brace Jovanovich, 1982), pp. 384–89.

late 1960s the FCC-approved service life for most central office communications equipment was about twenty years, and forty for electronic switching systems.[50]

But are regulation and monopoly the best explanations for the tepid rate of technical progress in communications? My estimated, quality-corrected prices for equipment supplied by the competitive fringe of equipment suppliers to small independent telephone companies in rural areas are remarkably close to Bellcore's estimates of prices for small exchanges procured within the Bell System. The similarities exist both in absolute price levels and in trends. If lack of competition were responsible for sluggish decreases in price, one might ask, shouldn't prices have fallen more quickly in the more competitive, albeit much smaller, rural switch market?

Perhaps not. The competitive fringe may not, after all, have been so competitive. Many of the equipment suppliers selling to this market may have taken the prevailing price as offered, with giant Western Electric setting that price. Also, equipment producers affiliated with service providers were sometimes barred from selling at lower prices than those they charged their large, regulated customers. Market structure and regulation might thus indeed be a reasonable explanation for the tepid rate of progress.

Another explanation for communications systems' slower rates of price decrease rests on fundamental technological differences with computers. It might be argued that computing systems mainly race against time; improvement means accomplishing the same task cheaper and faster. But improvement in voice communications is a battle waged primarily against distance; the challenge is to move information farther and more cheaply. Progress in the conquest of distance may not technically be highly correlated with advances in speed. Certainly, improvements in transmission technology rather than switching speed have been responsible for most progress in communications.

Finally, the return on relatively fixed investments in research and development is limited by the size of the market in which improvements can be sold. Communications switching systems constitute a small and specialized market. In the early 1960s the market for central office equipment was about one-half the billion-dollar market for computers. Today it is perhaps one-tenth. The justification for huge R&D investments in

50. Shepherd, "Competitive Margin in Communications," p. 108. The logic sketched out here seems confirmed by recent events. In December of 1988, AT&T announced a $6.7 billion charge against earnings to pay for the replacement of obsolete equipment. See Calvin Sims, "Big Change in Updating at AT&T," *New York Times*, December 2, 1988.

switch development is limited by that very specialized small market. Computers, however, are general-purpose tools whose market has been constantly expanding. And although specialized computers with small markets may require some unique R&D expenditures, such computers capitalize on existing, much larger investments in R&D to improve the technology of general-purpose computers.

Fundamentally, the communications industry is a user of computers, much as the machine tool, auto, chemical, or aircraft industries. In the past the special requirements of computers used to control switching networks may have argued for the development of special-purpose systems.[51] Focusing on such machines, however, limits the R&D investments that might be justified and may have slowed technical progress relative to that for general-purpose computers. Limited opportunities to develop technology for a much larger market—in part the product of regulatory constraints—may have slowed down the rate of progress.

Thus an absence of competition, the details of the regulatory process, the regulatory limitations placed on markets, and the details of technology itself stand out as the most likely explanations for differences in the rates of improvement of computer and communications systems. In computers, technological improvements in either system design or components clearly dominate any change in price derived from labor, capital, or commodity inputs. If computers and communications systems were as tightly linked by a common technology as popularly thought, one would expect to observe analogous rates of technological advance reflected in similar decreases in prices. Since the decreases were not similar, either the technological bases are not as intimately intertwined as thought or market structure and regulation restricted the rate of innovation in communications.

The regulatory isolation of communications producers in small, specialized markets where economies of scope could not be realized, or the producers' insulation from competitive pressures to introduce new technology quickly, may thus be root causes of sluggish price decreases. The apparent acceleration of the decreases since 1983, despite the relatively limited entry by communications equipment producers into new markets, argues for regulatory structure as the main factor.

51. Extreme reliability and fault-tolerance requirements may explain why the absolute price for switching processors is much higher than for commercial computers, but they do not explain the slow rate of decrease in price, since these requirements are generally thought not to have changed a great deal over the years.

Policy Issues

The deregulation of the communications network has left three major questions unresolved. First, to what extent should firms occupying the regulated niches in the system be allowed to compete in unregulated markets? Second, how will divestiture affect research activities at the Bell Telephone Laboratories, historically a major and quasi-public source of U.S. advances in information technology? Third, how will future technological and regulatory constraints interact with economic forces to determine the architecture of the U.S. communications network?

The past and future pace of technological progress is connected to all three questions. Arguments for allowing regulated service providers to enter into unregulated markets are often based on supposed economies of scope flowing from investments in technology. Changes in the regulatory structure may fundamentally alter the mechanisms producing R&D investment and alter the speed of innovation in the network. Those who see ongoing upheavals in the structure of the network base their vision on differences in the pace of technological innovation in different components of the network.

Entry by Regulated Firms into Unregulated Markets

Without considering markets other than computers (such as information services), this paper raises doubts about the significance of economies of scope across markets for computer and communications systems. Certainly AT&T's lack of success in marketing general-purpose computers would seem to confirm these doubts.[52] Nor have computer firms such as IBM been conspicuously successful in the communications equipment marketplace. And large software firms rarely enter the hardware business. The burden of proof would thus seem to be on those who argue that the potential for economies of scope justifies allowing the local operating companies to enter general-purpose computer hardware and software markets (as opposed to special-purpose communications products directly linked to their regulated businesses).

Other rationales—for example, potential benefits to managerial efficiency in regulated lines of business from the fresh air of competition in

52. Though one might argue that the natural entry point for AT&T would have been to branch out into the specialized computer applications that most resemble the particulars of its switch expertise—fault-tolerant systems oriented toward real-time control of large numbers of inputs and outputs, digital-analog signal processing, or transactions processing—rather than the general-purpose computing emphasis actually chosen.

unregulated markets—may well prove more convincing. But technological convergence and associated economies of scope do not seem to leap out at the observer from the historical record.[53]

Monopoly and Innovation

Historically, communications equipment manufacturers—particularly AT&T through the Bell Laboratories and Western Electric—have spent large amounts on R&D, especially basic research. The usual presumption is that basic research is less easily appropriated for private gain than more applied types of R&D. But the existence of a regulated monopoly may have encouraged both the high levels of R&D expenditure—through fixed percentage levies on revenues of the operating companies—and the emphasis on basic research, because intraindustry returns would certainly be captured by the monopolist. In a more competitive environment, the results of such basic research might be appropriated without cost by a competitor.

There have, however, been suggestions that both an unchallenged monopoly and regulation of rates of return might slow the diffusion of innovations. Paradoxically, then, the old market structure might actually have both increased basic research and slowed innovation. And deregulation and increased competition might step up the pace of innovation yet reduce spending on basic research. This may be happening. The share of basic research in all R&D in the communications equipment industry seems to have declined sharply between 1979 and 1985, after divestiture began to take shape (figure 17). Basic research by communications equipment producers as a share of all basic research performed in American industry registered a similar decline.

The Future of the Network

One point firmly established here is that to identify digital switches with computers is misleading. The otherwise sober Huber Report makes this leap when it argues that "in the past fifteen years . . . revolutionary developments in electronics have slashed the costs of switching and other forms of network intelligence."[54] Huber argues that the fundamental net-

53. Ideally, one would like to estimate empirically the extent of economies of scope using an econometric model. Unfortunately, the history of regulation and the consequent separation of the producers of computers and the producers of communications systems eliminate variation in the data needed to produce reliable estimates. Nonetheless, it may be possible to undertake such an exercise examining the relation between computers and data communications products.

54. Huber, *Geodesic Network*, p. 1.3.

Figure 17. *Basic Research in Communications Equipment as a Share of All Communications R&D and of All Industrial Basic Research, Selected Years, 1974–85*

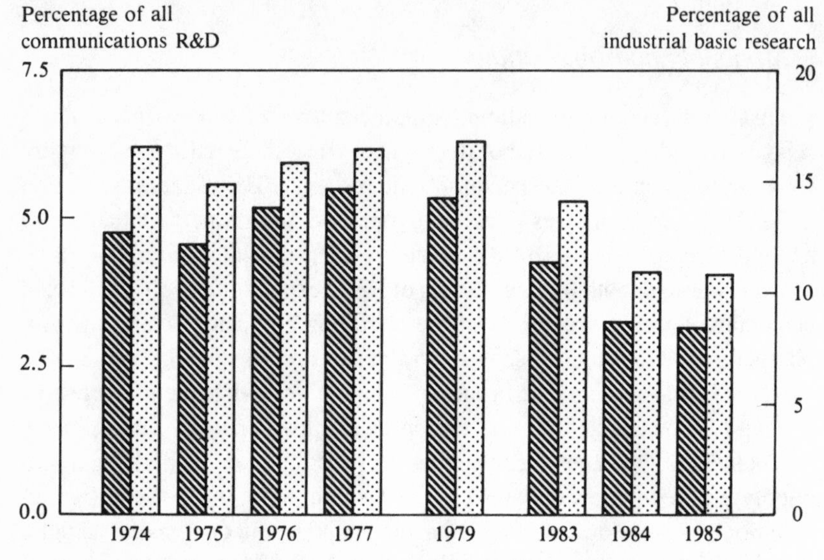

Source: Data supplied by National Science Foundation, Division of Science Resources Studies.

work trade-off between trunking and switching has been tilted toward switching by this revolutionary cost decline. His observation that local, low-density transmission costs have not much changed and are not likely to do so leads him to conclude that as switch costs decline, switching will move out toward the user so that communications firms can economize on relatively more expensive transmission.

But for the network as a whole, the historical record refutes these assumptions. (Indeed, even if true, they are not sufficient to guarantee cost advantages for the more decentralized network.)[55] Transmission costs have dropped much faster than switching costs and, with the advent of fiber optics, have continued to do so.[56] Rather than shifting toward a much

55. If there are no economies of cross section in transmission, a geodesic ring architecture will always be more expensive than a centralized star architecture, regardless of the relative costs of switching and trunking. Even with substantial economies of cross section, superiority of the geodesic architecture is not guaranteed. See appendix D at the end of this volume.

56. This point is also made by Lee L. Selwyn and W. Page Montgomery in their consulting report *Factual Predicates to the MFJ Business Restrictions: A Critical Analysis of the Huber*

less hierarchical, decentralized network model as Huber envisions (the geodesic network), intermediate layers of switching may instead be displaced by cheap transmission.

One might argue that this is currently happening, that so-called local bypass essentially means cheap transmission systems linking clusters of users directly to the national network, eliminating the layers of expensive switches controlled by the local operating companies. Cheaper high-capacity local transmission links may also ultimately replace today's relatively expensive low-density local distribution system. In a pattern more typical of growth in the computer industry, users might begin to consume cheap bandwidth in prodigious quantities, finding new applications for the video, graphics, voice, and data that can be piped in to their local terminals. Since real-time transmission of high-resolution graphics can absorb huge amounts of bandwidth (a billion bits per second are needed for high-resolution color images), today's excess fiber-optics capacity could easily be absorbed as services designed to use low-cost bandwidth come on line.

In short, AT&T may not have "chosen to surrender the heart of its old network for permission to participate fully in building the new one."[57] Instead, it may have recognized that the local switching systems that once fed its central toll system were endangered by the continued cheapening of transmission and trunking. The core toll network it retained may eventually reach the user through fewer intermediaries. AT&T may in fact have decided that the services sold through the network of tomorrow will be accessed through the center, not the periphery. Today that center remains firmly within its grasp.

Report (Boston: Economics and Technology, 1987), pp. 16–18. Selwyn and Montgomery, however, base their conclusions on the Bell System TPI data.

The reply by the Department of Justice to these comments noted that the new higher-capacity transmission technology has not yet reached out to the lowest-level local links in the system. But this does not mean that it will not eventually. See "Response of the U.S. to Comments on Its Report and Recommendations concerning the Line-of-Business Restrictions Imposed on the Bell Operating Companies by the Modification of Final Judgment," U.S. District Court for the District of Columbia, civil action 82-0192, April 27, 1987, pp. 65–66.

57. Huber, *Geodesic Network*, p. 1.6.

U.S. Interexchange Competition

Leonard Waverman

THIS PAPER focuses on the emerging structure of the U.S. inter-LATA (local access and transport area) interexchange market and on the degree of regulatory supervision over four types of interexchange competitors: AT&T Communications; other facilities owning interexchange common carriers (OCCs); other market participants, such as firms that lease bulk services from AT&T Communications (or the OCCs) and resell this capacity; and potential entrants currently not allowed into the interexchange market (the Bell operating companies, or BOCs).

With 70–75 percent of interstate domestic revenue but only 40 percent of interexchange capacity, AT&T Communications is regulated by the Federal Communications Commission (FCC) on a rate-of-return basis, with tariff filings, required approval for construction, hearings, and the full panoply of regulatory strictures.[1] The FCC proposed in August 1987 to replace strict rate-of-return regulation of AT&T Communications with a set of "price caps." The FCC forbears from regulating AT&T Communications' competitors, including the OCCs; such forbearance will presumably continue if AT&T Communications is subject to price caps. The BOCs are forbidden from offering interexchange service according to the 1982 terms of the modified final judgment, or MFJ, to the 1956 AT&T consent decree,[2] but the BOCs currently have a virtual monopoly of intra-LATA interexchange.

I thank Melvyn Fuss, Mike Denny, Robert Crandall, Jeff Rholfs, and Michael Crew for their comments and Arthur Wilson and Stephen Murphy for their able research assistance. I am also grateful to David Evans for pointing out a misinterpretation present in an earlier version.

1. Before the recent proposed changes to price caps, AT&T Communications had been regulated on a stricter basis than the old AT&T: its maximum rate of return was by service category rather than overall interstate earnings; if over a two-year period its earnings exceeded the maximum return, its prices were automatically reduced.

2. *United States* v. *American Telephone & Telegraph Co.*, 552 F. Supp. 131 (D.D.C. 1982), *aff'd sub nom Maryland* v. *United States*, 460 U.S. 1001 (1983). On September 10, 1987, Judge Greene held that these restrictions would continue to hold for the foreseeable future. *United States* v. *Western Electric Co.*, 673 F. Supp. 525 (D.D.C. 1987).

Substantial capacity additions have taken place in the interexchange market, principally through large investments in fiber-optic routes and networks by recent entrants into the market. Interexchange capacity in 1988 is some three to four times that which existed seven years ago, and it may double in several years because of technological advances in signaling in existing fiber-optic channels.[3] Besides capacity additions by AT&T Communications and the OCCs, several new additions to capacity have been made by firms that are "carrier's carriers"; that is, firms that provide interexchange capacity but do not retail that capacity to final customers.[4] Besides these market participants, all of which have "sunk" investments, a growing number of firms lease capacity or services from a facility carrier (normally AT&T Communications) and repackage or resell these services or provide value-added services in addition to capacity.[5]

As stated, the seven Bell operating companies are expressly forbidden from offering interstate interexchange service. It is thus fair to say that disequilibrium exists in the interexchange market, with fundamental changes under way in the means of connecting distant cities and with asymmetric regulation of market participants.

This paper addresses the appropriate form of future regulation, if any, over competing firms in the interexchange market. The need for regulations such as entry or exit restrictions and pricing limits depends on the underlying technological conditions of providing interexchange service—the economies of scale in interexchange, and the economies of scope between interexchange and other telecommunications services. The second section of the paper therefore examines the technological conditions of production, with focus on the meaning, evidence, and interpretation of economies of scale and scope and on the measurement of appropriate standards for pricing rules, standards such as incremental costs and stand-alone costs. That section is followed by a discussion of the needs for regulation and the alternatives available. The last section provides a list of conclusions.

3. Porter states that "with current electronics technology, volume in 1990 will fill less than 25% of available [fiber] capacity. With newly available 1.7 gigabit-per-second technology it will fill no more than 6–7%." See Michael E. Porter, "Competition in the Long Distance Telecommunications Market: An Industry Structure Analysis," Monitor Company, Inc., Cambridge, Mass., October 1987, p. 12.

4. These include Fibretrak, Lightnet, and Electra. See William B. Johnston, "The Coming Glut of Phone Lines," *Fortune*, January 7, 1985, pp. 96–100.

5. The FCC allowed the resale and sharing of all AT&T services except message telecommunications services and wide area telecommunications services in 1976. In 1981 MTS and WATS resale and sharing were permitted.

Before these issues are addressed, however, it is useful to describe briefly the structure of the telecommunications market just before and after divestiture.[6]

The Interexchange Market before and after Divestiture

In 1981 the telecommunications market in the United States consisted of several large, vertically integrated common carriers, the largest of which, AT&T, had 83 percent of industry revenue (local and toll). AT&T served 80 percent of the telephones in the United States through a series of operating companies that offered both exchange and toll services (as well as ancillary services, such as the Yellow Pages). AT&T's Long Lines was the interstate toll arm of the system; Western Electric was the manufacturer of equipment. Through various consent decrees with federal antitrust authorities, constraints were placed on the operations of various pieces of the Bell system.

Although entry by OCCs into the long-distance market was allowed, AT&T had shown both a willingness to use its monopoly over local access to frustrate entry and the ability to price discriminate in order to compete with entrants.[7]

The 1982 modification to the 1956 consent decree (the MFJ), agreed to by AT&T and the Department of Justice, was designed to eliminate many of the constraints on AT&T's activities while satisfying Justice's two major concerns: eliminating cross subsidies from monopoly services and preventing the use of bottleneck facilities to retard competitive entry (vertical foreclosure). The intent of the MFJ was simple—to separate the interexchange portion of AT&T from the local-access monopolies. The BOCs were divested from AT&T Communications, which retained its Long Lines and manufacturing arm, Western Electric.[8]

6. A description of the history of the interexchange market is given in Leonard Waverman, "Domestic Competition in the U.S. Interexchange Market: 1900–1986," paper presented at the Sixth International Conference on Forecasting and Analysis for Business Planning in the Information Age, Tokyo, December 2, 1986.

7. See Gerald W. Brock, *The Telecommunications Industry: The Dynamics of Market Structure* (Harvard University Press, 1981), chap. 8; Gerald R. Faulhaber, *Telecommunications in Turmoil: Technology and Public Policy* (Ballinger, 1987); and Waverman, "Domestic Competition."

8. For a discussion of the Department of Justice's position and details on the exact division of AT&T, see Timothy J. Brennan, "Regulated Firms in Unregulated Markets: Understanding the Divestiture in *U.S.* v. *AT&T*," Economic Analysis Group Discussion Paper 86-5 (Washington, D.C.: U.S. Department of Justice, Antitrust Division, April 1986). See also Faulhaber, *Telecommunications in Turmoil.*

The judgment required the BOCs to offer equal access to all interexchange carriers by 1987 and to refrain from offering interexchange service. Four percent of the BOCs' lines had been converted to equal access as of December 31, 1984; by December 31, 1985, the figure was 50 percent; by December 31, 1986, 74 percent had been converted; and by the end of 1987 the figure was 84 percent.[9]

Table 1 presents data on AT&T's Long Lines revenue (nominal) in selected years from 1951 through 1983, and scattered revenue data for the OCCs from 1975 through 1983. In 1951 AT&T offered six basic services and received $602 million from message telecommunications service (including revenue from international calls) and $82.3 million from the other five services—MTS teletype, private-line telephone and teletype, and audio and video program services. In 1958, the year before the "Above 890" decision,[10] the same six services were offered—with MTS earning $1.195 billion (85 percent of total Long Lines revenue) and the other five services earning $219 million.

In 1961 three new "services" were offered by the BOCs—WATS, Telpak, and "other." These new services did not represent technological changes or advances but tariff filings—price discrimination to compete with the newly licensed private microwave systems. Many of AT&T's "competitive" responses in the postwar period were changes in prices.[11] For example, in 1965 MTS revenues were $2.8 billion, or 83 percent of total Long Lines revenue; WATS was the second-largest revenue earner ($163 million); and Telpak (bulk private lines) was the third-largest revenue source ($115 million), some 30 percent of all private-line revenues.

By 1983 AT&T's private-line revenue was $3.4 billion, WATS revenues were $5.9 billion, and domestic MTS was $15.4 billion. In that year all the other interexchange carriers had combined service revenues of just $2.2 billion.[12]

Table 2 provides more complete data on the relative size of interexchange carriers in the 1981–87 period. These data are not definitive on

9. Common Carrier Bureau, Industry Analysis Division, "Trends in Telephone Service," Federal Communications Commission, Washington, D.C., January 1988, p. 14.

10. *Allocation of Frequencies in the Bands above 890 Mc.*, 27 FCC 359 (1959).

11. This point should be kept in mind when considering the concept of "contestability" and its applicability in describing the incumbent's (AT&T's) behavior.

12. The OCCs argued successfully that it was unfair that customers who made no choice of carrier were automatically routed to AT&T. Therefore, on "election day" in May 1985, customers were allowed to choose an interexchange carrier of first choice, one whose services would automatically be used. On election day, 30 percent of customers failed to make a choice; of those choosing, 80 percent picked AT&T, 10 percent MCI, and 4 percent Sprint.

Table 1. *AT&T's Long Lines Nominal Revenues, Selected Years, 1951–83*

Millions of current dollars

	Message telecommunications service			*Private-line services*								*Other (non-AT&T) common carriers (OCCs)*
						Program						
Year	*Phone*		*Teletype*	*Phone*	*Teletype*	*Audio*	*Video*	*Telpak*	*Other services*	*DP digital*	*WATS*	
1951	602.2		18.1	17.0	27.6	11.7	7.9	...	...	...	...	...
1955	917.4		33.5	33.3	43.1	12.5	29.2	...	...	...	...	...
1960	1,435.3		48.7	103.3	95.5	13.2	36.2	...	...	...	...	...
1965	2,763.8		68.3	106.0	96.7	21.9	43.5	115.0	16.6	...	163.2	...
1970	4,918.2		66.6	208.6	95.9	21.6	76.5	331.3	55.4	...	387.2	...
1971	5,526.7		16.5	219.0	87.9	23.0	77.2	341.9	65.6	...	460.6	...
1972	6,156.9		n.a.	255.4	86.0	22.8	78.8	349.0	85.6	...	588.5	...
1973	7,143.0		n.a.	286.2	75.6	23.1	69.2	372.8	108.8	...	757.6	...
	Domestic	*International*										
1974	6,582.7	1,308.4	n.a.	280.6	67.0	23.2	65.6	409.8	122.7	...	928.5	...
1975	7,309.2	1,505.0	n.a.	295.5	62.3	23.4	64.0	450.3	145.8	0.7	1,139.4	30
1976	8,164.4	1,810.7	n.a.	331.9	59.7	24.4	64.1	464.5	174.6	5.1	1,490.0	...
1977	8,981.7	2,108.1	n.a.	373.2	50.1	25.4	60.3	458.0	204.3	13.5	1,861.8	...
1978	10,200.3	2,598.9	n.a.	443.5	43.1	27.2	63.8	448.2	295.6	22.9	2,236.5	...
1979	11,448.9	3,281.4	n.a.	505.8	37.4	29.5	66.3	477.6	373.5	43.2	2,611.5	50
1980	12,709.4	4,037.7	n.a.		1,803.9[a]						3,082.1	...
1981	14,680.4	4,596.1	n.a.		2,705.2[a]						3,803.5	...
1982	16,019.7	4,223.0	n.a.		3,258.8[a]						4,820.1	...
1983	15,434.8	4,881.8	n.a.		3,376.2[a]						5,942.1	2,200

Sources: AT&T data are from AT&T's 1952–83 December monthly 1B reports to the Federal Communications Commission. Other common carriers' data are author's estimates from industry sources. Figures are rounded.

a. Includes all private-line services (phone, teletype, audio, video, Telpak, DP, and others).

the relative sizes of the firms because somewhat different (larger) markets are included for the OCCs' revenues than for AT&T's revenues. These data indicate that AT&T had 96.7 percent of the interexchange market in 1981 and 91.8 percent two years later. There is a definite break in the data in 1984, when AT&T Communications was formed, after divestiture; the break makes it difficult to compare market share before and after divestiture. These data suggest that in 1984 AT&T Communications' share of the interexchange market was near 89 percent, with MCI having a 6.5 percent market share, and that between 1984 and 1985 AT&T Communications' market share fell 4.3 percentage points (to 84.6 percent of the interexchange market). The major gains in market share were made by MCI and four new entrants. Between 1985 and 1987 these data indicate that AT&T Communications' market share (of revenues) fell 8.6 percentage points, to 76 percent of the interexchange market.

Table 3 gives market data on inter-LATA traffic volume (hours of service), a different basis of judging market share than that used in tables 1 and 2. The 1984 data are actual statistics and show AT&T with 87.3 percent of traffic volume; table 2 indicates AT&T's market share to be 88.9 percent of revenues. Booz-Allen and Hamilton, the source of the data in table 3, forecast traffic volumes and market shares for 1985 and 1986 based on a certain assumption of aggregate market growth and on the hypothesis that the OCCs would capture 100 percent of this aggregate market growth. In table 3, I have altered the Booz-Allen and Hamilton projections to divide market growth equally between AT&T Communications on the one hand and the OCCs on the other. These calculations suggest AT&T Communications' market share of inter-LATA traffic to have been 84.7 percent in 1985 and 82.3 percent in 1986. An October 1987 report on market shares by the FCC shows that AT&T averaged 77 percent of all interstate minutes (including international) in 1986 and 73.7 percent of that market in the second quarter of 1987. In that report, the FCC estimated AT&T's 1986 market share of switched revenues to be 78 percent and about 76 percent if international calls were eliminated.[13]

Table 4 provides two of AT&T's estimates of its own market share, based on interstate conversation minutes (column 1) and on interstate long-distance revenue (column 2). No reconciliation has been attempted between table 4 and tables 2 and 3. These data suggest lower market shares for AT&T than in tables 2 and 3: namely, 71 percent of minutes and 74

13. Peyton L. Wynns, "AT&T's Share of the Interstate Switched Market," Federal Communications Commission, Common Carrier Bureau, Industry Analysis Division, October 21, 1987, pp. 7, 12, 23.

Table 2. *Total Operating Revenues of Interexchange Carriers, 1981–87*
Amounts in millions of current dollars; market shares in percents

	1981		*1982*		*1983*		*1984*		*1985*		*1986*		*1987*	
Company	*Dollars*	*Percent*	*Dollars*	*Percent*	*Dollars*	*Percent*	*Dollars*	*Percent*	*Dollars*	*Percent*	*Dollars*	*Percent*	*Dollars*	*Percent*
Other common carriers														
MCI	413.1	1.9	802.0	3.1	1,326.3	4.9	1,761.0	6.5	2,330.8	7.7	3,371.7	10.7	3,938.4	12.5
GTE-Sprint[a]	230.9	1.1	393.4	1.5	740.0	2.7	1,052.0	3.9	1,121.9	3.7	2,132.0	6.8	2,591.5	8.2
Allnet[b]	...	...	...	...	...	...	...	...	309.2	1.0	449.7	1.4	394.6	1.3
Lexitel[b]	...	...	...	...	...	...	...	...	127.2	0.4	...	...	...	...
Cable and Wireless[c]	...	...	...	...	...	...	...	...	145.8	0.5	170.8	0.5	180.3	0.6
USTS	82.6	0.4	128.3	0.5	163.3	0.6	161.5	0.6	241.0	0.8	282.1	0.9	268.8	0.9
US Telecom[a]	...	...	...	...	...	...	...	...	387.1	1.3	...	...	...	...
Teleconnect	...	...	...	...	...	...	...	...	...	...	...	...	167.6	0.5
OCC total	726.6	3.3	1,323.7	5.2	2,229.6	8.2	2,974.5	11.0	4,663.0	15.4	6,406.3	20.4	7,541.6	24.0

AT&T Long Lines Domestic	21,252.0	96.7	24,182.4	94.8	24,833.5	91.8	...	...	...	...	...	...	...	
AT&T Communications, domestic	...	...	...	...	...	...	23,943.0	88.9	25,598.3	84.6	25,031.8	79.6	23,906.7	76.0
AT&T Communications, total	...	...	...	...	...	...	35,035.8	...	37,458.0	...	36,629.1	...	34,982.7	...
Total, all companies and revenue[d]	...	...	...	...	...	...	38,010.3	...	42,121.0	...	43,035.4	...	42,524.3	...
Total, all companies, domestic	21,978.6	...	25,506.1	...	27,063.1	...	26,917.5[e]	100.0	30,261.3[e]	100.0	31,438.3	100.0	31,448.3[e]	100.0

Sources: Data for specialized common carriers and AT&T Communications, total, are from Federal Communications Commission, Common Carrier Bureau, Industry Analysis Division, ''Interexchange Carriers, Total Operating Revenues,'' compilation of information submitted by the companies in FCC Form P. For AT&T Long Lines, domestic, see table 1. For AT&T Communications, domestic, 1984, see Booz-Allen and Hamilton Inc., ''Long Distance Telephone Market Price and Size Forecast,'' New York, N.Y., March 28, 1985, tables 1 and 2, which calculate AT&T Communications gross service revenues less ''estimated international'' as $23,943 million, a value I chose to accept, though Booz-Allen total gross revenues are well below AT&T Communications revenues as reported by the FCC. Data for AT&T Communications, domestic, 1985, 1986, were estimated as 23,943/35,036 × the AT&T Communications value reported by the FCC. Figures are rounded.

a. GTE-Sprint and U.S. Telecom merged and became known as U.S. Sprint in July 1986; the data here are for each firm before July 1986 and for U.S. Sprint from July 1986 on.

b. Allnet and Lexital merged at the end of 1985.

c. Formerly TDX Systems Inc.

d. Sum of specialized common carriers' total and AT&T Communications total.

e. Not comparable to 1981–83 values because of differences before and after divestiture and differing revenues; estimates for 1984–87 depend on assumptions made above and ignore any international revenues of the specialized common carriers.

Table 3. *Inter-LATA Traffic Volume by Carrier*
Hours in millions; market shares in percent

Company	1984[a]		1985[b]		1986[b]	
	Hours	*Percent*	*Hours*	*Percent*	*Hours*	*Percent*
MCI	127	7.5	162	8.9	201	10.3
Sprint	73	4.3	95	5.2	118	6.0
Allnet	16	0.9	21	1.2	26	1.3
OCC total	216	12.7	278	15.3	345	17.7
AT&T Communications	1,481	87.3	1,540	84.7	1,607	82.3
Total	1,697	100.0	1,818	100.0	1,952	100.0

Sources: For 1984 data: information supplied by the Federal Communications Commission; for 1985–86 data: estimates based on Booz-Allen and Hamilton, "Prospects for Major-Facilities Based Other Common Carriers," Supplement, August 22, 1985. In this report Booz-Allen forecast the total market for 1985 and 1986. The forecast values were altered to divide the total forecast market growth 50:50 between AT&T Communications and the other three firms. Figures are rounded.
a. Actual.
b. Forecast.

percent of revenues in 1986. In 1987 AT&T estimated its own market share to be in the range of 70 percent.

Another description of market structure is offered in table 5. For an interexchange carrier to offer service, it must have some form of access with a BOC. Once that access is made, the OCC receives a "carrier identification code" (CIC). Table 5 lists the number of CICs outstanding at the end of each quarter from January 1982 through December 1987 and the number of separate entities holding CICs. In January 1982, CICs had been assigned to 7 entities. In December 1987, 573 CICs had been assigned to 451 entities. Most of these CICs are held by firms that engage in resale and sharing rather than by firms that own interexchange facilities.

Data on *capacity* of interexchange facilities are difficult to acquire. AT&T Communications has estimated that in 1984 its *in-service* capacity was 987 million circuit miles, whereas the *engineered* capacity of other systems was somewhat larger, at 1,190 million circuit miles.[14] This same analysis for 1985 showed AT&T Communications with 1,029 million circuit miles of interexchange capacity and "others" with 1,689 million circuit miles.[15] These comparisons of AT&T Communications' in-service capacity with its competitors' engineered capacity is designed to show the readily available "supply" of competition to AT&T Communications; data on AT&T Communications' engineered capacity (rather than in-

14. "Reply Comments of American Telephone & Telegraph," FCC docket 83-1147, June 4, 1984, app. B, p. 3-1.

15. Comments of American Telephone & Telegraph for the National Telecommunications and Information Administration (NTIA) Study on Structure and Regulation of the Telecommunications Industry, March 29, 1985, notes 17–18.

Table 4. *AT&T's Measures of Its Market Share, 1979–87*
Percent

Year	Interstate long-distance conversation minutes	Interstate long-distance revenues	Year	Interstate long-distance conversation minutes	Interstate long-distance revenues
1979	94.2	96	1984	77.3	81
1980	93.4	96	1985	75.4	79
1981	91.5	94	1986	71.3	74
1982	87.6	91	1987[a]	68.2	70
1983	82.1	85			

Source: Michael E. Porter, "Competition in the Long Distance Telecommunications Market: An Industry Structure Analysis," Monitor Company, Inc., Cambridge, Mass., October 1987, figs. 2, 3.
a. Estimate.

Table 5. *Carrier Identification Codes Assigned by Bell Communications Research, 1982–87*

Year	Number of CICs	Number of entities
1982		
January	7	7
April	11	11
August	13	13
December	11	11
1983		
January	11	11
April	16	16
August	31	31
December	42	42
1984		
January	46	46
April	71	54
August	108	82
December	155	123
1985		
January	165	130
April	199	169
August	227	190
December	256	217
1986		
January	265	224
April	298	249
August	352	295
December	413	334
1987		
January	422	343
April	463	374
August	522	415
December	573	451

Source: Common Carrier Bureau, Industry Analysis Division, "Number of Carriers Holding Carrier Identification Codes," Federal Communications Commission, Washington, D.C., January 1988, pp. 4–8.

service capacity) would have been more valuable for analysis of market structure.

Tables 6 and 7 provide statistics on the market shares of fiber-optic interexchange facilities, a technology used only since the early 1980s for interexchange service. The data in table 6 show that AT&T Communications in 1985 had less than *one-third* of the actual capacity in fiber-optic plants. The FCC conservatively estimated a near doubling of fiber-optic route miles and of miles of fiber between 1985 and 1986 (table 6). Data for 1987 show a slowing down of the installation of new fiber, but a substantial improvement is forecast in the carrying capacity of existing fiber plant because of significant advances in the technology of signaling in fiber-optic systems. According to Siperko, technological advances will *double* existing capacity by 1988, and further advances will continue to be made.[16]

In summary, as of December 1987, AT&T Communications' market share of inter-LATA revenues was probably in the 70–75 percent range, its share of conversation minutes was somewhat less, and its share of inter-LATA capacity was substantially less (40 percent). These market shares have decreased markedly in the past ten years. AT&T Communications' competitors consist of two major facilities-owning carriers (neither of which are experiencing robust health) and a series of regional facilities carriers, resellers, and value-added networks (VANs). Other carriers (the BOCs) are prevented from entering the market, and the FCC continues to regulate AT&T Communications but not its competitors.

Economies of Scale and Scope: Telecommunications Services

An economist's view of the appropriateness of regulation or of the efficiency of multiproduct offerings by a single firm is conditioned by the underlying technological conditions of production. Where economies of scale (how unit costs for x change as the output of x changes) and scope (how unit costs for x change as some other output, y, changes) exist over a wide range of output levels for two or more services, restrictions on entry and regulated prices may be socially desirable so that a single firm exists.[17] The concept of a multiproduct natural monopoly rests on the

16. Charles M. Siperko, "LaserNet—A Fiber Optic Intrastate Network (Planning and Engineering Considerations)," *IEEE Communications Magazine*, vol. 23 (May 1985), p. 38.

17. The reference here is to the literature on sustainability. See Leonard J. Mirman, Yair Tauman, and Israel Zang, "Supportability, Sustainability, and Subsidy-Free Prices," *Rand Journal of Economics*, vol. 16 (Spring 1985), pp. 114–26.

Table 6. *Estimated Fiber Deployment by Major Interexchange Carriers, 1985–87*

Company	Route miles		Fiber miles		Average fiber cross section (number of fibers)[a]
	Number	Percent	Number	Percent	
1985					
NTN[b]	3,340	16.6	48,094	10.5	14.4
RCI	580	2.9	6,960	1.5	12.0
Electra	493	2.4	10,194	2.2	20.7
MCI	2,560	12.7	79,200	17.4	30.9
AT&T	5,677	28.2	136,248	29.9	24.0
GTE-Sprint[c]	1,200	6.0	24,000	5.3	20.0
U.S. Telecom[c]	4,100	20.3	98,400	21.6	24.0
U.S. Sprint[c]	. . .	. . .	. . .	. . .	. . .
Lightnet	2,200	10.9	52,800	11.6	24.0
Norlight	. . .	. . .	. . .	. . .	. . .
Total	20,150	100.0	455,896	100.0	22.6
1986					
NTN[b]	7,642	19.0	133,171	15.0	17.4
RCI	580	1.4	6,960	0.8	12.0
Electra	493	1.2	10,194	1.1	20.7
MCI	5,580	13.9	167,400	18.8	30.0
AT&T	10,893	27.1	261,432	29.4	24.0
GTE-Sprint[c]	. . .	. . .	. . .	. . .	. . .
U.S. Telecom[c]	. . .	. . .	. . .	. . .	. . .
U.S. Sprint[c]	10,000	24.9	190,000	21.4	19.0
Lightnet	5,000	12.4	120,000	13.5	24.0
Norlight	. . .	. . .	. . .	. . .	. . .
Total	40,188	100.0	889,157	100.0	22.1
1987					
NTN[b]	7,979	14.1	156,421	11.9	19.6
RCI	620	1.1	7,440	0.6	12.0
Electra	493	0.9	10,194	0.8	20.7
MCI	6,317	11.1	223,894	17.1	35.4
AT&T	19,000	33.5	456,000	34.8	24.0
GTE-Sprint[c]	. . .	. . .	. . .	. . .	. . .
U.S. Telecom[c]	. . .	. . .	. . .	. . .	. . .
U.S. Sprint[c]	17,000	30.0	323,000	24.6	19.0
Lightnet	5,300	9.3	127,200	9.7	24.0
Norlight	670	1.2	8,040	0.6	12.0
Total	56,709	100.0	1,312,189	100.0	23.1

Source: Jonathan M. Kraushaar, "Fiber Deployment Update: End of Year, 1987," Federal Communications Commission, Common Carrier Bureau, Industry Analysis Division, Washington, D.C., January 1988, table 1; and author's calculations. Figures may not add because of rounding.

a. The number of circuits per fiber has been steadily increasing.

b. National Telecommunications Network.

c. See note a to table 2.

Table 7. *Estimated Fiber Deployment by Local Bell Operating Companies, 1985–87*[a]

Company	Route miles	Fiber miles	Average fiber cross section (number of fibers)[b]
1985			
Ameritech	3,200	77,700	24.3
BellSouth	3,830	50,807	13.3
Bell Atlantic[c]	1,240	83,085	67.0
Nynex	1,606	83,384	51.9
Pacific Telesis[c]	3,847	100,022	26.0
Southwestern Bell[c]	1,913	70,490	36.8
U.S. West	3,527	47,341	13.4
Total	19,163	512,829	26.8
1986			
Ameritech	5,200	111,100	21.4
BellSouth	8,694	170,092	19.6
Bell Atlantic[c]	4,374	150,847	34.5
Nynex	3,209	129,743	40.4
Pacific Telesis[c]	2,484	64,740	26.1
Southwestern Bell[c]	4,374	151,043	34.5
U.S. West	5,017	70,082	14.0
Total	33,352	847,647	25.4
1987			
Ameritech	6,500	143,100	22.0
BellSouth	11,593	212,623	18.3
Bell Atlantic[c]	6,482	225,128	34.7
Nynex	4,346	177,694	40.9
Pacific Telesis[c]	2,976	81,291	27.3
Southwestern Bell[c]	6,105	185,672	30.4
U.S. West	6,451	100,622	15.6
Total	44,453	1,126,130	25.3

Source: Kraushaar, "Fiber Deployment," table 3.
a. Aggregated to the regional holding company level.
b. The number of circuits per fiber has been steadily increasing.
c. Differences in reporting by the regional holding companies, though not expected, may have resulted in underestimates of the total amount of fiber in the ground. Significant variation in average fiber cross section between 1985 and 1986, as can be seen in the Bell Atlantic data, is possibly a result of this problem.

notion that it is less costly for a single firm to offer a range of services than for those services to be divided among two or more firms. Natural monopoly is technically measured by whether the cost function is *subadditive*. The alternatives to such socially desirable regulation would be either unconstrained monopoly or mandated multiple firms (say, as a result of antitrust divestiture), each of suboptimal size.

Before the modified final judgment in 1982, both sides in the AT&T antitrust suit were preparing evidence on the degree of economies of scale and scope in the production of telecommunications services. In this section I review the evidence on scale and scope economies and provide some new results. The purpose of this discussion is not to reanalyze the judgment but to shed light on economies of scale in the interexchange market, thus highlighting one possible source of market dominance. Second, the findings of these studies will also indicate if there are any cost inefficiencies from preventing firms (AT&T Communications, OCCs, or the BOCs) from offering both interexchange and local telecommunications services. The literature on scale and scope is also valuable for a third purpose. Several economists have proposed replacing rate-of-return regulation of AT&T Communications with floor-and-ceiling constraints on pricing, a proposal now pending at the FCC. These proposed constraints are "incremental costs" and "stand-alone costs," respectively. These two concepts are measurable, at least in theory, by parameters estimated in econometric studies of scale and scope.

The empirical evidence on economies of scale and scope in AT&T (or in other telecommunications firms) by and large does not rely on precise process modeling of the actual technologies used in providing services.[18] Instead, the studies estimate quite general cost functions with firm-specific time-series data. In these studies, technology enters as some exogenous factor augmenting technological progress and is represented by the introduction of electronic switching, direct toll dialing, or research and development expenses incurred by the firm.[19] Rather than incorporating engineering estimates of the economies of scale in microwave, cable, and fiber-optic interexchange transmission and estimating the optimal network given various output levels, econometric studies use data on the changing output levels and output mix as well as data on changing input levels and factor prices to estimate both product-specific (for example, interexchange or local) and overall economies of scale (and scope). By correlating cost and output changes, the econometric models attempt to measure the "incremental costs" attributable to changes in one output (say interexchange) while other output levels (say local) are held constant. Theoretically, these studies can identify the changes (if any)—in equipment, transmission, switching, and even access—that accompanied changes in interexchange services. These product-specific incremental costs represent the floor in some economists' suggested pricing rules for improved regulation of AT&T Communications. Several of these econometric studies also attempt to

18. Such models exist in the engineering and economics literature. For a theoretical model using process analysis, see S. C. Littlechild, "Peak-Load Pricing of Telephone Calls," *Bell Journal of Economics and Management Science*, vol. 1 (Autumn 1970), pp. 191–210.

19. Process modeling would examine, in contrast, the technological choices involved in each type of switch.

measure stand-alone costs—the costs of a separate service, such as interexchange, offered by an independent interexchange firm, costs also designated as important for ceiling pricing constraints.

Cost Function Models

The basic cost model used in these studies of scale and scope is the translog, a second-order Taylor series expansion in logarithms.

I postulate that the firm minimizes total costs (C) by using three inputs (labor, capital, and materials) purchased from competitive factor markets at prices W_i; two outputs (Q_1 and Q_2—local and toll, respectively) are produced; and t represents the exogenous change in technology. This translog cost function may be written

$$
\begin{aligned}
\log C(W_1, Q_1, Q_2, t) = a_0 &+ \sum_{i=1}^{3} a_i \log W_i \\
&+ \sum_{j=1}^{2} b_j \log Q_j + a_t \log t + \tfrac{1}{2} \sum_{i=1}^{3} c_i (\log W_i)^2 \\
&+ \tfrac{1}{2} \sum_{j=1}^{2} d_j (\log Q_j)^2 + \tfrac{1}{2} a_t \log t^2 + \sum_{i \neq j}\sum c_{ij} \log W_i \log W_j \qquad (1) \\
&+ b_{12} \log Q_1 \log Q_2 + \sum_{i=1}^{3} b_{it} \log W_i \log t \\
&+ \sum_{j=1}^{2} C_{jt} \log Q_j \log t + \sum_{i=1}^{3}\sum_{j=1}^{2} d_{ij} \log W_i \log Q_j .
\end{aligned}
$$

Using Shephard's lemma[20] adds two equations to the set to be estimated, one for each of two factor shares (S_i); the estimates for the third factor share are calculated by using the fact that the three factor shares must equal unity):

$$
(2) \quad S_i = a_i + c_i \log W_i + \sum_{j=1}^{2} c_{ij} \log W_j + b_{it} \log t + \sum_{j} d_{ij} \log Q_j .
$$

Because the translog model is obviously not well defined in regions near zero output (the log of zero is infinity), a generalized translog model

20. The derivative of the cost function (equation 1) with respect to an input price is equal to that input's share of total input costs.

has been proposed[21] and has been used by other authors as well.[22] This generalization involves applying a Box-Tidwell transformation to the variables, here shown for output:

$$Q_j^* = \frac{Q_{j-1}^{\alpha}}{\alpha}, \tag{2a}$$

where Q_j is one of the two outputs, and α is a parameter to be estimated. As α approaches zero, the Box-Tidwell transformation approaches the translog model. The authors' approaches vary; some use the Box-Tidwell transformation on all variables, others transform only the output variables.

The authors estimating these cost functions usually impose conditions from the economic theory of production *symmetry,*

$$b_{ij} = b_{ji};\ c_{ij} = c_{ji}, \tag{3}$$

and *homogeneity of degree one in input prices,*

$$\sum_i a_i = 1; \quad \sum_i c_{ij} = \sum_i b_{it} = \sum_i d_{ij} = 0. \tag{4}$$

The parameters of interest to those concerned with the appropriate degree of competition or pricing constraints in the telecommunications industry deal with estimates of scale and scope and the subadditivity of the cost function. For a firm producing only one output, economies of scale is an unambiguous measure; obviously economies of scope cannot exist. For a multiproduct firm, even one with only two outputs, measures of scale and scope become complex depending on whether these measures are product-specific or for the firm's overall operations, and on whether the measures are local (in the neighborhood of the level of operations) or global (for all possible output levels). These measures therefore depend on the combinations of output actually produced or hypothesized as produced by the firm(s).

21. Melvyn Fuss and Leonard Waverman, "The Regulation of Telecommunications in Canada," Technical Report 7, Economic Council of Canada, Ottawa, 1981.

22. See Ferenc Kiss, Seta Karabadjian, and Bernard J. Lefebvre, "Economies of Scale and Scope in Bell Canada," in Léon Courville, Alain de Fontenay, and Rodney Dobell, eds., *Economic Analysis of Telecommunications: Theory and Applications* (Amsterdam: North-Holland, 1983), pp. 55–82; and David S. Evans and James J. Heckman, "Multiproduct Cost Function Estimates and Natural Monopoly Tests for the Bell System," in Evans, ed., *Breaking Up Bell: Essays on Industrial Organization and Regulation* (New York: North-Holland, 1983), pp. 253–82.

OVERALL ECONOMIES OF SCALE (LOCAL MEASURE ONLY). The inverse of the scale elasticity (*SE*)—the cost elasticity (*CE*)—is calculated in a cost function analysis, with the property that

$$SE = \frac{1}{CE}. \tag{5}$$

As the derivative of the cost function with respect to all outputs, cost elasticity obviously depends on the levels of all outputs; the overall cost elasticity is then a *local* measure and is given in the translog format as

$$CE = \sum_{j=1}^{2}(b_j + d_j \log Q_j + b_{12} \log Q_j + c_{jt} \log t + \sum_i d_{ij} \log W_i). \tag{6}$$

OVERALL ECONOMIES OF SCOPE, STAND-ALONE COSTS, AND INCREMENTAL COSTS. *Global* economies of scope are said to exist if producing local and toll services together costs less than producing each service separately:

$$C(Q_1, Q_2) \leq C(Q_1,0) + C(0, Q_2), \tag{7}$$

where $C(Q_1,0)$ and $C(0, Q_2)$ are measures of *stand-alone costs*, the costs of having each of the two services produced independently.

Measures of *incremental costs* also depend on these estimates of stand-alone costs. The estimated *average incremental costs* for some output level Q_1 (with the second output level held constant at $\overline{Q}_2$) is

$$C(Q_1, \overline{Q}_2) - C(0, \overline{Q}_2). \tag{7a}$$

The *local* measure of overall economies of scope is the estimate of cost complementarity (*CC*) at the observed (or postulated) output levels:

$$CC = \frac{\partial^2 C}{\partial Q_1 \partial Q_2} < 0. \tag{7b}$$

If *CC* is negative, an increase in the output of one of the services reduces the marginal costs of producing the second service.[23]

23. Note that *CC* depends on the specific output levels and is a local or neighborhood condition, whereas scope (equation 7) is a global condition.

PRODUCT-SPECIFIC ECONOMIES OF SCALE. Panzar and Willig[24] have suggested measuring *product-specific economies of scale* for some service k (here called *PSSE*) as the ratio of average incremental cost (equation 7a) to marginal cost, or

$$\frac{IC_k/Q_k}{\partial C/\partial Q_k} = \frac{IC_k/C}{CE}. \tag{8}$$

When the ratio of average incremental cost to marginal cost (equation 8) is greater than one ($IC_k > MC$), product-specific economies of scale exist. Note that to measure *PSSE* for service k requires measuring the stand-alone costs for the other service, j.

NATURAL MONOPOLY. A *natural monopoly* is said to exist if the cost function is subadditive. Subadditivity exists if the two outputs can be produced at lower total cost by one firm than by two or more firms, or if

$$\sum_i \sum_j C(\alpha_i Q_1, B_j Q_2) > C(Q_1, Q_2),^{25} \tag{9}$$

where α_i and B_j are the proportions of services i and j produced by each firm, and where

$$\sum_i \alpha_i = 1, \sum_j B_j = 1. \tag{9a}$$

Baumol, Panzar, and Willig have shown that, for subadditivity to hold, economies of scope (equation 7b) is a necessary condition (otherwise each of the two outputs could be produced by a single firm) while economies of scope (equation 7b) and product-specific economies of scale (equation 8) are sufficient conditions.[26]

Clearly, however, product-specific economies of scale *are not* necessary conditions for subadditivity of the cost function, since economies of scope could be so large that even with product-specific diseconomies of scale, or overall diseconomies of scale, total costs could still be lower with both services produced together.

24. John C. Panzar and Robert D. Willig, "Economies of Scope, Product Specific Economies of Scale, and the Multiproduct Competitive Firm," Bell Laboratories Economics Discussion Paper 152 (Murray Hill, N.J.: Bell Laboratories, August 1979).

25. If the left-hand side of equation 9 is less than the right-hand side, the cost function is superadditive, and lower costs of service would exist if the firm were "broken up."

26. William J. Baumol, John C. Panzar, and Robert D. Willig, *Contestable Markets and the Theory of Industry Structure* (Harcourt Brace Jovanovich, 1982).

A problem with tests for natural monopoly is that the concepts of global economies of scope and product-specific economies of scale imply knowledge of *stand-alone costs* for each of toll and local services (implications for the pricing rules that will be examined later). If the data set contains firms that offer only local service (these do exist) or only toll service (there are several of these), as well as firms that offer both services, one could have high confidence in the econometric estimates of stand-alone costs. The data used in the published studies are data for either AT&T or predivestiture Bell Canada; these data do not contain any observations of stand-alone production and are thus suspect for measuring stand-alone costs, incremental costs, product-specific economies of scale, and economies of scope.[27]

Evans and Heckman have proposed a direct test of subadditivity using only the actual data over which the firm had operated.[28] They calculate, for each time period sub_t such that

$$(10) \qquad sub_t\,(\alpha, B) = \frac{C(Q_1, Q_2) - \sum_i \sum_j C(\alpha_i Q_1, B_j Q_2)}{C(Q_1, Q_2)},$$

where α, B are constrained to be within the boundaries of the *actual* observed minima and maxima of toll and local outputs. They calculate $sub_t\,(\alpha, B)$ for all time periods (over which the constraints on minima and maxima hold) for a large possible set of combinations of firm size (for two firms). If the maximum of their measured sub_t coefficients is negative, then the cost function is locally subadditive.

The Concepts of Scale and Scope in Telecommunications

Before turning to the evidence on economies of scope and scale and on subadditivity, it is useful to examine what these concepts mean in the specific context of a telecommunications firm such as AT&T.

Economies of scope (a global condition), or cost complementarity (a local or neighborhood condition)—necessary conditions for subadditivity or the existence of a natural monopoly—require some *jointness in supply* for the two products, local and toll service.[29]

27. The simple translog model cannot be used to study these concepts because, as mentioned, it is undefined in the region of zero output (the log of zero is infinity).

28. Evans and Heckman, "Multiproduct Cost Function Estimates," pp. 265–71.

29. Ronald Braeutigam, "Comment" on Melvyn Fuss and Leonard Waverman, "Regulation and the Multiproduct Firm: The Case of Telecommunications in Canada," in Gary Fromm, ed.,

Imagine a world with no local service; that is, neither local loops nor telephones (these are "local" service in the pre-1981 AT&T and Bell Canada worlds). In this imaginary world the provision of toll service would require toll firms to build "local" facilities—that is, the local loops and telephones. In a different imaginary world, where local telecommunications facilities already existed, it would be far more costly for toll firms to use *their own* local facilities than for those toll firms to pay for the use of the already existing local facilities. These two worlds illustrate the jointness of local and toll telecommunications facilities. *Global economies of scope must then exist between local and toll communications*, especially over the data sets (1947–78) used in the econometric analyses. *This existence of global economies of scope in facilities does not, however, necessarily imply that local and toll services should be supplied by the same firm*. Facilities (or production) and services (or operation) are separate. One can, as in the second hypothetical example, envision the *sharing* of local facilities by toll firms contracting to pay for usage. Cost savings of having a single firm offer both services (economies of scope) would exist only if either facilities or operating costs could be reduced with common service provision, compared with multiple firm operation.

Cost complementarity (the neighborhood analogue to global economies of scope) may or may not exist for local and toll output. Although the local loop, the telephone, and the local switch all handle toll services, toll switches and toll lines do not handle local service. An *increase* in toll service, given the existence of a local network with excess capacity, creates no additional need for local facilities—thus seeming to imply cost complementarity (at least in the short run), hence economies of scope. An increase in local service does not press on toll facilities; this cost complementarity is then not symmetric. These facts are, however, in large part irrelevant to the optimal market structure. Unless there are lower costs from having one firm offer both the existing local and the incremental toll services, no economies of scope in *operation* exist.

The cost complementarity important for regulatory decisions is not simply the sharing of some common facility (the local loop) by two services (local calls and toll calls), but some cost saving by having one firm provide both services. Trucks, buses, and private automobiles all use highways

Studies in Public Regulation (MIT Press, 1981), p. 317, suggests that economies of scope can exist without subadditivity. If the cost function (CQ_1, Q_2) is of the form $F + Q_1 + Q_2$, "the cost function does exhibit economies of scope, even though production is not joint." The F, or fixed costs, in Braeutigam's cost function, however, are *used* for both Q_1 and Q_2 (otherwise the cost function would be separable), and thus some jointness exists.

and local streets, but there is no cost complementarity in operation either in having the same firm provide the highways as well as truck, bus, and automobile service or in having the same operator for all truck, bus, and auto trips.[30]

Imagine yet another hypothetical world, in which one carrier provides interexchange service between two nodes, with local service at each node being provided by different local-exchange carriers (LECs). The interexchange carrier must then engage in bilateral monopoly negotiations with two LECs; in addition, each LEC can attempt to increase its share of the interexchange revenue pie at the expense of the other LEC—''hold-ups'' can occur. Such bilateral monopoly negotiations lead to inefficient outcomes because pricing either LEC or interexchange service above its marginal costs will diminish the use of that service, thus reducing the potential profits of the three firms.[31] One possible way out of this inefficiency is to have multiple nodes and multiple interexchange carriers between nodes. With many potential paths between LECs, hold-ups can be avoided, but again with some inefficiency in the form of higher costs of indirect paths (increased switching, transmission, and noise).

The textbook answer to this problem of bilateral monopoly negotiation is vertical integration—or AT&T before divestiture. But that integration led to abuses of power against interexchange competitors (vertical foreclosure) as well as to the present inefficient telecommunication system, with the price of access held well above the marginal costs of access.[32]

Regulation is a *potential* answer to the problem of bilateral negotiations. The regulator can ensure equal access (what the modified final judgment orders) *and* set the price for access (again, at present done by regulators, but done incorrectly). Once that price is set by the regulator at the unbundled cost of access, no inefficiencies of bilateral monopoly negotiations

30. In addition to these notions of scope, toll systems can place investment demands on the local loop; that is, cause incremental local costs. Trucks cause heavier damage to roads than do private cars and should be charged for that use. If toll users cause costs of local loops, they should pay those costs; cost causality is not an argument for joint supply.

31. See Leonard Waverman, ''The Regulation of Intercity Telecommunications,'' in Almarin Phillips, ed., *Promoting Competition in Regulated Markets* (Brookings, 1975), pp. 201–39. This may also be a description of bilateral negotiations for international telecommunications services (or airline services) with each country as an LEC.

32. Wenders and Egan have suggested that the access charge and the recovery of non-traffic-sensitive costs in usage formulas are the source of major inefficiency of the present system. That inefficiency arose in the era of monopoly provision of telecommunications services because of asymmetric regulation. See John T. Wenders and Bruce L. Egan, ''The Implications of Economic Efficiency for U.S. Telecommunications Policy,'' *Telecommunications Policy*, vol. 10 (March 1986), pp. 33–40.

remain. The need for regulatory supervision of the price of access remains as long as market power exists on the part of the LECs.

The concepts of scale and scope must therefore be carefully defined before the empirical evidence is examined. One would like to know whether the larger the telephone exchange, the lower the unit costs of providing service to an individual incremental customer (local plant economies of scale). One would also like to know, for public policy purposes, whether combining several local exchanges of given size leads to lower unit costs compared with having those same exchanges separately owned and operated (local firm economies of scale). At the interexchange level, one would want to know whether increasing the scale of a link joining two points lowers unit interexchange costs (toll plant economies of scale) and whether there are economies in a single interexchange firm's owning the links between multiple points (firm interexchange economies of scale). Finally, there could be overall firm economies of scale in the ownership and operation of local and interexchange facilities—this requires economies of scope between local and interexchange facilities and perhaps economies of scale: the larger the firm offering both interexchange and local services, the lower the unit costs of providing both.

Survey of Economies of Scale and Scope

Fuss has surveyed the econometric evidence on economies of scale and scope as given in the academic literature through 1981. Table 8 summarizes Fuss's work and extends the survey to papers produced through early 1986.[33]

The table summarizes twenty econometric studies of economies of scale and scope (some of these studies involve different cost functions or aggregations published in the same paper, and the survey may not be complete). Of the twenty, fourteen used the translog cost function, two used a Box-Tidwell transformation of the output variables, and four used a Box-Tidwell transformation of all variables. Of the twenty, seven (all for Bell Canada) examined three outputs (local, MTS, and competitive toll or private line); seven examined two outputs (local and toll for six, local plus MTS and competitive toll in a seventh), and six examined a single

33. M. A. Fuss, "A Survey of Recent Results in the Analysis of Production Conditions in Telecommunications," in Courville and others, eds., *Economic Analysis of Telecommunications*, pp. 3–26. See also the more recent survey undertaken by Ferenc Kiss and Bernard Lefebvre, "Econometric Models of Telecommunications Firms: A Survey," *Revue Economique*, vol. 38 (March 1987), pp. 307–73.

Table 8. *Estimates of Economies of Scale and Scope, 1977–86*

Study and year	*Firm*	*Years used in estimation*	*Cost model used*	*Outputs*	*Economies of scale*	*Economies of scope*
Fuss-Waverman, 1977[a]	Bell Canada	1952–75	Translog	Local, MTS; competitive, toll	0.89–1.15	No[b,c]
Fuss-Waverman, 1979[a]	Bell Canada	1952–76	Translog	Local, MTS; competitive, toll	1.45	No[b,c]
Denny and others, 1979[d]	Bell Canada	1952–76	Translog	Local, MTS; competitive, toll	1.37–1.59	Yes[b,c]
Nadiri-Schankerman, 1979[e]	AT&T	1947–76	Translog	Single aggregate	1.75–2.69	n.a.
Breslaw-Smith, 1980[f]	Bell Canada	1952–78	Translog	Local, MTS; other toll	1.29	"Small"[b]
Denny-Fuss, 1980[g]	Bell Canada	1952–72	Translog	Single aggregate	1.10–2.15	n.a.
Christensen and others, 1980[h]	AT&T	1947–77	Translog	Single aggregate	1.56–1.94	n.a.
Fuss-Waverman, 1981[i]						
1	Bell Canada	1957–77	Translog	Local, MTS; competitive, toll	1.26–1.63	Yes[b,c]
2	Bell Canada	1957–77	Hybrid translog	Local, MTS; competitive, toll	0.83–1.09	Some diseconomies of scope
Christensen and others, 1983[h]	AT&T	1947–77	Translog	Single aggregate	1.3–1.7	n.a.
Kiss and others, 1983[j]						
1	Bell Canada	1952–78	Translog	Local, MTS; competitive, toll	1.67[k]	Unreported[c]
2	Bell Canada	1952–78	Translog	Local, toll	1.66[l]	[m]
3	Bell Canada	1952–78	Translog	Single, aggregate	1.62	n.a.
4	Bell Canada	1952–78	Generalized translog	Local, MTS; competitive toll	1.43[k]	Unreported
5	Bell Canada	1952–78	Generalized translog	Local, toll	1.62[l]	[n]
6	Bell Canada	1952–78	Generalized translog	Single aggregate	1.61–1.73	n.a.

Evans-Heckman, 1983–84[o] 1986 revision[n]	AT&T	1947–77	Translog Hybrig translog Generalized translog	Local, toll Local, toll Local, toll	1.39–1.57	No, diseconomies of scope based on direct calculation of subadditivity
Charnes-Cooper-Sueyoshi, 1986[p]	AT&T	1947–77	Translog	Local, toll	Unreported	Yes, economies of scope based on dire calculation of subadditivity

Sources: The studies undertaken between 1977 and 1981 are summarized in M. A. Fuss, "A Survey of Recent Results in the Analysis of Production Conditions in Telecommunications," in Léon Courville, Alain de Fontenay, and Rodney Dobell, eds., *Economic Analysis of Telecommunications: Theory and Applications* (Amsterdam: North-Holland, 1983), pp. 6–7, 19. Published versions of these and other individual papers are noted below.

n.a. Not available.

a. Melvyn Fuss and Leonard Waverman, "Regulation and the Multiproduct Firm: The Case of Telecommunications in Canada," in Gary Fromm, ed., *Studies in Public Regulation* (MIT Press, 1981), pp. 277–313; updated and reported in M. Denny and others, "Estimating the Effects of Diffusion of Technological Innovations in Telecommunications: The Production Structure of Bell Canada," *Canadian Journal of Economics*, vol. 14 (February 1981), pp. 24–43.

b. Cost complementarity (neighborhood test); where a single value is given under economies of scale, it denotes the estimate at the expansion point, typically middle to late 1960s.

c. Global economies of scope cannot be measured in the translog format.

d. Denny and others, "Estimating the Effects of Diffusion of Technological Innovations in Telecommunications."

e. M. Ishaq Nadiri and Mark A. Schankerman, "The Structure of Production, Technological Change, and the Rate of Growth of Total Factor Productivity in the U.S. Bell System," in Thomas G. Cowing and Rodney E. Stevenson, eds., *Productivity Measurement in Regulated Industries* (Academic Press, 1981), pp. 219–47.

f. J. Breslaw and J. B. Smith, "Efficiency, Equity and Regulation: An Econometric Model of Bell Canada," Final Report to the Canadian Department of Communications, Ottawa, March 1980.

g. M. Denny and M. Fuss, "The Effects of Factor Prices and Technological Change on the Occupational Demand for Labour: Evidence from Canadian Telecommunications," Working Paper 8014 (University of Toronto, Institute for Policy Analysis, July 1980).

h. L. R. Christensen, D. Cummings, and P. E. Schoech, "Econometric Estimation of Scale Economies in Telecommunications," paper presented at a meeting of the Econometric Society in Denver, September 1980; revised and published under the same title in Courville, de Fontenay, and Dobell, eds., *Economic Analysis of Telecommunications*, pp. 27–53.

i. Melvyn Fuss and Leonard Waverman, "The Regulation of Telecommunications in Canada," Technical Report 7 (Ottawa: Economic Council of Canada, 1981).

j. Ferenc Kiss, Seta Karabadjian, and Bernard J. Lefebvre, "Economies of Scale and Scope in Bell Canada," in Courville, de Fontenay, and Dobell, eds., *Economic Analysis of Telecommunications*, pp. 55–82.

k. Significantly greater than one at the .05 level, but "the 3-output TL [translog]/ GTL [generalized translog] model proved to be too general. Many estimated parameters exhibited a high degree of instability." Kiss and others, "Economies of Scale and Scope in Bell Canada," p. 63.

l. Significantly greater than one at the .05 level.

m. The authors commented that the test produced "meaningless" results.

n. The authors commented that the generalized translog had no cost complementarity and "meaningless" scope; restricting some of the second-order parameters to zero led to evidence (not significantly different from zero) of cost complementarity and scope.

o. David S. Evans and James J. Heckman, "Multiproduct Cost Function Estimates and Natural Monopoly Tests for the Bell System," in David S. Evans, ed., *Breaking Up Bell: Essays on Industrial Organization and Regulation* (North-Holland, 1983), pp. 253–82; revised as David S. Evans and James J. Heckman, "A Test for Subadditivity of the Cost Function with an Application to the Bell System," *American Economic Review*, vol. 74 (September 1984), pp. 615–23, and David S. Evans and James J. Heckman, "Erratum: A Test for Subadditivity of the Cost Function with an Application to the Bell System," *American Economic Review*, vol. 76 (September 1986), pp. 856–58.

p. A. Charnes, W. W. Cooper, and T. Sueyoshi, "A Goal Programming/Constrained Regression Review of the Bell System Breakup," Center for Cybernetic Studies, University of Texas, Austin, January 1986; revised and published in *Management Science*, vol. 34 (January 1988), pp. 1–26.

aggregate measure of output. The studies using a single output aggregate shed no light on economies of scope. A more serious limitation is that the studies using two or more outputs tested for and rejected the existence of a single aggregate measure of output. The six studies using a single aggregate output therefore will not be discussed in detail in the following analysis.

I begin with the evidence on economies of scope and cost complementarity. Fourteen studies used more than one measure of output and thus ostensibly could shed light on economies of scope. Of the seven cases in which three outputs were examined, five used the translog format and thus cannot provide evidence of economies of scope. Of these five, two found significant cost complementarities between local and toll. One of the other two studies examining three outputs (and using a generalized translog model) found some evidence of diseconomies of scope.[34]

Of the seven studies using two outputs, one (for Bell Canada) found ''small'' economies of scope;[35] five found no evidence of scope economies (but between the aggregate of local and toll compared with competitive toll) and one of the five (for AT&T by Evans and Heckman) found significant diseconomies of scope; and finally the seventh estimated substantial economies of scope for AT&T (Charnes, Cooper, and Sueyoshi).[36]

The contrast between the results of Evans and Heckman and those of Charnes, Cooper, and Sueyoshi is most interesting because the same data (AT&T, 1947–77) and the same functional form (translog) were used by the two sets of authors. Charnes, Cooper, and Sueyoshi argued that the difference in the studies is the objective function being maximized; Evans and Heckman maximized the logarithm of the likelihood function, implicitly assuming both positive and negative residuals between realized

34. Ferenc and Lefebvre have suggested that this result for this one study (Fuss and Waverman, ''Regulation of Telecommunications in Canada'') is due to those authors' constraining the Box-Tidwell parameter to be the same across all output variables. Kiss and Lefebvre suggested that removing this constraint would reverse the 1981 results. Kiss and Lefebvre did not, however, estimate the exact system used by Fuss and Waverman but a simplified version. It is unclear how this simplification has affected reestimation.

35. J. Breslaw and J. B. Smith, ''Efficiency, Equity and Regulation: An Econometric Model of Bell Canada,'' Final Report to the Canadian Department of Communications, Ottawa, March 1980.

36. A. Charnes, W. W. Cooper, and T. Sueyoshi, ''A Goal Programming/Constrained Regression Review of the Bell System Breakup,'' Center for Cybernetic Studies, University of Texas, Austin, January 1986; revised and published in *Management Science*, vol. 34 (January 1988). References are to the 1986 version. The authors did not test directly for economies of scope. Instead, they found local subadditivity of the cost function by using the tests developed by Evans and Heckman in ''Multiproduct Cost Function Estimates.'' Economies of scope, however, is a necessary condition for subadditivity.

costs and those represented by cost-minimizing behavior. Charnes, Cooper, and Sueyoshi argued that the assumption of cost minimization implies that realized costs cannot be less than those predicted by cost minimization. This, they suggested, requires the modeler to use one-sided residual estimation techniques. Charnes, Cooper, and Sueyoshi used one such technique, goal-constrained programming, and in so doing suggested that they reversed the most important results (most important, at least, for public policy) of Evans and Heckman. A recent paper by Evans and Heckman has demonstrated that the reversal of results is due to more than just a change in objective function. Evans and Heckman have shown that Charnes, Cooper, and Sueyoshi used a different data set and a different model from their own. Evans and Heckman concluded that "the CCS model most comparable to our 1983 model gives very similar results on the natural monopoly test."[37]

The evidence on overall economies of scale as summarized in table 8 would appear to favor the presence of such economies. In only two cases does the lower-bound estimate of overall economies of scale (95 percent confidence region below the mean estimates) fall below unity.

On the basis of this evidence, one would not, however, rush to the FCC or the Canadian Radio, Television, and Telecommunications Commission and argue that this econometric evidence supports divestiture between local and toll services but further supports the existence of a single firm in the toll market. The weight of the evidence of all these studies is simply not strong enough, since changing the level of aggregation, the functional form, the constraints imposed, or the objective function dramatically alters the results. The message is simply that the data available are insufficient to enable researchers to discriminate between alternative hypotheses.[38] Moreover, I also share the doubts of Fuss, who argued that the evidence cannot adequately disentangle the effects of two factors that both reduce unit costs—economies of scale and technological advance.[39]

To demonstrate these assertions, let me undertake a more complete analysis of the two papers—Evans and Heckman, and Charnes, Cooper, and Sueyoshi—that give diametrically opposed views of the optimal degree of integration and competition in the U.S. interexchange market. The

37. David S. Evans and James J. Heckman, "Natural Monopoly and the Bell System: A Response to Charnes, Cooper and Sueyoshi, *Management Science*, vol. 34 (January 1988), pp. 27–38.

38. A letter to the author from David S. Evans, Fordham University, dated August 3, 1987, shares this view.

39. Fuss, "Survey of Recent Results."

data used in both studies consist of input levels, input prices, outputs, and output price data for AT&T, both Long Lines and BOCs, from 1947 through 1977. These data, however, are too aggregated to provide the evidence crucial to the question of the optimal design of the U.S. telecommunications market.[40]

AGGREGATION BIAS. The unit of operation in the analysis is the firm—AT&T. It is simple to see that *aggregation bias* must plague any estimate of product-specific scale economies (*PSSE*) at this firm level. The concept of *PSSE* at, say, the local-exchange level involves the calculation across exchange sizes of the effect on costs of adding a single subscriber, other things being equal. What this econometric evidence measures is the effect *across* all local exchanges, in the Bell System before divestiture, of increasing local output somewhere. If all local exchanges in the United States expanded at the same rate over the 1947–77 period, the econometrically estimated *PSSE* would indicate the unit cost changes from a 1 percent increase in local output. Because local exchanges did not increase at the same rate, aggregation bias affects the results and the interpretation of the results. If economies of scale were observed with the local aggregate used, moreover, this finding would not imply that *one firm* should provide all local output.

Aggregation bias exists as well at the level of all interstate toll output for AT&T. An increase of 1 percent in the aggregate toll output, if accompanied by less than a 1 percent increase in unit costs, suggests toll output *PSSE*. Toll service provision increased unevenly, however, across AT&T's territory over the 1947–77 period. Hence the "measured" scale elasticity is biased. Therefore, empirical evidence on "product-specific scale economies" for interexchange does not suggest that all routes be owned by one firm.

TECHNOLOGICAL CHANGE. The characterization of technological change is vital to accurate estimation of econometric cost functions. During 1947–77 technical advance benefited interexchange plant and switching but not the provision of local loops. Microwave exchange plant was first put in place in 1947. Crossbar switches began to replace step-by-step switches in the 1950s and were replaced themselves by electronic switches beginning in the 1960s. The inherent economies of scale of each of these specific pieces of equipment differ. Microwave plant on a route-by-route basis has few scale economies; cable and especially fiber optics have far

40. Kiss and Lefebvre, "Econometric Model of Telecommunications Firms," compare and contrast these studies as well.

greater economies of scale. Electronic switches, on an installation-by-installation basis, provide far greater economies of scale than do crossbar switches.

Evans and Heckman, and Charnes, Cooper, and Sueyoshi represented technological advance for AT&T's toll and local outputs by a proxy—the yearly research and development expenditures of Bell Labs. This proxy is not related to the actual pace of technological advance of specific components of telecommunications plant and thus its use in the empirical investigations added unknown bias to the results.

ECONOMIES OF SCOPE. The lack of evidence of economies of scope is *theoretically* surprising in that, as suggested earlier, global economies of scope must exist in telecommunications networks. The lack of evidence on economies of scope (cost complementarity) as *measured*, however, is not surprising—the data are too limited. While it is possible to hypothesize cost complementarities at the individual exchange level, again, aggregation bias compounds the actual results surveyed. The aggregate data in effect may correlate the actual increase in local service for some year in, say, New England with the decreased marginal cost of toll in that same year in, say, New Mexico. With the aggregate data, one cannot control the experiment so that changes in toll service are correlated with changes in exchange service *over the same production unit*. Aggregate data are just too large a production unit for the analysis. The aggregate data cannot answer the question whether cost complementarities exist between toll service and local service.

SUBADDITIVITY. Finally, what is one to make of the evidence on subadditivity, the test of the existence of natural monopoly? Aside from the papers by Evans and Heckman, and Charnes, Cooper, and Sueyoshi, such tests have been applied indirectly by measuring economies of scope and scale; these tests are then flawed because of the inherent difficulty of extrapolating from AT&T or Bell Canada actual data to a hypothetical world in which only one of the services is produced. This extrapolation requires entering the shadowy world where AT&T (or Bell Canada) is hypothesized as producing either zero toll or zero local output. Extrapolation this far beyond the boundaries of the actual data cannot produce reliable results.[41]

The subadditivity tests reported by Evans and Heckman, and Charnes, Cooper, and Sueyoshi directly examined the shape of the cost function in

41. See Fuss and Waverman, "Regulation of Telecommunications in Canada." In that paper, the authors attempted to measure stand-alone costs only for competitive toll services, an output that was actually zero at the beginning of the estimation period.

output and cost space. The tests as formulated by Evans and Heckman involved dividing the actual outputs between two hypothetical firms in such a way that the minimum output levels actually produced over the period become floor constraints to each hypothetical firm's output levels. Duopolists of various size having between 10 percent and 90 percent of output were examined. This test of subadditivity does not have the fundamental problems of the tests for global economies of scope. Evans and Heckman found substantial significant cost *savings* from dividing output between two firms.[42] Charnes, Cooper, and Sueyoshi found substantial significant *increases* in cost from "breaking up Bell." My view is that the subadditivity test for aggregate AT&T data is so sensitive to data and to econometric technique that it cannot be relied on for making policy. Evidence to back up this assertion is provided below.

Additional Evidence on Economies of Scale and Scope

The studies summarized in table 8 used data for either AT&T or Bell Canada and are thus limited in the number of data points available (thirty or twenty-eight observations, respectively). In this section, I report the results of estimating the cost function given in equation 1 and the factor share equations given in equation 2 with cross-sectional time-series data for both AT&T and Bell Canada (fifty-eight observations).

Several changes had to be made to equations 1 and 2 for this cross-sectional time-series analysis. First, a dummy variable, *DC*, was introduced to allow for the differences in the relative size of the two firms.[43] Second, the proxies for technical change differ between the U.S. and Canadian data series. In the U.S. data, technical change is represented by Bell Labs' research and development expenditures. In the Canadian data, two representations of specific technical advances in telecommunications were used—the rate of introduction of direct distance dialing and the rate of introduction of electronic switching. It is impossible to estimate equations with cross-sectional data that measure technical change in different ways. Therefore, two approaches were used. In the first, technical change (t in equations 1, 2, and 6) was represented by research and development data in the United States and by the rate of introduction of

42. Divestiture was not the subadditivity test applied by Evans and Heckman because it requires zero local output for the firm. Evans and Heckman, and Charnes, Cooper, and Sueyoshi provide little guidance on the effects of divestiture.

43. The change is to introduce a parameter that will compensate for the large difference in total and factor costs between the two firms given their relative size.

direct distance dialing in Canada. Any second-order terms in t were set to zero, however, with technical change constrained to be Hicks neutral. The second approach was to measure technical change in both countries by the passage of time, with second-order terms allowed to be nonzero.

The results of this analysis were unsatisfactory but no worse than the results of Evans and Heckman, and Charnes, Cooper, and Sueyoshi. Before examining these data, one should note that the emphasis in Charnes, Cooper, and Sueyoshi is the superiority of their approach to that employed by Evans and Heckman. Charnes, Cooper, and Sueyoshi castigated Evans and Heckman for insufficient and incomplete testing for bias and instability of regression coefficients. In addition, they lamented that other authors have not criticized Evans and Heckman because they are "all members of a discipline [economics] where the checks we are using are not commonly employed—in which case the advisability of security cross checks from other disciplines again becomes apparent."[44] Neither set of authors appears to have used the simplest cross-check of all—the calculation of marginal costs, since neither set reported them. These marginal costs can be calculated from the data and the respective parameter estimates. The overall scale elasticity is simply the obverse of the sum of weighted marginal costs (see equations 5 and 6).

Examination of the results shows that the estimated marginal costs are meaningless.[45] The estimates by Charnes, Cooper, and Sueyoshi of the marginal costs of interexchange service fall monotonically during 1956–66 (by nearly 90 percent) and then turn *negative* for nine of the last eleven years. Charnes, Cooper, and Sueyoshi's estimated marginal costs of local service are positive. Evans and Heckman estimated negative marginal costs for thirteen years of local service and for six years of interexchange service. These results of Charnes, Cooper, and Sueyoshi are no better or no worse than the results of Evans and Heckman—negative marginal costs for interexchange service.[46]

I had hoped that pooling the data on both Bell Canada and AT&T would allow estimation of a "sensible" cost function. This was not the case. All the estimates yielded negative marginal costs for some output levels. These analyses together with those of Evans and Heckman, and Charnes, Cooper, and Sueyoshi show that the available aggregate data in

44. Charnes and others, "Goal Programming/Constrained Regression Review," pp. 27–28, 29. This language was modified somewhat in the published version.

45. Detailed results are available from the author on request.

46. A method (one neither recommended nor used here) to "correct" the negative marginal costs is to constrain them to be nonnegative.

telecommunications are insufficient, without further constraints, to allow estimation of a cost function and its characteristics such as scale and scope. To measure such technological characteristics, either disaggregated data are necessary or additional information and constraints must be included. Fuss and Waverman reported that the translog cost function and its variants produced negative estimated marginal costs for Bell Canada local service unless additional information on the characteristics of demand and additional constraints on the behavior of the firm (profit maximization) were imposed.[47] I have not attempted to estimate this extended model here because of lack of data.

Some discussion of the results of the cross-sectional time-series analyses, however, is warranted. First, analysis of variance tests indicated that, at the 95 percent confidence level (but not at the 99 percent level), the parameter estimates for AT&T and for Bell Canada were taken from the same population. This analysis then indicated that pooling the data could be done—the two firms operated under similar technological conditions.

Neither of the two approaches to incorporating technical change yielded superior results. Table 9 reports the parameter values obtained from using the explicit proxies for technical change (rather than from using time as a proxy) and setting second-order terms involving T equal to zero. Results are reported for the translog and for a generalized Box-Tidwell translog function using a different Box-Tidwell parameter (BT) for each variable.[48] Serial correlation (SC) was corrected for by using a Zellner iterative procedure. Estimating the BT parameters and the degree of SC jointly requires additional information to identify the BT and SC parameters separately. A grid-search method was used whereby the BT parameter was set at levels between -2.5 and 2.5, and the SC value was estimated for each value of BT. The results reported here are those which maximized the logarithm of the system likelihood function ($BT(\alpha) = -0.5; SC = 0.8545$).

I calculated the estimated marginal costs for local and interexchange service from the parameter estimates given in table 9 for the translog model.[49] These results also are unsatisfactory. Although positive marginal costs were observed at all Canadian data points, marginal costs were

47. Fuss and Waverman, "Regulation of Telecommunications in Canada."

48. Evans and Heckman constrained the Box-Tidwell parameter to be the same for all variables. This constraint was likely due to the few degrees of freedom available to them. The parameter estimates in the cross-sectional time-series analysis of both AT&T and Bell Canada data demonstrated that the Box-Tidwell parameters significantly differed from each other. See Kiss and others, "Economies of Scale and Scope in Bell Canada," for the same point.

49. Detailed results are available from the author on request.

Table 9. *Cross-Sectional Time-Series Analysis of AT&T and Bell Canada Data*

	Translog cost function[a]		*Generalized translog cost function*[a,b]	
Parameter	*Estimate*	*Standard error*	*Estimate*	*Standard error*
Constant	9.467	0.010	9.465	0.021
Capital	0.535	0.005	0.541	0.006
Capital^2	0.222	0.015	0.242	0.012
Capital and labor	−0.152	0.012	−0.169	0.011
Capital and local	0.354	0.064	0.197	0.049
Capital and toll	−0.156	0.039	−0.066	0.030
Capital and technology	0.036	0.012	0.045	0.012
Labor	0.348	0.004	0.343	0.005
Labor^2	0.146	0.021	0.157	0.020
Labor and local	−0.336	0.050	−0.171	0.040
Labor and toll	0.148	0.031	0.050	0.024
Labor and technology	−0.042	0.010	−0.042	0.009
Toll	0.292	0.123	0.128	0.145
Local	0.155	0.219	0.728	0.237
Technology	0.068	0.070	−0.099	0.051
Local and toll	−0.547	1.352	−1.970	0.958
Local^2	0.676	2.075	3.568	1.472
Toll^2	0.418	0.884	1.279	0.613
Dummy (Canada)	−3.002	0.015	−2.894	0.034
Local and dummy (Canada)	0.237	0.241	−0.478	0.222
Toll and dummy (Canada)	−0.154	0.133	0.050	0.119
Technology and dummy (Canada)	−0.027	0.095	0.036	0.059
Capital and dummy (Canada)	−0.043	0.007	−0.049	0.008
Labor and dummy (Canada)	−0.016	0.006	−0.011	0.006
Rho	0.624	0.002	0.855	0.001

Source: Author's calculations.
a. Corrected for serial correlation.
b. $\alpha = -0.5$.

negative for local service in the United States for the last seven years of the sample. Sixteen of the estimated annual marginal costs of U.S. toll service calculated from the Box-Tidwell generalized translog model are negative, as are five of the estimated annual marginal costs of local service in Canada. Moreover, the *size* of the estimated marginal costs of each service differs substantially depending on which measure of technical change was used, indicating that even the level of costs is dependent on the estimation technique used.

Summary

My view is that neither scale nor scope was significant in the 1947–77 period at the level of a firm such as AT&T before divestiture. First, AT&T was one of the largest firms in the world in the pre-1981 period. It provided local telephone service to 150 million telephones and toll service over the entire United States. The "firm" consisted of twenty-seven subsidiaries, twenty-four at the state telecommunications operating level, one providing interstate interexchange service, one research and development firm, and one supplier. It is unlikely that this mammoth organization experienced economies of scale and scope. The more fundamental questions are, first, whether at the "plant" level scale and scope in telecommunications services existed and, second, if there were overall networking economies of scale.

It is unlikely that significant economies of scale existed in interexchange service between 1950 and 1980. Microwave was conducive to multifirm transmission facilities over many routes, with demand levels for several thousand voice circuits. Beginning in 1947 AT&T battled to prevent others from using that technology; the interexchange monopoly that lasted until 1968 was not driven by technology but was regulation driven. It is important to note that optical fiber has greatly increased the plant economies of scale in interexchange service at the very time that regulation (and antitrust enforcement) ensured that competition could occur.

I would suggest that cost complementarities were probably minimal between interexchange and local service as well. In an important sense, local facilities have been quite separate from toll facilities.[50] The local loop and local switch route all calls, toll or local (the global economies of scope). The toll switch and interexchange facilities have up to this point been separate and connected to the local switch by a dedicated line. Whether the customer's telecommunications services are handled by its own switch and routed to interexchange facilities or sent by the switched

50. See Peter W. Huber, *The Geodesic Network: 1987 Report on Competition in the Telephone Industry* (Washington, D.C.: U.S. Department of Justice, Antitrust Division, January 1987), pp. 1.2, 1.3, for a description of the "pyramid" with toll switches at the apex and local switching distinct and forming the foundation.

local system by way of a toll switch, interexchange facilities will continue to be separate from local systems. Ignoring network planning aspects, I see no good reasons why joining these quite separate facilities in one firm will lower costs; that is, there are few benefits of vertical integration between upstream (local) and downstream (toll) facilities. There is likely some minimal maintenance sharing—the local and toll switch can be serviced by the same person—but these savings are probably trivial.

Network planning and call routing would require offering some "jointness" of service such as number identification, billing, and the like. More intelligent networks (such as the integrated services digital network, ISDN, or open network architecture, ONA) will allow service providers to signal, bill, and design customer-specific services rather than have the facilities perform these tasks. As long as clear access rules and clear, regulation-supervised pricing exists, it is not necessary for a single firm to provide all services for costs to be minimized.

These views are not necessarily shared by all observers. In his 1987 report to the Department of Justice, Peter Huber said: "Making the parts work together requires a high level of coordination. The companies best able to provide that level of coordination are those that build, own, and operate all the necessary pieces under a single corporate umbrella. And the information age brings with it extraordinary new economies of scope. . . . Vertical integration is the future of the telecommunications market, just as it is the past."[51]

In other parts of his report, Huber suggested that vertical integration of local service, interexchange service, and switching are necessary to provide optimal network services. Without this integration, he argued, technological advance will be retarded and the incorrect mix of transmission and switching will be supplied. Huber stated that AT&T, GTE-United Telecom, MCI-IBM, and Wang are all attempting to provide switching, interexchange service, and local service (at least to very large customers) and that to prevent such integration is futile. He suggested that various vertically integrated firms can survive and provide competitive services but did not prove that vertical integration of facilities is necessary for efficient operation. In my view, where access rules and correct pricing signals exist, the optimal mix of switching and transmission can be provided *without* the need to reverse the modified final judgment.

51. Ibid., pp. 1.7, 1.30.

The Interexchange Market—Market Shares and Regulatory Alternatives

Several pieces of evidence are clear: AT&T's 1986 share of the interstate interexchange market was some 70–75 percent of revenues, 70–75 percent of minute miles, but only 40 percent of capacity. AT&T's shares of revenues and capacity have been falling, with the latter falling much more rapidly than the former as several entrants have built extensive fiber-optic systems. Several observers, including AT&T, view the long-distance market as having many characteristics of a competitive market. Baumol, for example, has argued that "many of the array of interexchange services will, quite properly, have to be judged to be effectively competitive."[52] The test for effective competition that Baumol would apply would involve determining whether competitive forces alone are sufficient to maintain market prices below *stand-alone costs*, the price a firm with "market dominance" would charge.[53]

AT&T has argued that "barriers to entry have virtually disappeared and the interexchange marketplace has become intensely competitive." Other participants have different views. U.S. Sprint has stated that "AT&T continues to have substantial market power. . . . In addition, there continue to exist substantial barriers to entry," but also "Any attempt by AT&T to extort monopoly rents through the use of its remaining monopoly power would be constrained, at least to some extent." Huber has stated that "entry barriers to the inter-LATA market remain high." Simnett argues that economies of scale in marketing and the continued charging for access via interexchange usage provides AT&T Communications with substantial advantages in the interexchange market.[54]

In his recent lengthy decision on removing various restrictions placed on the BOCs by the modified final judgment, Judge Greene devoted but half a page to the degree of competition in the interexchange market. Judge Greene did not issue an opinion on the degree of competition in interexchange but did agree with the Department of Justice that "AT&T's rivals appear to be making sufficient progress [but] that it would be at

52. William J. Baumol, "Modified Regulation of Telecommunications, and the Public Interest Standard." Princeton University, August 25, 1986, submitted as Comments in NTIA docket 61091-6191, December 15, 1986, p. 31.

53. This is not always the correct test, as discussed below.

54. Comments of AT&T in NTIA docket 61091-6191, December 15, 1986, p. 12; comments of U.S. Sprint in ibid., pp. 2, 4; Huber, *Geodesic Network*, p. 3.4; and Richard E. Simnett, "The Infeasibility of Full Competition in the U.S. Long-Distance Telephone Industry under the Present Regulatory Regime," Bell Communications Research, Inc., October 23, 1986.

least premature to view the entry of the Regional Companies as necessary to preserve interexchange competition.''[55]

It is clear that substantial entry barriers exist in the interexchange market, principally the sunk capital that must be precommitted to service markets;[56] the interexchange market is not even imperfectly contestable, or susceptible to hit-and-run entry.[57] Take as an example the National Telecommunications Network (NTN), a partnership of seven companies presently joining a number of cities in the Southeast, Midwest, and West Coast. At 1987 year-end, NTN had 7,979 route miles and 156,421 fiber miles in operation (up 139 percent and 225 percent respectively since 1985).[58] The FCC estimates that this network cost $487.5 million, or $61,097 per route mile, somewhat less than the FCC's general estimated investment costs of $75,000 per route mile.[59] Most of the costs of the NTN or any other fiber-optic system are the cost of cable installation (a sunk cost if there ever was one). The salvage value of these systems is likely to be small—the value of signaling equipment, other transportable equipment, and multiuse buildings.

Besides AT&T, MCI, and U.S. Sprint, several regional networks capable of joining in national networks exist. As already mentioned, most of the expansion of these entrants has occurred since 1980. These investments were predicated either on being the first in the market with low-cost, high-quality fiber-optic systems or on a surge in telecommunications demand.[60] Whatever the reason or the rationality behind these investments, they now represent resources that clearly diminish the market power that AT&T may once have had in some interexchange markets.

These fiber-optic networks can provide effective checks on the use of market dominance precisely *because* they are sunk, long lasting, and have little salvage value. When an oil producer goes bankrupt, the oil resources are not lost, but remain in the ground and are revalued. Bankruptcy of a

55. *United States* v. *Western Electric Co.*, 673 F. Supp. 525, 550 (D.D.C. 1987).

56. Some observers would disagree with this statement. For an opposing view, see Porter, ''Competition in the Long Distance Telecommunications Market,'' pp. 19–23.

57. Some argue that satellites provide the kind of hit-and-run entry the contestable market theory requires because satellite capacity is easily switchable between markets. Satellites are not well suited for large-volume, point-to-point data services and therefore provide little effective competition to terrestrial systems. See Siperko, ''LaserNet.''

58. Jonathan M. Kraushaar, ''Fiber Deployment Update: End of Year, 1987,'' Federal Communications Commission, Common Carrier Bureau, Industry Analysis Division, Washington, D.C., January 1988, table 1.

59. Ibid., table 2, note 2.

60. Porter estimates that at the end of 1986, AT&T, U.S. Sprint, and MCI had fiber-optic capacity almost four times as great as actual total 1986 inter-LATA conversation minutes. Porter, ''Competition in the Long Distance Telecommunications Market,'' fig. 26.

fiber-optic system owner is then equivalent to a lowering of the asset value of the system. The plant remains in place in the ground; it may not be used, or it may be used by different owners. Whatever its ownership status, it represents competitive pressure on the largest carrier—AT&T Communications.

These competitive pressures through ownership of facilities are not ubiquitous; many small interexchange markets have little or no competition. Moreover, in many areas interexchange competition is not through facilities but through the resale and sharing of AT&T Communications' facilities. The ability to provide national interexchange competition to AT&T Communications requires competitors to purchase services (such as WATS or foreign exchange, FX, lines) from AT&T Communications, leaving the *potential* for that company to act strategically to harm its competitors; after all AT&T Communications' share of interexchange revenues is still close to 75 percent.

The view that AT&T Communications' present market share and any residual market power are not based on the facilities it owns is not unique. Both Wenders and Simnett have viewed the present access charge system as a barrier to entry.[61] Simnett argued that, because of this system and AT&T Communications' 80 percent market share (as Simnett estimates it), that enterprise has lower marginal costs.[62] Changing the procedures of recouping fixed costs through interexchange usage charges will remove this distortion, but such a change does not seem imminent.

A second advantage to AT&T Communications may be the fixed costs, and thus economies of scale, of marketing.[63] An advantage to AT&T Communications exists because of "brand recognition," the difficulty new interexchange firms have in persuading customers to change from the firm they know, AT&T Communications. Customers will switch interexchange firms only if they are offered real advantages—lower prices or "better" service. If all firms have similar production costs (there are

61. John T. Wenders, "Local Telephone Competition," University of Idaho, January 1987; and Simnett, "Infeasibility of Full Competition."

62. With a common carrier line (CCL) charge of 7.4 cents per interexchange usage minute, a stimulation of demand by AT&T Communications through lowering its tariff will reduce the CCL charge for everyone; however, AT&T Communications will capture 80 percent of the "new" demand. Thus the effective marginal cost of access to a firm with a large market share is lower than for a firm with a small market share. Simnett, "Infeasibility of Full Competition," pp. 5–6.

63. Porter estimated that AT&T spent some $100 million on advertising in 1986, or 0.5 percent of AT&T revenue. Sprint spent some $40 million on advertising, or 2.0 percent of revenues. Porter, "Competition in the Long Distance Telecommunications Market," figs. 23, 24.

no advantages in pure production of interexchange for anyone), prices tend, as in competitive markets, to be similar. Firms then offer better service—hence the rush to offer full fiber-optic systems. Because AT&T Communications is already known and once had 100 percent of the customers, it need only market to keep them—new firms must market to capture customers. If there are different costs of marketing (capture costs exceed retention costs) and if AT&T Communications has certain advantages—brand recognition, ubiquitous service offerings, operator services, and "800" services—it is not surprising that AT&T Communications' market share is still near 75 percent.

In my view, if there are marketing advantages to AT&T Communications, these are real and substantial. (Far more research is required into any advantages of AT&T Communications that are not facilities driven.) Combining marketing advantages with consumer preferences provides AT&T Communications with an advantage over competitors (similar to the "first mover" or "incumbent" advantage in much of the literature in industrial organization). The advantage is neither artificial nor anticompetitive, but one that limits the ability of *new* entrants to compete for large segments of the interexchange market. Yet it is the competitive pressure of these OCCs that provides AT&T Communications with the incentive to be efficient.

Regulatory Alternatives

The interexchange market is made up of a diverse set of geographic areas, a diverse set of buyers, and an equally diverse set of service providers. There are actually numerous interexchange markets—some composed of connections between large urban areas used mainly by large businesses, others composed of connections between small rural communities used mainly by individual households.

In this new, potentially competitive environment, what rules if any should be established to constrain AT&T Communications' behavior? In addition, what role, if any, should the BOCs play? In all this, it is clear that the minimal set of rules consistent with encouraging, not hampering, competition is required. The best rules are those which are "incentive compatible"—that is, induce the firm to choose prices that are in the social interest. The potential set of rules is as follows: (1) the status quo—that is, rate-of-return regulation of AT&T Communications with the full panoply of tariffing procedures now in place, forbearance on all OCCs, BOCs not allowed to offer interexchange service; (2) some form of liberalized regulations on AT&T Communications (several to be discussed below), no BOC entry; (3) liberalized regulation of AT&T Communications and BOCs, BOCs' entry into interexchange allowed; and (4) abolition of the FCC—that is, no regulation. I discuss each of these in turn.

STATUS QUO. The status quo is unlikely to remain. There is little merit in engaging in natural monopoly rate-of-return price setting for one firm in an industry consisting of several competitors, even when some residual market power exists. The reasons for dropping rate-of-return regulation are legion and are well covered in several publications.[64]

Some states (eleven as of January 1987) have abandoned traditional rate-of-return regulation for inter-LATA interexchange service within states, and others (thirteen) have introduced flexible pricing. These state commissions in general agree that inter-LATA interexchange services are competitive, hence that AT&T Communications has little market power. All of these same state commissions continue to regulate *intra-LATA* interexchange, which in most cases is subject to total entry restrictions, many authorizing the BOC as the monopoly carrier.

This schizophrenic view of the degree and benefits of competition, which appears to be distinguished by the border of an artificial region (the LATA), is fairly easy to explain. Many state regulators view inter-LATA competition as "safe" because access charges are fixed regardless of the price of an interexchange call. Competition therefore cannot diminish the subsidy from inter-LATA to local service. Intra-LATA interexchange service does not bear some known and fixed access charge. Therefore, competition in intra-LATA is viewed by regulators as decreasing the subsidy to local service, and intra-LATA competition is to be feared.[65]

64. See Wynns, "AT&T's Share of the Interstate Switched Market"; NTIA, "Comprehensive Review of Rate of Return Regulation of the U.S. Telecommunications Industry," notice and request for comments, docket 61091-6191, *Federal Register*, vol. 51, October 16, 1986, pp. 36837–41; and Baumol, "Modified Regulation."

65. For example, on April 15, 1985, the Public Service Commission of Nevada endorsed inter-LATA interexchange competition and the abandonment of rate-of-return regulation on interexchange carriers but disallowed intra-LATA competition: "The Commission is persuaded that the current state of the intrastate telecommunications environment is such that effective competition can develop. For the following reasons, the Commission finds that the public interest will best be served by implementation of rules which provide for the equal and streamlined regulation of all intrastate inter-LATA carriers." *In Re Petition of AT&T Communications of Nevada, Inc.*, docket 84-758, slip opinion (Public Service Commission of Nevada, April 15, 1985), p. 13. Yet "the Commission must conclude that it is not prudent at this time to allow intra-LATA competition." Ibid., p. 19. The commission went so far as to disallow intra-LATA resale, even though at least one firm had been providing service. One means of opening up the intra-LATA market to competition might be to fix an access charge, as in inter-LATA service.

The reader should be warned that states' abandonment of rate-of-return regulation over inter-LATA interexchange within states is not synonymous with deregulation. No state has deregulated interexchange. In place of rate-of-return regulation, most states impose strict reporting requirements and other forms of tests that are close to rate-of-return regulation. For example, the Georgia Public Service Commission stated: "It is further found that competitive forces make strict rate of return regulation unnecessary. Rather, it should be sufficient to protect the public interest for carriers to demonstrate as to rate decreases that the service is not priced below cost and as to rate increases that present cost exceed expenses."[66] Having said that the days of rate-of-return regulation are numbered does little to establish alternatives. I do not discuss alternatives more suited for the BOCs—the "social contract" phenomenon in Vermont and Nebraska. What I do consider are the needs, if any, in maintaining tariffing procedures and the types of price rules possible in the interexchange market.

LIBERALIZED REGULATION OF AT&T COMMUNICATIONS (NO BOC ENTRY). There are multiple reasons for continuing some form of regulation over AT&T Communications (and over access): to prevent overall monopoly profits; to prevent predation and strategic abuse of competitors; to prevent monopolistic abuse of some customers; to maintain political concerns (geographic averaging of rates; it is politically difficult to totally deregulate); and to prevent inefficient bilateral monopoly negotiations.

In the past, AT&T has attempted, as all firms would, to limit competitive entry. AT&T used or abused its pricing structure and the regulatory process to frustrate entry.[67] AT&T Communications still has some market dominance—over competitors that are dependent on it for the lease of facilities and over customers who have no or few competitive alternatives—and an advantage in marketing. AT&T Communications and others stress the degree of competition that has arisen in interexchange.[68] The FCC's data indicate that more than 80 percent of U.S. telephone lines have been converted to equal access.[69] Porter has stated that AT&T Communications competes with at least one facilities-based interexchange car-

66. *In Re Interexchange Telephone Carrier Regulation and Proposed Rule Making*, docket 3522-U, slip opinion (Georgia Public Service Commission, January 8, 1986), p. 11.

67. See Faulhaber, *Telecommunications in Turmoil*.

68. For example, John R. Haring and Evan R. Kwerel, "Competition Policy in the Post-Equal Access Market," Working Paper 22 (Washington, D.C.: Federal Communications Commission, Office of Plans and Policy, February 1987).

69. Common Carrier Bureau, "Trends in Telephone Service," p. 14.

rier in 87 percent of all LATAs.[70] In most markets, two is not a number synonymous with a great deal of competition, especially when one firm has 75 percent of the market, another 12 percent, and the rest consists of a competitive fringe.[71] FCC data (table 5) also show the presence of many interexchange firms that have received a carrier identification code for access payment purposes. These data show the number of resellers of AT&T Communications' facilities or the degree of arbitrage possible in AT&T Communications' price schedule but not the degree of competition in the interexchange market. These resellers are true hit-and-run entrants that have little sunk capital and lease their "equipment" from AT&T Communications at prices decided by AT&T Communications. In a truly contestable market hit-and-run firms are not dependent on the incumbent firm for their capacity and its costs.

The existence of a large number of "competitors" to AT&T Communications is then partly a function of the degree of price discrimination the company uses—a characteristic of entry under its control.[72] AT&T's major competitive weapon has always been pricing—private-line tariffs, Telpak, WATS—to deter entry of private microwave systems or to discourage the growth of OCCs. Although there is little evidence to judge whether these tariffs were predatory, their existence does provide a basis for judging the value of regulatory alternatives. First, AT&T always responded to entry threats by reducing prices (suggesting another reason why the contestable market story is not appropriate here). Second, in the past AT&T had the potential to cross-subsidize in these competitive markets—not from local service where losses were being made, but from monopoly MTS, then and now an area of profits (or "contributions").

One would then expect price competition from AT&T Communications—the question being whether that entity, in the future, will be "foolish" enough or, more important, able to cross-subsidize. As Baumol has stated, competition "prevents cross subsidy because, by definition, such an arrangement entails the overpricing of the products that are the sources of the cross subsidies and that, in turn, is an invitation to entrants to take

70. Porter, "Competition in the Long-Distance Telecommunications Market," p. 11.

71. At the turn of the century, business users in many communities had two possible vendors of local telephone service. This brief period of competition did lower rates to most users, business and residential.

72. A similar point is made by Simnett, "Infeasibility of Full Competition," pp. 15–17. Porter, "Competition in the Long Distance Telecommunications Market," pp. 19–23, recognizes this point by stressing the ability of resellers and VANs to lease facilities from AT&T's competitors.

over the supply of the overpriced products.''[73] The issue then becomes whether AT&T serves any areas where entrants will not respond to higher prices. If such areas exist, then AT&T has some market power—the ability to raise price over costs.

Robson and Levinson of Multinational Business Services have attempted to measure the areas where AT&T Communications is ''uncontested'' (that is, faces little or no competition). They suggested that four areas of uncontested service exist: (1) certain segments of intra-LATA service, (2) low-density interexchange markets, (3) small-volume customers (85 percent of all business customers and 75 percent of all residential customers), and (4) short-haul, private-line service.[74] Robson and Levinson suggested that these ''uncontested'' markets, rather than being the source of further monopoly profits, are areas of losses or are uncontested because they are unprofitable. *They found that these uncontested markets account for 46 percent of AT&T's long-distance revenues and 80 percent of the number of customers.*[75] Their calculations therefore suggest that competition in the interexchange market exists for 54 percent of AT&T's revenues, but for only 20 percent of its customers. They calculate that AT&T Communications has a 68 percent market share in the ''contested'' market where AT&T Communications makes all its profits.

Although toll prices have fallen since 1983, at the margin toll prices are still well above their costs.[76] Because 80 percent of AT&T's customers are uncontested, there may be a substantial market where AT&T Communications could exercise power in the future if totally unregulated.[77] The potential to cross-subsidize other services may exist. The natural question to ask, therefore, is what services might AT&T Communications want or be able to cross-subsidize in the future if unregulated.

One hypothesis is that the competition for large-volume business users would drive prices below marginal cost for the toll firm with deep pockets. This hypothesis does not withstand careful analysis. In the past AT&T

73. Baumol, ''Modified Regulation of Telecommunications,'' p. 6.

74. Ernest S. Robson III and Bruce Scott Levinson, ''Competition in the Telecommunications Marketplace,'' Multinational Business Services, October 1986.

75. Ibid., pp. 13, 17. Note that Porter, while relying on the 1986 Multinational Business Services report for some of his facts, did not advance this argument and in fact argued the opposite, that AT&T faces competition in nearly all of its markets.

76. Wenders and Egan show that in 1983 AT&T's interstate toll prices were $0.42 and $0.27, with marginal costs of $0.066 and $0.086 for business and residential service, respectively. Wenders and Egan, ''Implications of Economic Efficiency,'' p. 36.

77. Viewing prices and costs at the margin suggests that in the future far less than 80 percent of the number of customers will be uncontested.

may have wished to forestall entry at the margin, reasoning correctly that the opening wedge would lead (and did lead) to the deluge. Now AT&T Communications has 40 percent of installed circuit miles, and its competitors' circuit miles sunk in the ground *cannot* be driven out of the market by low prices. The firms (temporarily) owning these circuit miles may be driven out, but the circuit miles will then be sold, and new owners will appear. The new owners will then have costs substantially lower than AT&T Communications because of AT&T Communications' strategic behavior against previous owners. Predation against lower-cost rivals is a good form of suicide, not a useful business tool. Moreover, attempts to gain market share against competitors in the interexchange market by pricing below cost have one obvious strategic response—the competitors could leave these large customers to AT&T Communications, saddling it with losses. If prices cannot be raised in uncontested markets (because some ancillary form of regulation exists) or if most markets become contested, AT&T Communications is only worse off by engaging in predation. Simple price controls (pricing schemes suggested several years ago by Linhart and Radner), such as preventing price increases in uncontested markets *if* other prices are lowered, could be sensible regulatory tools because it is in uncontested markets, not in the connections between large urban areas, that AT&T Communications has some residual market power.[78] Price "caps" or ceilings are another suggestion, and one dealt with below. Maintaining geographic rate averaging, although inefficient, would also prevent increases in prices in low-density areas.

AT&T Communications may be able to constrain competitors in other ways. As already stated, it can determine the number of resellers or competitors by altering its price schedule. In addition, its charges for facilities needed by competitors in thin markets (WATS, FX, and the like), if above its marginal cost but below the competitors' stand-alone costs, will diminish competition.[79] Some regulatory controls may therefore be required to prevent AT&T Communications' pricing services or facilities so as to raise rivals' costs while it enjoys these same services or facilities at marginal cost. One means of controlling this possibility is to

78. P. B. Linhart and R. Radner, "Deregulation of Long-Distance Telecommunications," in Vincent Mosco, ed., *Policy Research in Telecommunications: Proceedings from the Eleventh Annual Telecommunications Policy Research Conference* (Norwood, N.J.: Ablex, 1984), pp. 102–14; and P. B. Linhart and R. Radner, "Relaxed Regulation of AT&T, Reconsidered," paper presented at the Fourteenth Annual Telecommunications Policy Research Conference, Airlie, Va., April 1986.

79. See Steven Salop and David T. Scheffman, "Cost-Raising Strategies," Georgetown University, July 1986.

force AT&T Communications to unbundle all charges and to act as a carrier's carrier—allowing all firms, users or competitors, to use any facility by paying the unbundled tariff. AT&T Communications would in its pricing of an end-to-end service have to include the tariffed amount: that is, it would have to charge its customer and itself the same as it would charge a competitor. Such a tariff would be incentive compatible.

If AT&T Communications has overall advantages in interexchange provision, price ceilings and floors (or alternatively, price caps) can also prevent abusive actions. It might then earn higher rates of return than if regulated; such a return, however, is a "reward" for its superior advantages, not evidence of monopolization.

BOC ENTRY INTO INTEREXCHANGE AND REGULATION. The modified final judgment and Judge Greene's 1987 review of it require that the BOCs not offer inter-LATA interexchange service. The BOCs suggest that they be allowed to serve this market; that the BOCs would be major competitors to AT&T; that customers want, and BOCs should be able to provide, end-to-end service.

The question of the entry (or reentry) of the BOCs into inter-LATA interexchange requires considering the *social* purposes, costs, and benefits that would be served by that entry; how that entry would or could occur; and the regulatory oversight, if any, that would be required. A potential benefit of BOC entry into the interexchange market would be the enhanced competition with AT&T Communications. The major argument against BOC entry into the interexchange market is that this renewed vertical integration revives the potentials for abuse of bottleneck facilities—the vertical foreclosure problem, the major reason for the Department of Justice's suit against AT&T and the major motivation for divestiture.

In my view, access to the local system can be maintained on a nondiscriminatory basis if BOCs are allowed to compete for interexchange service. First, equal access as mandated by the MFJ ensures that each interexchange carrier receives the same technical quality of connection at the same trunk-end point in the local system. There would then be little potential for the BOCs to misuse the technical conditions for access to prevent competition. Huber suggests that LECs would delay or refuse to provide access services and that the LECs have substantial control over network architecture and information.[80] If these advantages are abused, however, antitrust remedies are available to competitors.[81]

80. Huber, *Geodesic Network*, p. 3.29.

81. It is odd that Judge Greene, in writing the MFJ to an antitrust case, dismissed future

The entry of all seven BOCs into interexchange services may prevent any single BOC from abusing its market dominance. Assume BOC *A* attempts to engage in strategic anticompetitive abuse of the technical conditions of access or of network architecture. Interexchange carriers AT&T, MCI, and U.S. Sprint, which are not in the business of local service, may be unaware of their disadvantage resulting from unequal technical access. However, BOCs *B*, *C*, *D*, . . . , *N* will not be so fooled. Because all BOCs offer access, each will be well aware of any gaming by BOC *A*. As long as there is some agency (perhaps antitrust) to complain to, the technical abuse of access will not be an issue.[82]

The principal benefit of permitting BOCs to offer interexchange service is that they would provide strong competition to any attempt at market dominance by a single interexchange carrier. Any advantages that AT&T Communications has in marketing or in uncontested markets could be matched by the BOCs.

Allowing unlimited entry into interexchange by the BOCs, however, places AT&T Communications at a distinct disadvantage. Most states have legislated BOC monopolies in intra-LATA interexchange, providing BOCs with pockets of market power. Combining this monopoly and a monopoly over local access provides the BOCs with a customer base and marketing and billing advantages. One is therefore left with potential harm from unlimited BOC entry into the interexchange market (tying these services to its monopoly provision of local services), as well as with potential benefits (increased competition to AT&T Communications, especially in ''thin'' markets).

There are possible ways of harnessing the benefits of BOC entry into interexchange and minimizing the potential for competitive abuses. One means of avoiding vertical foreclosure is to allow BOCs to compete for interexchange service in all markets except their own (the suggestion of Peter Huber).[83] Each can then use its expertise to market end-to-end service

---

antitrust action as an effective constraint on potential BOCs' abuses in the interexchange market. See *United States* v. *Western Electric*, 673 F. Supp. at 550.

82. One reason that equal access was not the sole goal of the MFJ is that an undivested AT&T may have been ready and able to thwart equal access. Divested BOCs have had no such incentives to favor particular interexchange carriers.

83. See Huber, *Geodesic Network*, chap. 3. This was the original position of the Department of Justice in the 1987 analysis of the restrictions on the BOCs; see U.S. Department of Justice, Antitrust Division, ''Report and Recommendations of the United States concerning the Line of Business Restrictions Imposed on the Bell Operating Companies by the Modification of Final Judgment,'' filed February 2, 1987, in *United States* v. *Western Electric*, civ. no. 82-0192. The department later reversed itself. Judge Greene dismissed the concept as unworkable. See *United States* v. *Western Electric*, 673 F. Supp. at 543-44.

in all local territories but its own. Because they cannot provide interexchange service in the market where they have a local monopoly, the BOCs cannot engage in vertical foreclosure. Although this suggestion solves the problem, it most likely does so by minimizing BOC entry into interexchange.

A second possible means of allowing BOC entry into interexchange, and one not raised in the Department of Justice's 1987 review, is to allow the BOCs in their own territories to enter into consortia with OCCs but not with AT&T Communications. Allowing direct contracting between the BOCs and AT&T Communications would rekindle fears of vertical integration or foreclosure. In this plan, BOC *A* can provide end-to-end service by contracting with OCC *B* (but not with AT&T Communications) for interexchange service to all BOC areas. AT&T Communications can contract with BOC *B* to provide interexchange service in BOC *A*'s geographic area (part of the Huber proposal). To make this competition real and symmetric, (1) any restrictions on intra-LATA service would have to be eliminated; (2) the interexchange service provided by any BOC or OCC combination would have to be unbundled and tariffed so that AT&T Communications or any other OCC could purchase the local segment at the same price as the BOC was charging itself (its customer); and (3) AT&T Communications or any OCC could bypass the local system by any means at any time. Under this scenario one could foresee BOC *A* advertising interexchange service and AT&T Communications countering that the interexchange carrier is not AT&T. Consumers would not know which part of the former Bell system to believe!

Significant BOC-OCC contracting may not take place. AT&T Communications will, however, be severely limited by the threat of BOC entry into interexchange. Similarly, although AT&T Communications might not directly offer intra-LATA service, its threatened entry will forestall BOC abuse.

DEREGULATION. Total deregulation of the interexchange market is an unlikely scenario politically. No carrier (including AT&T Communications) is calling for total deregulation. The advantages of deregulation are the reductions in the inefficiencies occasioned by regulation. Sufficient potentials of abuse of market dominance, however, still exist—abuses by AT&T Communications or the BOCs directed at rivals and at customers. An important additional reason for regulation has been suggested—regulation of access rules and prices so as to prevent bilateral monopoly bargaining, which would lead "back to the future," to vertical integration.

There equally would appear to be no good reason to maintain the present form of asymmetric regulation or to prevent totally the BOCs' providing interexchange service. I now turn to a discussion of the types of possible price regulation over interexchange competitors.

Reasonable Price Regulation

Pricing is a major form of competition—consider the Telpak, WATS, or Reachout America pricing packages. Each of these is not a "new" service but simply a form of price discrimination. Although it is unclear (and probably untrue) that any of these pricing packages or services was predatory, none of them was justified on the grounds that the *difference* in price (between, say, switched MTS and Reachout America) was equivalent to the *difference* in the costs of providing the two services. The discount explicit in the Reachout America tariff, then, is not equal to the reduction in costs between offering MTS and Reachout America. Indeed, there are no cost savings in Reachout America compared with MTS—the same facilities and metering costs are incurred whether a customer uses an hour of toll or an hour under Reachout America. Reachout America, like Telpak and WATS, is purely and simply price discrimination. Equating these services with price discrimination is not condemning them (although as evidence of price discrimination, they may be evidence of market power)—price discrimination in most cases is welfare improving,[84] and in the presence of economies of scale, price discrimination or pricing above marginal cost is necessary.[85] The problem for regulation is whether and how AT&T Communications should be controlled in its offering of new services through price-discriminatory tariffs. What, if any, bounds should regulation impose on the tariffs of AT&T Communications?

Baumol has proposed two price controls—*incremental cost* as a price floor and *stand-alone cost* as a price ceiling. He has argued that any price above incremental costs is not predatory and that any price below stand-alone cost is not abusive of market dominance. There are two problems with these (and other) price tests.[86] First, it is not obvious that stand-alone

84. See Louis Phlips, *The Economics of Price Discrimination* (Cambridge University Press, 1983).

85. See Baumol, "Modified Regulation of Telecommunications"; or Stephen J. Brown and David S. Sibley, *The Theory of Public Utility Pricing* (Cambridge University Press, 1986).

86. It is clear that marginal costs are the "correct" floor rate. Measuring marginal costs is not an easy proposition, as was demonstrated in the section "Economies of Scale and Scope," above.

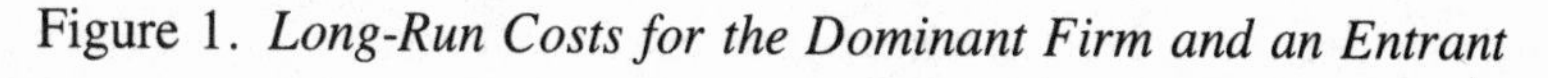

Figure 1. *Long-Run Costs for the Dominant Firm and an Entrant*

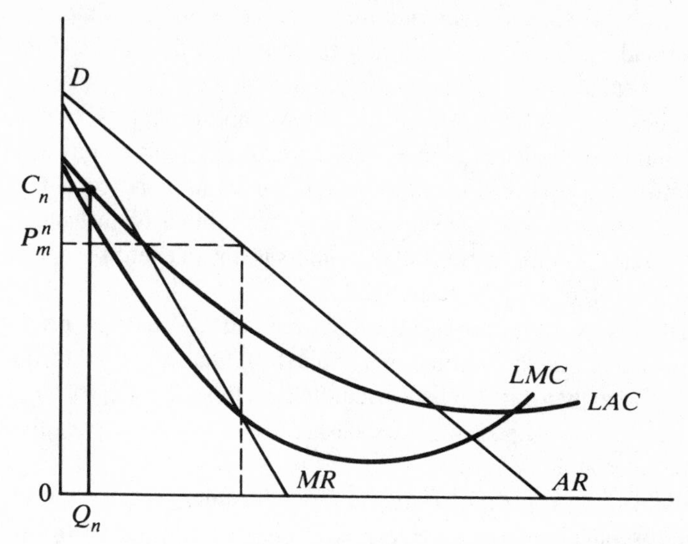

costs are a limit on monopoly pricing; second, there are substantial empirical problems in providing correct measures of these theoretical concepts.

In figure 1, *LAC* and *LMC* represent the dominant firm's long-run average and marginal costs, respectively, of providing the quantity of service Q. The individuals, Q_n, a subset of Q, can construct their own facilities at stand-alone costs C_n; C_n then limits the monopoly price for this n subset, P^n_m, if and only if $C_n < P^n_m$. If there are economies of scale in providing interexchange service to a number of customers, $C^n_m < C_n$. It is thus possible for $C^n_m < P^n_m < C_n$—in effect, blockaded limit pricing.

Whether stand-alone cost as measured for a single customer or a set of customers is less than the monopoly price is an empirical, not a theoretical, question. An ancillary reason why stand-alone cost for a single customer or set of customers is an inappropriate regulatory constraint is as follows. In the theoretical literature it is the minimum *set* of all stand-alone costs measured for all possible sets of customers that represents the appropriate pricing signal.[87] The correct regulatory signal then requires measuring *not* the stand-alone cost for AT&T Communications or for

87. See Mirman and others, "Supportability, Sustainability, and Subsidy-Free Prices"; and Brown and Sibley, *Theory of Public Utility Pricing*.

some particular service or customer, but the minima of the stand-alone costs for all the various *combinations* of services and customers.

A second major problem in using these tests is the great difficulty in measuring stand-alone costs for a regulatory proceeding. As already discussed, econometric evidence at this point is not terribly useful. Further, this econometric evidence produces (if it could) the stand-alone costs using AT&T Communications technology, not the stand-alone costs of an efficient competitor. The measurement of the minima of the stand-alone costs for various combinations of customers is not an exercise that is likely to appear in regulatory proceedings.

I therefore am not convinced of the realism of ''banded pricing''—setting price floors and ceilings while allowing market prices to fluctuate freely between them. It will be difficult to develop theoretically valid price rules, and nearly impossible to measure these prices, especially in an adversarial regulatory setting.

I would impose a simpler solution—constraints on price increases in the interexchange market. Price is well above marginal cost for MTS services; it is therefore not necessary to worry about the need to raise prices to ensure that no subsidies exist. I would therefore deregulate AT&T Communications from all rate-of-return or pricing rules and require for it that (1) tariffs be published; (2) all services be unbundled; (3) any new service price not be allowed to be *raised* for two years without a hearing; (4) prices in ''uncontested'' interexchange markets not be allowed to rise—more important, their rate of price decrease must be no less than the average rate of price decrease for prices in contested markets (continuing geographic averaging would assist as well);[88] (5) any services used by OCCs or competitors must not be available on superior terms or conditions to AT&T Communications; and (6) services that exist only as differentiated by the quantity taken (that is, WATS or bulk user discounts) cannot be changed without several months' lead time—it is these services that are essential to resellers and OCCs to function, and these entities should at least have adequate warning of changes in their costs of operation.

Price Caps

The FCC has proposed the replacement of traditional rate-of-return regulation of AT&T Communications with a set of price caps. The price-cap model has been used in England for the regulation of the privatized

88. If these prices were demonstrated to be below ''costs,'' price rises would occur.

British Telecom. It is important to understand that the price-cap model was intended to cap the prices of *monopoly* services (local exchange) and to deregulate other services. As implemented, the scheme put caps on the overall index of exchange services as well as on local and domestic toll services. The benefits of price caps and the details of their implementation have been discussed elsewhere.[89] The issue here is their appropriateness for interexchange service and for AT&T Communications alone.

Given the state of competition in interexchange, it would not be sensible to define it as a monopoly. Yet the price-cap model has been developed for monopoly service. The FCC notice of rulemaking appears to suggest a set of constraints for AT&T Communications that are hardly a decrease in regulation. The FCC, besides being concerned about proper overall price indexes, measures of productivity increase, correct time periods for review, and so forth, is also concerned about the various sets of services to be combined under various price caps, the need to control new service offerings, and the need to monitor whether price caps are effective—namely, is AT&T Communications earning monopoly returns? These latter concerns, although perhaps politically motivated, make little economic sense. To replace present rate-of-return regulation by more complex rate-of-return regulation is misguided. In my view, a price-cap approach to regulation of interexchange should consider the areas where AT&T Communications may have some market power (that is, uncontested markets). A cap on the average price for these services and deregulation of all other services could prove effective and relatively simple to administer. If all AT&T services are to fall under price caps, one would want to keep "uncontested services" under a separate price cap to prevent any potential abuse of prices to benefit AT&T Communications.

Summary and Conclusions

—The inter-LATA interexchange market is competitive—not perfectly, since AT&T Communications has real advantages.

—Certain segments of the inter-LATA interexchange market are more competitive than others. Large-volume business users that wish to send traffic between two large population centers will have several facilities carriers and interexchange competitors offering service. Low-volume residential customers sited in areas of low population density may not have

89. See Linhart and Radner, "Relaxed Regulation of AT&T, Reconsidered"; and Haring and Kwerel, "Competition Policy in the Post–Equal Access Market."

many alternatives to AT&T Communications. These facts are not reasons to regulate AT&T Communications or to prevent entry into interexchange but are similar to facts for other services (retail trade, airlines, newspapers, and the like).

—It is unlikely that predation against any rival will prove to be an advantageous business proposition for AT&T Communications. One important factor is the excess capacity already present in fiber-optic systems. These sunk investments act as backstops preventing attempts at predatory abuse.

—Certain customers (up to 80 percent of AT&T Communications' customers) might be "abused" by a totally unconstrained AT&T Communications. Simple price rules (a cap on the average price of such services) can prevent such abuse.

—Rate-of-return regulation is an anachronistic mechanism to set prices in the interexchange market and should be replaced.

—There are benefits to BOC entry into interexchange, but overall disadvantages if BOCs are able to reintegrate with AT&T Communications.

—There are unlikely to be higher operating costs emanating from divestiture. This statement has no real backup because the econometric studies summarized provide no substantive evidence on this point.

—The purpose of interexchange regulation should be to determine the rules and price of access and to establish simple price rules that maximize competition and minimize interference with the market.

—Stand-alone costs and average incremental costs are not workable price rules.

—*Simple* price caps are workable if they are imposed on services for which ideal competition does not exist; in the long term, with sensible regulators, these price caps can be akin to economists' notions of correct prices. Complex price cap rules could turn out to be complex rate-of-return regulation.

—BOCs should be allowed to offer inter-LATA interexchange service if (1) they do not contract with AT&T Communications (that is, new forms of the old vertical integration should be disallowed), (2) intra-LATA interexchange is opened to competition, and (3) local bypass is permitted.

—AT&T Communications should not be permitted to offer customers "better" prices than competitors; that is, it should unbundle all end-to-end tariffs and offer any piece as a service to competitors.

—An important and continuing reason for regulation is to prevent inefficient bilateral monopoly bargaining, in essence controlling the conditions and prices for access.

—Finally, regulators should turn their attention from profits (higher profits to AT&T Communications, if they exist, would be a reward for superior service or advantages) to "correct" prices for access and to simple rules for price caps.

The Role of the U.S. Local Operating Companies

Robert W. Crandall

THE AT&T DIVESTITURE was designed to prevent the Bell operating companies (BOCs) from using their assumed monopoly power in local-exchange and access services to reduce competition elsewhere, especially in long-distance services. To prevent a recurrence of AT&T's alleged abuse of its monopoly power, the U.S. Justice Department sought and obtained in 1982 a decree that separated the Bell operating companies from AT&T's long-distance service, Long Lines, and from its manufacturing arm, Western Electric.[1] Since January 1, 1984, these operating companies have been limited largely to local-exchange and intrastate (intra-LATA) toll services. They are forbidden to offer interexchange (inter-LATA) service and information services or to manufacture communications equipment.[2] At first, other nontelephone services were open to the divested BOCs only on a case-by-case basis under a formal waiver system enforced by the court, but the court has now removed all limits on BOC entry into noncommunications markets.

In some ways the 1982 decree (often referred to as the modified final judgment, or MFJ) resembles a 1956 decree entered into by AT&T to settle an earlier antitrust suit.[3] In this earlier consent order, AT&T agreed to limit its activity to regulated telecommunications services in order to keep its Western Electric manufacturing subsidiary and remain intact. This limitation kept AT&T from entering a variety of new computer-related markets and from exploiting any joint economies from combining computer and communications technologies. Ironically, AT&T's desire to escape from the strictures of the 1956 decree was at least partly responsible for its acceptance of the 1982 decree, through which it was forced to shed all its local operating companies.

1. *United States* v. *American Telephone and Telegraph Co.*, 552 F. Supp. 131 (D.D.C. 1982) *aff'd sub nom. Maryland* v. *United States*, 460 U.S. 1001 (1983).

2. For an explanation of these terms, see the "Overview" at the opening of this volume.

3. *United States* v. *Western Electric Co.*, 1956 Trade Cases (CCH) para. 68,246 (D.N.J. January 24, 1956).

Even though the current decree has been in existence for only seven years, and the restrictions on what lines of business the operating companies may enter have been operative for only five years, strong pressure has already built up to reconsider the wisdom of constraining the divested BOCs. In 1987 the Department of Justice released its required triennial report on the subject, recommending that many of these restrictions be lifted or eased.[4] The court, however, rebuffed the recommendation, arguing that structural conditions in the industry had not changed.[5] In this paper, I examine the theoretical and empirical bases for continuing the policy of strict structural separations required by the decree.

Boundaries between Regulated and Competitive Markets

In every exercise of government regulation, policymakers must draw boundaries between regulated and unregulated markets. For example, regulation of natural gas was first imposed on its transport through interstate pipelines and on its retail distribution. Later, regulation was extended to the extraction and sale of the gas to pipeline companies.[6] The manufacture of drilling equipment and the processing of natural gas into chemical feedstocks remain unregulated. But should pipeline companies be allowed to own chemical plants? If they are, how can the regulators be sure that the pipeline companies are not cross-subsidizing the production of chemicals from the sales of natural gas to other downstream buyers of natural gas? Or should retail utilities own natural gas reserves? If so, how can regulators police the retail price of gas sold to consumers?

The market boundary problems encountered in regulation are of two kinds: vertical and horizontal. Vertical boundaries are those between successive stages of production. Horizontal boundaries are those between substitute products or services. In telecommunications, important vertical boundaries exist between equipment manufacture and local-exchange or access service, between local-exchange or access service and interexchange service, and between local-exchange or access service and enhanced information services. Examples of horizontal boundaries are those

4. U.S. Department of Justice, Antitrust Division, "Report and Recommendations of the United States concerning the Line of Business Restrictions Imposed on the Bell Operating Companies by the Modification of Final Judgment," filed February 2, 1987, in *United States* v. *American Telephone and Telegraph Co.*, civ. no. 82-0192.

5. *United States* v. *Western Electric Co.*, 673 F. Supp. 525 (D.D.C. 1987). In March 1988 the court relaxed this order for some information services.

6. See Stephen G. Breyer and Paul W. MacAvoy, *Energy Regulation by the Federal Power Commission* (Brookings, 1974), chap. 1.

between leased private-line services, private-microwave services, and switched message services, but these are less relevant to an analysis of the current decree.

In telecommunications, local-exchange and access services and interexchange services have traditionally been regulated, but the manufacture and sale of telephone equipment have not. Nor are such hybrid communications services as videotext formally under rate-of-return regulation. Only "common carriage" is regulated, but even the vertical limits of common carriage are constantly disputed. For example, there is no natural division between local-exchange and interexchange services, nor between exchange services and such enhanced services as call forwarding or protocol conversion. Nor is it clear where the division between the manufacture and the sale of terminal equipment exists.

The issues raised by the structural separation provisions in the modified final judgment are largely those of vertical market boundaries. The divested BOCs are prohibited from venturing backward from local-exchange or access service to equipment manufacture, or forward into (inter-LATA) interexchange or information services. Under the judgment the role of the BOCs is simply to offer the local-exchange or access service that is used as an input in the delivery of interexchange and information services.[7]

Vertical Integration in Economic Theory

The theory of vertical integration is really the theory of the extent of the firm.[8] Successive stages of production are combined to achieve efficiency or market power. The most obvious example of efficiencies within the firm are technological economies—the joint economies of, say, melting and forming metals. Vertical integration also allows activities to be better coordinated than they might be through a market—for example, investment or research and development decisions can be better timed.[9] Vertical integration may also be a means of avoiding the price-output distortions that can arise because of bilateral monopoly. Integration may

7. This is true in theory. In practice, however, the divested Bell operating companies continue to offer intra-LATA interexchange services, often protected by state regulatory authorities from competition from the interexchange carriers.

8. Two major contributions to this literature are R. H. Coase, "The Nature of the Firm," *Economica*, vol. 4 (November 1937), pp. 386–405; and Oliver E. Williamson, *Markets and Hierarchies: Analysis and Antitrust Implications* (New York: Free Press, 1975).

9. Williamson, *Markets and Hierarchies*, pp. 82–105.

also be required if an upstream firm with monopoly power is to prevent inefficient downstream substitution against its output.[10]

It is the relation between vertical integration and market power that is important to an understanding of the modified final judgment and the 1956 consent decree. A simple and often invalid notion that pervades the early antitrust literature is that vertical integration allows a firm to exercise "foreclosure" of suppliers or customers.[11] But as long as newcomers can enter into these upstream or downstream markets, vertical integration by merger or internal growth is unlikely to be the route to monopoly profits.

A related theory is that vertical integration may increase the capital requirements for another firm to enter the market, thereby reducing competition. But such capital barriers are hard to substantiate. In the last analysis, one must usually demonstrate the control of a "bottleneck" facility to find a theory that links vertical integration with market power.[12] In telephony, the local-exchange network is obviously such a bottleneck. But why is integration forward or backward from this bottleneck profitable to a monopolist—regulated or unregulated?

The standard Chicago school explanation of vertical integration demonstrates the specific dangers of integration in telephony. Assume that a monopolist controls all production in the market for access and sells its output to competitive firms in the interexchange market. For simplicity assume that access and interexchange services are produced at constant unit costs and that one unit of access is required per unit of interexchange service. The monopolist of access can extract all of the profit from the sale of interexchange services by setting the price of access at a level equal to the monopoly price of interexchange service less the sum of the costs of access and interexchange service. This assumes that there is no possibility for substituting against access in the production of interexchange services and that price discrimination is impossible in the interexchange service market.

If downstream production admits of factor substitution, integration will be profitable to the upstream monopolist and may enhance economic welfare.[13] Similarly, if price discrimination is possible in the downstream

10. Frederick R. Warren-Boulton, "Vertical Control with Variable Proportions," *Journal of Political Economy*, vol. 82 (July–August 1974), pp. 783–802.

11. For a critical discussion of this doctrine, see Robert Bork, "Vertical Integration and the Sherman Act: The Legal History of an Economic Misconception," *University of Chicago Law Review*, vol. 22 (Autumn 1954), pp. 157–201.

12. *United States* v. *Terminal Railroad Association of St. Louis*, 224 U.S. 383 (1912).

13. See Warren-Boulton, "Vertical Control," pp. 796–99, for a discussion of this point.

market, integration will also be profitable and is likely to enhance economic welfare. If, however, the intermediate access market is regulated but the downstream (or the upstream) market is not, integration may simply be a means of avoiding the regulatory constraint.

The monopolist of access in the example may also wish to integrate forward even if there is no regulation of either market. If the downstream interexchange market is subject to economies of scale such that there are few firms or perhaps even a single firm offering this service, the access monopolist may be faced with a situation of bilateral monopoly. This fact may make efficient pricing of access or interexchange service or both impossible and lead to sellers of interexchange service appropriating some of the monopoly rents from the market. Forward integration by the access monopolist would prevent this outcome.

The Theory of the Modified Final Judgment

In the preceding section, I discussed vertical integration in *unregulated* markets. The presence of regulation complicates the analysis.

The theory that underlies the modified final judgment is straightforward. AT&T owned bottleneck facilities in a regulated industry. This control gave it the power to exclude competitors in upstream and downstream markets through denial of access or patronage and allowed it to subsidize unregulated activities from its regulated markets. Because its operating companies controlled access to more than 80 percent of the country's telephones, it could use its franchised monopoly of local-exchange and access services to monopolize equipment supply and interexchange.

Since no decision was reached in the antitrust case, there was no formal finding that AT&T had engaged in these practices. However, in his decision on September 11, 1981, dismissing AT&T's motion for summary judgment, Judge Harold Greene concluded that the government had presented enough evidence to sustain a rebuttable conclusion that AT&T had

—denied access to competitors in the markets for long-distance and information services and customer premises equipment (CPE);

—cross-subsidized its long-distance and other businesses from the regulated local services; and

—engaged in preferential procurement policies, favoring its Western Electric subsidiary.[14]

14. *United States* v. *American Telephone and Telegraph Co.*, 524 F. Supp. 1336 (D.D.C. 1981).

DENIAL OF ACCESS. By 1982, when the judgment was entered, competition in the market for customer premises equipment had been well established. Nonetheless the judgment required that the manufacture of CPE be separated from the provision of local-exchange and access services partly because operating companies might use their power over network standards to exclude competitors. Similarly the court opined that the operating companies "would have substantial incentives to subvert these equal access requirements" that were part of the decree if they were permitted to offer long-distance services.[15]

Because the government produced evidence that AT&T had used its control over local-exchange and access facilities to frustrate the entry of competitors into CPE and long-distance services, the judgment decisively bars the divested Bell operating companies from long-distance services. Its separation of CPE from local-exchange and access services is more equivocal, as the divested operating companies are allowed to market, but not manufacture, such equipment. Finally, the judgment bars operating companies from information services because they "could discriminate by providing more favorable access to the local network for their own information services than to the information services provided by competitors" even though the government's case failed to provide evidence of competitive problems in these services.[16]

CROSS SUBSIDIZATION. Key to the argument for separating various telecommunications functions is the fear that a regulated monopoly could choose to cross-subsidize unregulated activities by charging the costs of competitive activities to regulated activities without the regulatory authorities' permission or detection. This shift would allow regulated carriers to raise regulated rates to subsidize the competitive ventures.

The government alleged that AT&T subsidized interexchange service by "pricing without regard to costs" once Microwave Communications, Inc. (MCI), and other competitors entered interexchange markets. Similar allegations were made for equipment markets—that Western Electric cross-subsidized competitive equipment from monopoly equipment by pricing the former "without regard to cost."

The judgment also raises the specter of subsidization of information services by the divested operating companies from their remaining local "monopoly" services. This fear, combined with the network access issue, described above, led to the prohibition on all information services.

15. *United States* v. *American Telephone and Telegraph Co.*, 552 F. Supp. at 188.
16. Ibid., p. 189.

MARKET FORECLOSURE. In his decision denying AT&T summary judgment, Judge Greene asserted the government had presented a strong but rebuttable case that AT&T had foreclosed competitors in the equipment market by favoring its Western Electric subsidiary. Seventeen separate episodes were presented by the government as evidence that AT&T's operating companies and Long Lines favored Western Electric by denying competitors access to technical information, by making it hard for competitors to obtain technical assessments of their products, and by instituting crash programs to copy competitors' products. While this "foreclosure" argument is standard antitrust reasoning, Judge Greene's decision fails to provide a theory of why AT&T should wish to monopolize the supply of its own equipment. Nonetheless, the judgment bars the BOCs from the manufacture or design of central office or customer premises equipment.

Each of these purported evils of vertically integrated monopoly applied to the pre-1982 AT&T and especially to the AT&T of the 1970s, when new competitors were appearing in the equipment and long-distance markets. None is necessarily relevant to the postdivestiture world. Before looking at the world as it now exists, however, it is useful to examine somewhat more closely the theory on which the modified final judgment is based.

The Theory Examined

Clearly the theory behind the judgment is at least superficially plausible. Because it controlled local telephone exchanges, AT&T could have engaged in predatory denial of access, cross subsidization, and market foreclosure. But why would it have wanted to do so? Assume that the interexchange, CPE, and central office equipment markets were potentially competitive—that is, that competition among actual and potential suppliers of these products and services would have driven prices to marginal cost. Then why would AT&T have wanted to extend its business from the simple offering of local-exchange and access services? If it had done so, would resource allocation have been affected?

Exclusionary Activity: Interexchange Services

It is by now common wisdom that AT&T used its local monopoly in telephone service to restrict access to its customers by rivals for long-distance business. MCI, the first of the new competitors, filed an antitrust suit against AT&T alleging such restraints and won in the courts.[17] The Justice Department suit was based in large part on similar evidence of exclusionary activity, and Judge Greene found this evidence persuasive enough to deny the defendants their motion for dismissal.

There is little doubt also that local operating companies enjoy some degree of monopoly power, although this power is constrained by state public utility regulation and increasingly by new local-access (''bypass'') technologies. But why should these franchised monopolists wish to extend their monopoly to another *regulated* service that uses their regulated local services as inputs? By setting the right price of access, the local monopolist should be able to extract *all* the monopoly rents from long-distance services. That familiar result derives from the Chicago school analysis of vertical integration. But if *both* services are regulated, extracting monopoly profits by the integrated or unintegrated carrier is inhibited.

The historical reason for AT&T's integration into long-distance services is well known.[18] After its major patents on basic telephone equipment expired in the 1890s, the company faced a wave of firms entering local-exchange services, to which it responded with price competition and mergers. To gain a competitive advantage over these independent companies, AT&T moved aggressively to consolidate its position in long-distance services, where it continued to enjoy patent protection, and to deny this service to the independents. A potential antitrust suit over this and related issues led AT&T to enter into the Kingsbury Commitment of 1913, which allowed independent carriers access to AT&T's long-distance network.

At virtually the same time, long-distance and local telephone services were brought under public utility regulation with the support of both AT&T and the independent companies. Congress gave the Interstate Commerce Commission regulatory authority over interstate telephone services in 1910, and the states slowly followed with public utility regulation of local services—arrangements that froze competitive entry into telephone services until the late 1960s.

Because it had consolidated its local-exchange monopolies after the merger wave around 1910 and after the Willis-Graham Act of 1921 exempting telephone mergers from the federal antitrust laws, AT&T could have chosen to concentrate on local service and to extract the rents that

17. *MCI Communications Corp.* v. *American Telephone and Telegraph Co.*, 708 F. 2d 1081 (7th Cir. 1983).

18. The following discussion is based largely on Gerald W. Brock, *The Telecommunications Industry: The Dynamics of Market Structure* (Harvard University Press, 1981).

the state regulatory bodies would permit while divesting itself of its long-distance operations. It rejected that strategy for two obvious reasons. First, the natural monopoly characteristics of the long-distance network during the interwar period would have created a troublesome situation of bilateral monopoly between its local companies (the sellers of access) and the long-distance supplier (the demander of access). Second, because the switched network of the time used rather primitive switches and simple transmission wires, it might have lost economies of scope.

Besides, long-distance services were serving another purpose for AT&T. As regulatory constraints became more tightly binding in some states, long-distance business could support local services, especially after the development of the vacuum-tube technology that allowed AT&T to extend quality long-distance service from coast to coast.

In the decades after World War II, more and more of non-traffic-sensitive plant costs were charged to long-distance services for rate-setting purposes (figure 1). This shift of revenues was engineered by a joint board of federal and state regulators, with whom AT&T might have been expected to wield some influence. Thus AT&T found itself, perhaps willingly, allowing the profits from long-distance service to defray the costs of its local-exchange operations. This strategy was politically attractive because after 1950 long-distance transmission costs fell dramatically. By simply not reducing long-distance rates as rapidly as costs were falling, AT&T could use the rents from long-distance to keep local charges from rising.

These distortions in telephone pricing through the "separations and settlements" process rose dramatically just when a new technology—microwave—enabled other companies to offer competitive long-distance services. Clearly AT&T could not have known as early as the 1950s that competitors would try to enter the switched long-distance market, but by the late 1960s it must have been extremely worried that such entry might occur. Despite this threat, the percentage of subscriber plant costs charged to the long-distance service nearly tripled between 1965 and 1975 (figure 1).

Why did AT&T not complain about the enormous rise in the regulatory distortion of telephone rates as it began to be threatened by entry? Perhaps it saw this distortion as its ultimate, political defense against competitive entry in long-distance services. No one in Washington would want to allow competitors to skim the cream that subsidized local telephone service, thereby endangering America's much valued "universal service." That is, AT&T could now argue that its long-distance monopoly was needed to preserve universal service through its local operating companies.

Figure 1. *Percentage of AT&T Subscriber Plant Costs Allocated to Interstate Service, 1943–82*

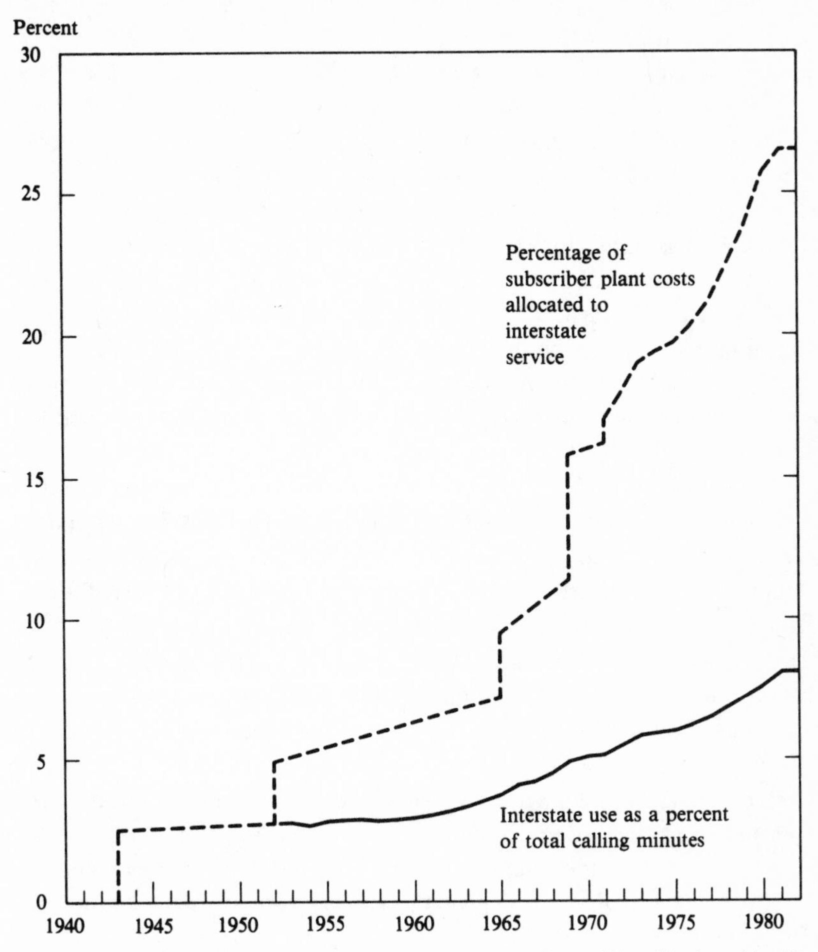

Source: Congressional Budget Office, *The Changing Telephone Industry: Access Charges, Universal Service, and Local Rates* (Washington, D.C., June 1984), fig. 1.

Indeed, the FCC and local regulators consistently opposed the entry of new (switched-message) long-distance carriers until MCI circumvented them in a maneuver that the courts upheld.[19]

AT&T might still have been content to allow newcomers into long-distance services in the 1960s and 1970s and to extract all the rents it was allowed simply by offering access and local-exchange services at regulated

19. *MCI Telecommunications Corp.* v. *Federal Communications Commission*, 561 F. 2d 365 (D.C. Cir. 1977), *cert. denied*, 434 U.S. 1040 (1978).

monopoly rates. If it was protected against entry into local services by monopoly franchises in many jurisdictions, it might have extracted the maximum rent simply by setting the right interstate-access charges, regulators permitting. Why did it fail to do so?

Unfortunately for AT&T, it *owned* the long-distance facilities. Once competitors were allowed into the long-distance market, AT&T's embedded plant made it vulnerable. And its operating companies could hardly expect to be free to set access charges to extract rents from these new competitors. Before divestiture, the Exchange Network Facilities for Interstate Access (ENFIA) tariff for competitors was set by FCC and industry negotiation at only 35 to 55 percent of the cost that AT&T's Long Lines bore for access to local loops.[20] The new long-distance carriers justified receiving this discount on the grounds it reflected the lower quality of the connections provided them, but it was low enough to encourage them to make a substantial (and perhaps uneconomic) investment in competitive facilities.

It was the distortion in rates and the large embedded capital base that AT&T had dedicated to interstate services that drove AT&T to defend its long-distance monopoly in the 1970s. Long-distance service was far from a "contestable" market.[21] Entry and exit were not easy, and AT&T's exit barriers were made worse by a depreciation policy that overvalued its capital assets in its accounts. If AT&T had suffered a major loss in long-distance market share, it would have been forced to take capital losses on redundant plant that could not be sold for its book value. More important, AT&T was unable to charge its new long-distance competitors an access charge for switching their calls to subscribers that extracted all the rents from the overpriced long-distance service. Thus AT&T had a powerful incentive to defend the long-distance market—an incentive exaggerated by a history of conservative depreciation policies and regulatory distortion of rates.

Note that this analysis points to regulatory ratemaking as part of the reason for AT&T's aggressive behavior in defending its long-distance business. Had regulators not tried to charge a large part of access costs to long-distance services, AT&T might not have been interested in monopolizing both markets. Even during its hegemony, General Motors did not try to extend its market power to the steel, tire, glass, or automobile retailing industries.

20. Brock, *Telecommunications Industry*, pp. 228–29.

21. See William J. Baumol, John C. Panzar, and Robert D. Willig, *Contestable Markets and the Theory of Industry Structure* (Harcourt Brace Jovanovich, 1982).

Exclusionary Activity: Customer Premises Equipment

Historically, AT&T's market position and market power derived from its patents on telephone technology. The handset was part of this technology. Over time, however, these patents expired, and AT&T's monopoly depended more on operating company franchises than on control of equipment supply. By the 1960s and 1970s, Western Electric no longer had a domestic monopoly on the technology for manufacturing handsets, key telephones, and PBXs (private branch exchanges). Why didn't AT&T at least begin to tap outside suppliers of customer premises equipment if others could manufacture it as efficiently or even more efficiently?

Undoubtedly, rate regulation and depreciation practices combined with inertia to keep CPE manufacture within AT&T (Western Electric). The Bell operating companies leased CPE to subscribers as part of their monthly access or local-exchange service package. The equipment was depreciated by the operating companies very slowly, resulting in a net book value far greater than its true economic value once competitive sources of equipment developed. Indeed, after AT&T lost its battle to preclude competition in this market, it was forced to take a write-off of $5.5 billion on a variety of equipment that it had depreciated too slowly before the divestiture.[22] It was hardly surprising that AT&T worked to prevent competition in CPE during the 1960s and 1970s.

The Anticompetitive Effects of Cross Subsidies

The fear of cross subsidy ("fooling the regulators" by shifting some of the costs of service in a new market to a regulated market) is a major reason for restricting regulated companies solely to their limited lines of business. Yet there are several problems with the cross-subsidy argument.[23] First, if regulators can be fooled, why must a regulated firm enter unregulated product or service markets in order to fool them? Other mechanisms may be available, such as the manipulation of various accounting conventions, to do the same thing. Second, how does one know that the regulatory constraint is actually binding? It is assumed that unregulated prices will be higher than regulated prices, but this assumption must always be questioned. For AT&T, much of its unexploited market power may

22. *AT&T 1983 Annual Report*, p. 2.

23. Some of the following points have been raised in William S. Reece, "Anticompetitive Cross Subsidy and the AT&T Divestiture," Loyola College, n.d.

have been absorbed by its unionized workers.[24] The voluminous empirical literature on the demand for telephone service suggests, however, that the demand for local-exchange or access services is extremely inelastic at regulated rate levels.[25] Thus the AT&T local-exchange carriers probably had substantial unexploited market power.

Third, even if a monopolist is successful in fooling its regulators, the resulting cross subsidies may not greatly affect competition in the *unregulated* markets. For these cross subsidies to affect the regulated monopolist's prices in unregulated markets, they must affect incremental production costs in these unregulated markets. But this outcome may not occur. The subsidies may be exhausted in the production of inframarginal units, or they may be unrelated to costs that vary with output.

In short, one must doubt the cross-subsidy argument in the absence of empirical evidence. But even if validated empirically, this argument would seem to have more serious implications for resource allocation in the *regulated* industry than in the markets that the regulated firm attempts to enter through cross subsidies.[26] In the antitrust arena, of course, the courts' concerns are reversed: they worry that competition is inhibited in the *unregulated* markets.

In the case of telephone services, regulators may have so distorted relative prices that any attempt by AT&T or other regulated carriers to cross-subsidize unregulated ventures by raising regulated rates may actually have improved consumer welfare. This outcome seems especially likely for local service, which regulators have been cross-subsidizing from long-distance rates since the 1950s. If entry by other exchange carriers leads to pressures to reverse this subsidy, such entry will probably improve economic welfare.

Market Foreclosure or Forbearance

Finally, some observers advocate separating regulated from unregulated activities because they fear the regulated firms will use vertical integration to reduce competition in upstream or downstream markets. This is the logic behind the antitrust suit that was settled by the 1956 decree and

24. Recent evidence suggests that labor receives as much as three-fourths of monopoly rents from industry. See Michael A. Salinger, "Tobin's *q*, Unionization, and the Concentration-Profits Relationship," *Rand Journal of Economics*, vol. 15 (Summer 1984), pp. 159–70.

25. Lester D. Taylor, *Telecommunications Demand: A Survey and Critique* (Ballinger, 1980), chap. 3.

26. See Kenneth C. Baseman, "Open Entry and Cross-Subsidization in Regulated Markets," in Gary Fromm, ed., *Studies in Public Regulation* (MIT Press, 1981), pp. 329–60.

behind the modified final judgment's prohibition on the manufacture of equipment by Bell operating companies.

One reason that a franchised monopoly supplier of local-exchange or access service may want to supply all its own equipment for its regulated service is that such integration allows the firm to evade rate regulation. AT&T's initial monopoly was based on patents on equipment that it developed and produced, but once AT&T settled into a world of delivering a regulated monopoly service, it might have chosen to move away from manufacturing its own equipment. However, as long as telephone service is regulated but equipment manufacture is not, integration into equipment manufacture by AT&T allows it to price the capital equipment in its rate base above cost and to evade rate-of-return regulation altogether. Manufacture of customer premises equipment may have served the same purpose, but the detariffing of this equipment and the ability of competitors to lease or sell to the operating company's customers have largely negated this effect. It is remarkable that the government's theory of the *AT&T* case focused less on this "avoidance of regulation" theory than on a conventional argument of vertical extension of monopoly.[27]

Empirical Evidence on the MFJ's Anticompetitive Practices Arguments

Even if integration by local-exchange carriers could create traditional bottleneck or cross-subsidy problems by integrating into upstream or downstream markets, it is far from clear that these anticompetitive possibilities were realized. Nor is it clear that AT&T succeeded in evading rate regulation through its ownership of Western Electric. In this section, I examine some evidence related to the three kinds of anticompetitive practices addressed in the modified final judgment.

Denial of Access

There is little doubt that AT&T behaved aggressively toward its new competitors in the interexchange and the CPE markets in the 1970s. The court heard voluminous evidence on devices AT&T insisted were needed to protect its network from the CPE supplied by competitors.[28] Apparently,

27. Judge Greene did mention the avoidance of regulation theory. See *United States* v. *American Telephone and Telegraph*, 552 F. Supp. at 190.

28. See Judge Greene's decision denying AT&T's motion for summary judgment, 524 F. Supp. at 1336, for a summary of the government's evidence.

even within AT&T many felt that such protective couplers were unnecessary and that AT&T could not deny access to competitors in the CPE market. Nonetheless, the company's top management tried to prevent competitive "interconnect" companies from gaining access to AT&T customers.

Today much of this evidence seems moot. However hard it may have tried, AT&T had certainly failed in its attempts to maintain its share of most CPE markets by the time of the 1982 decree. In the late 1970s, the Federal Communications Commission forced state public utility commissions to allow competition in all terminal equipment because such equipment is used for both intrastate and interstate calls. Table 1 shows that by 1983 AT&T had lost an enormous share of the market for key telephones and PBXs and that it has failed to recover this lost share since divestiture. It seems unlikely that AT&T's ownership of Bell operating companies would alter its competitive standing in these markets.

The interexchange market is different. Even though rival interexchange service providers had begun to enter the market in the mid-1970s, AT&T had not lost much market share in most long-distance markets by 1982. The data in table 2 suggest that the new competitors failed in particular to divert many customers for switched interexchange services from AT&T before the divestiture despite favorable access rates under the ENFIA tariff. Their total market shares of all interexchange revenues were only 5.5 percent in 1982. With the divestiture and its requirement for equal access, these competitors' shares grew by 1986 to 14.4 percent of total interex-

Table 1. *U.S. Market Shares for AT&T and Others in the Manufacture of Customer Premises Equipment, Selected Years, 1976–86*
Percent

Company	*1976*	*1981*	*1982*	*1983*	*1984*	*1985*	*1986*
			Key and hybrid telephones				
AT&T	n.a.	53	48	28	26	27	27
Others	n.a.	47	53	72	74	73	74
			Private branch exchanges (PBXs)				
AT&T	50	36	27	23	19	23	21
Northern Telecom	13	8	13	17	22	22	23
Rolm	4	12	13	15	17	14	16
Mitel	0	9	12	11	10	9	9
Others	33	35	35	34	32	32	31

Sources: For key and hybrid telephones: North American Telecommunications Association, *1987 Statistical Review: Annual Market Study of the Telecommunications Equipment Industry* (Washington, D.C., 1987), table 1; for PBXs: Richard H. K. Vietor and Davis Dyer, *Telecommunications in Transition* (Boston: Harvard Business School, 1986), p. 62; and Peter W. Huber, *The Geodesic Network: 1987 Report on Competition in the Telephone Industry* (Washington, D.C.: U.S. Department of Justice, Antitrust Division, January 1987), p. 16.4.

n.a. Not available.

Table 2. *Market Shares of Interexchange Service Revenues, 1978, 1982, 1986*
Percent

Company	*1978*	*1982*	*1986*	*Non-LEC share 1986*
AT&T	99.0	94.4	56.7	79.6
LECs	. . .	. . .	28.8	. . .
MCI	0.6	3.1	7.6	10.7
U.S. Sprint	0.4	1.5	4.3	6.1
Others	<0.1	0.9	2.5	3.5
Total	100.0	99.9	99.9	99.8

Sources: Author's calculations based on Federal Communications Commission, *Statistics of Communications Common Carriers* (Washington, D.C., annual issues); letters from other common carriers to FCC in FCC common carrier docket 83-1291; and U.S. Telephone Association, *Telephone Statistics* (Washington, D.C., annual issues).
LEC Local-exchange carrier.

change revenues, or 20.3 percent of the non-local-exchange carrier (LEC) interexchange market—reflecting a steady erosion of AT&T's market position.

Obviously the failure of competitors to obtain a large market share does not prove that local-exchange carriers could or did exercise bottleneck monopoly power in denying competitors access to customers. But the ease with which CPE suppliers found customers certainly suggests the converse—that local-exchange carriers could not easily deny customers access to competitive CPE vendors once the FCC succeeded in declaring unlawful those tariffs that denied such interconnection.[29]

Most interexchange traffic continues to move to and from customers on access lines controlled by the local-exchange carriers. Much of this traffic is distributed by switched access lines at a tariffed total rate, for origination and termination of the call, of about 11 cents per conversation minute.[30] Many large customers bypass the switched service, either by leasing private lines or by using facilities not owned by the local-exchange carriers. Most of this bypass is done by leasing; at present only a small share of total interexchange traffic avoids the local-exchange carriers' facilities altogether. The Huber Report, submitted by the Department of Justice to the court as part of its first triennial review of the line-of-business restrictions in the modified final judgment,[31] estimates that perhaps 5.25

29. It must be admitted that AT&T's behavior between 1974 and 1982 was undoubtedly constrained by the ongoing antitrust case. Had AT&T not been charged with monopolization by the government in 1974, it might well have been more aggressive in attempting to prevent entrants from succeeding in the CPE market.

30. "Monitoring Report of the Federal-State Joint Board," in Federal Communications Commission Common Carrier docket 87-339, December 1987, p. 101.

31. Peter W. Huber, *The Geodesic Network: 1987 Report on Competition in the Telephone Industry* (Washington, D.C.: U.S. Department of Justice, Antitrust Division, January 1987).

Table 3. *Estimated Routing of Current Inter-LATA Traffic at Originating and Terminating Ends*

Route	Number of customers (thousands)	Number of minutes[a] (billions)
LEC switched access to interexchange carrier	100,000	150
LEC private lines to interexchange carrier	100	20
Non-LEC facility to interexchange carrier	0.1	0.25
End-to-end private line	0.1[b]	5[b]

Source: Huber, *Geodesic Network*, p. 3.9.
LEC Local-exchange carrier.
a. Originating and terminating minutes divided by 2.
b. Approximate.

billion of 175.25 billion minutes of annual inter-LATA interexchange traffic are routed to customers on lines other than those provided by local-exchange carriers (table 3).

The modified final judgment has addressed the problem of access to interexchange service by requiring local-exchange carriers to provide equal access to long-distance carriers for their larger exchanges. This requirement, phased in through September 1986, has led to more than 80 percent of local-exchange carriers' access lines being converted to equal access. The court, however, does not view this protection as sufficient by itself to permit the divested BOCs to reenter the long-distance market.

Cross Subsidies

More has been written about cross subsidies and less has been proven than for either of the other two alleged AT&T abuses. The FCC spent two decades searching for an accounting system to detect and quantify such subsidies accurately, without notable success.[32] The government's case against AT&T relied on a showing that AT&T priced "without regard to cost," not on estimates of actual cross subsidies.[33]

It is possible to construct at least a crude test for cross subsidies. All exchange carriers report their annual operating results to the FCC, which

32. See Walter G. Bolter, "The FCC's Selection of a 'Proper' Costing Standard after Fifteen Years—What Can We Learn from Docket 18128?" in *Assessing New Pricing Concepts in Public Utilities: Proceedings of the Institute of Public Utilities Ninth Annual Conference* (Michigan State University Graduate School of Business Administration, 1978), pp. 333–72.

33. *United States* v. *American Telephone and Telegraph*, 524 F. Supp. at 1364–70.

publishes them in the *Statistics of Communications Common Carriers*. From these data, one can estimate a cost function of the form

$$(1) \qquad C = f(Q_1, \ldots, Q_n; w_1, \ldots, w_j),$$

where C is some measure of annual costs, the Q's are the carrier's various outputs, and the w's are the input prices faced by the carrier. If the carrier is cross-subsidizing regulated services from forays into competitive communications markets, C should vary directly with the carrier's share of communications activities outside basic local-exchange or access service (non-LEC communications activities).[34]

The "fooling of the regulators" should be directly proportional to the share of the communications activity that is unregulated at the state level, since state regulators probably will not be easily fooled by attempts to charge the costs of noncommunications activities to the regulated carriers' accounts. To capture the cross subsidies, I enter a variable, *DIVERSIFICATION,* the ratio of all other communications revenues to local-exchange and access revenues (including intrastate toll) for the parent (holding) company into equation 1 as a separate variable.

Because there may be a difference between the ability of carriers to hide capital and noncapital costs in regulated accounts, equation 1 is estimated separately for noncapital costs and for capital costs. For noncapital costs, the measures of Q are local and toll calls handled (*CALLS*). For capital costs, the measure of Q is main telephones plus PBXs served (*PHONES*). In addition, a separate equation is estimated using net plant as the dependent variable, rather than capital costs, to capture the possible shifting of assets into the regulated telephone rate base.

Because all telephone carriers face a national capital market, it is assumed that the cost of capital does not vary across carriers.[35] The wage rate, however, varies regionally and by collective bargaining circumstances. Hence the total compensation per employee (*WAGE*) is entered as the sole measure of w to estimate equation 1.

The cost of providing telephone service varies with the density of the carrier's exchange area and with its rate of growth (because of historical

34. Baseman, "Open Entry and Cross-Subsidization," points out that it is difficult to determine if a nonregulated service is being subsidized by comparing the price charged for that service with the carrier's incremental cost of delivering it. The carrier could have chosen a technology that is inefficient for delivering the entire array of services but provides the nonregulated service at a low incremental cost.

35. Since most of the local-exchange carriers in the sample are part of large holding companies, estimating each one's cost of capital would present formidable problems. For this reason, I use the Cobb-Douglas rather than the translog formulation of the cost function.

cost accounting and the costs of installing new subscribers). Therefore measures of subscribers per square mile (*DENSITY*) and of the five-year growth in main telephones served (*GROWTH*) are included in the cost function. Finally, to capture any possible effects of differences in technology, the share of electronic switches (*ELECTRONIC*) is included as a separate regressor.

The cost function estimated is Cobb-Douglas with the insertion of the technological and diversification variables. It takes the form

$$\begin{aligned} \log C_i = {} & a_0 + a_1 \log Q_i + a_2 \log WAGE_i \\ & + a_3 \log DENSITY_i + a_4 \log DIVERSIFICATION \\ & + a_5\, GROWTH_i + a_6\, ELECTRONIC_i + u_i. \end{aligned} \tag{2}$$

The results of estimating the cost function for the years 1971 and 1981 appear in tables 4 and 5. These years were chosen because they were relatively tranquil times in the business cycle and were a decade apart, bridging the period of competitive entry into CPE and interexchange services. All reporting carriers for which complete data were available are included—forty-two carriers in 1971 and fifty-eight in 1981.

Noncapital costs clearly vary with the rate of calling—the elasticities for local and toll calls add to roughly one (table 4). The wage elasticity in noncapital costs is implausibly large in the 1971 estimate of noncapital costs, but much more reasonable for 1981. (Incidentally, wages are substantially higher for BOCs than for other operating companies.) The elasticity of capital costs with respect to main telephones plus PBXs is very close to unity (table 5). Electronic switching does not seem to have much effect on noncapital costs, capital costs, or net plant per telephone. Costs and net plant are generally inversely related to density.[36] As expected, the five-year growth rate is positively related to capital costs, noncapital costs, and net plant.

The coefficients of the diversification variables are of the greatest interest. Diversification assumes statistically significant coefficients in only the net-plant and capital-cost equations. When the equations are estimated with unit costs or net plant per phone as the dependent variable to eliminate heteroskedasticity, the estimated coefficient of diversification for capital costs is 0.04 in 1981, but not statistically significant for the 1971 sample (table 5). The diversification variable in the net-plant equation is significant

36. The sample consists of the larger holding-company entities as well as the larger independents. It does not include small, rural telephone companies.

Table 4. *Estimates of the Determinants of Noncapital Expenses for Telephone Operating Companies, 1971 and 1981*[a]

	Dependent variable			
	Noncapital expenses		*Noncapital expenses per call*	
Independent variable	*1971*	*1981*	*1971*	*1981*
Constant	−5.59	−3.70	−6.35	−4.43
	(6.99)	(6.40)	(9.82)	(6.74)
TOLL CALLS	0.164	0.413	. . .	. . .
	(2.08)	(6.68)		
LOCAL CALLS	0.807	0.612	. . .	. . .
	(9.26)	(10.72)		
DIVERSIFICATION	−0.041	0.044	−0.073	0.088
	(0.52)	(0.78)	(0.97)	(1.42)
DENSITY	−0.010	−0.057	−0.023	−0.054
	(0.27)	(2.43)	(0.69)	(2.21)
WAGE	1.60	0.589	1.50	0.637
	(4.92)	(3.18)	(4.93)	(3.15)
GROWTH	0.005	0.023	0.011	0.032
	(0.12)	(1.56)	(0.32)	(1.93)
ELECTRONIC	0.037	0.050	0.045	0.066
	(1.59)	(1.47)	(2.14)	(1.70)
Number of cases	42	58	42	58
$\bar{R}^2$	0.976	0.991	0.421	0.317

a. Numbers in parentheses are *t*-statistics.

in 1971 but not in 1981 (table 6). In short, there is no convincing, consistent evidence that telecommunications diversification raises telephone company operating costs per call or per phone served. Where this evidence is strongest (capital costs in table 5), it is small and declines between 1971 and 1981.

Because the diversification variable is at best a crude approximation for telecommunications diversification by local carriers, and because it is necessarily the same across all carriers for a given holding company (AT&T, GTE, Continental), I also examined the effects of diversification by using separate dummy variables for the BOCs and three other holding companies. The results are rather surprising (table 7). In 1971 none of the coefficients of the dummy variables is statistically significant in the operating-cost regression, and only GTE has a statistically significant (positive) coefficient for net plant per phone. This result suggests that the accounting costs across these various holding companies were generally not significantly different from one another after adjusting for differences

Table 5. *Estimates of the Determinants of Capital Expenses (Depreciation plus Interest) for Telephone Operating Companies, 1971 and 1981*[a]

	Dependent variable			
	Capital expenses		*Capital expenses per phone*	
Independent variable	*1971*	*1981*	*1971*	*1981*
Constant	3.67	4.76	3.65	4.75
	(7.84)	(9.11)	(7.42)	(9.11)
PHONES	0.941	1.02	. . .	. . .
	(35.41)	(56.95)	. . .	. . .
DIVERSIFICATION	0.118	0.036	0.098	0.041
	(2.28)	(2.29)	(1.83)	(2.91)
DENSITY	−0.044	−0.077	−0.062	−0.068
	(1.87)	(3.85)	(2.67)	(4.00)
WAGE	0.334	−0.083	0.140	−0.037
	(1.49)	(0.49)	(0.64)	(0.23)
GROWTH	0.074	0.240	0.076	0.245
	(3.33)	(1.85)	(3.29)	(1.90)
ELECTRONIC	0.005	−0.023	0.017	−0.025
	(0.29)	(0.73)	(1.17)	(0.81)
Number of cases	42	58	42	58
$\bar{R}^2$	0.985	0.991	0.478	0.349

a. All variables except *GROWTH* in natural logarithms. Numbers in parentheses are *t*-statistics.

in wages, density, and growth. In 1981, however, the BOCs show significantly higher noncapital operating costs and net plant per main phone than the other companies in the sample after accounting for the other variables. Operating costs per call and net plant per phone are 28 and 23 percent above those of the other companies, other things being equal.

A further examination of the mean values for noncapital operating costs and net plant per main phone and PBX provides some insight into the rise in BOC costs between 1971 and 1981. In 1971 the BOCs had substantially lower net plant per phone than Continental or GTE and marginally lower net investment per phone than United and the independents in the sample (table 8). By 1981 the BOCs had virtually the same net plant invested per phone as the others (except Continental). This fact suggests that the BOCs added plant in the 1971–81 period at a cost per telephone that far exceeded the rate for the other major telephone operating companies. Surprisingly the detariffing of CPE and inside wiring and the write-offs at divestiture failed to reduce net plant per subscriber much between 1981 and 1986, perhaps because of the cost of achieving equal access.

Table 6. *Estimates of the Determinants of Net Plant for Telephone Operating Companies, 1971 and 1981*[a]

	Dependent variable			
	Net plant		*Net plant per phone*	
Independent variable	*1971*	*1981*	*1971*	*1981*
Constant	5.84	6.69	5.83	6.53
	(13.67)	(12.29)	(13.55)	(10.49)
PHONES	0.970	1.05	. . .	. . .
	(13.67)	(56.57)	. . .	. . .
DIVERSIFICATION	0.131	−0.015	0.121	0.007
	(2.77)	(0.92)	(2.58)	(0.39)
DENSITY	−0.051	−0.073	−0.060	0.003
	(2.36)	(3.48)	(2.95)	(0.13)
WAGE	0.392	−0.063	0.292	0.225
	(1.91)	(0.36)	(1.54)	(1.17)
GROWTH	0.073	0.341	0.074	0.440
	(3.61)	(2.53)	(3.66)	(2.86)
ELECTRONIC	0.014	0.004	0.021	0.011
	(1.00)	(0.14)	(1.58)	(0.29)
Number of cases	42	58	42	58
$\bar{R}^2$	0.988	0.990	0.550	0.092

a. Numbers in parentheses are *t*-statistics.

Table 7. *Coefficients of Dummy Variables in Operating-Cost and Net-Plant Regressions for Telephone Operating Companies, 1971 and 1981*

	Noncapital costs		*Noncapital costs per call*[a]		*Net plant*		*Net plant per phone*[b]	
Group	*1971*	*1981*	*1971*	*1981*	*1971*	*1981*	*1971*	*1981*
BOCs	0.163	0.351*	0.091	0.278*	−0.014	−0.062	−0.022	0.227*
Continental	0.817	0.241*	0.888	0.341*	0.305	0.253*	0.320	0.180
GTE	−0.089	0.093	−0.135	0.020	0.208*	−0.009	0.207*	0.082
United	−0.065	0.035	0.002	0.051	0.002	0.055	0.007	0.086

*Statistically significant at 95 percent confidence level.
a. Calls are local toll calls.
b. Phones are main telephones plus PBXs.

The story for operating costs is a little different. The BOCs' rate of increase in operating costs between 1971 and 1981 was roughly equivalent to those of GTE and United but somewhat greater than Continental's (table 8).[37] After 1981, however, the other operating companies' costs continued to rise, but the BOCs' noncapital operating costs actually fell in nominal

37. The Continental comparison is of limited value, however, because there was only one Continental operating company in the sample in 1971.

Table 8. *Selected Expense Categories, Telephone Operating Companies, 1971, 1981, 1986*
Current dollars

Group	*1971*	*1981*	*1986*
		Noncapital operating costs per call	
BOCs	0.051	0.101	0.083[b]
Continental	0.080	0.130	0.143
GTE	0.036	0.082	0.096
United	0.042	0.081	0.109
Others	0.037	0.082	0.074
		Net plant per main telephone and PBX line	
BOCs	621	1,197	1,184[a,b]
Continental	1,460	1,414	1,438[a]
GTE	788	1,233	1,393[a]
United	683	1,227	1,181[a]
Others	647	1,167	1,130[a]

Source: Author's calculations based on data from FCC, *Statistics of Communications Common Carriers* (Washington, D.C., annual issues).
a. Net plant per access line.
b. After divestiture from AT&T.

terms, perhaps because some costs were transferred to AT&T in the January 1984 divestiture.

One is thus left with something of a puzzle. If AT&T did shift some of the costs of interexchange service to its operating companies during the 1970s, why did it not do so earlier? And why should the investment per main telephone in the BOCs have risen so rapidly between 1971 and 1981?

None of these results prove that there was or was not any cross subsidization in AT&T's pricing strategy before divestiture. They are hostage to the accounting conventions employed in the data submitted to the FCC. I have no way of knowing if these conventions were similar across all companies. Alternatively, there could have been substantial cross subsidies among various categories of services that did not lead to costs being transferred from other categories to local-exchange or access activities. The seven-way cost studies conducted by AT&T certainly suggest such cross subsidies among interstate services, but no one can be sure that such embedded-cost analyses offer much information about true opportunity costs.

When similar equations are estimated for the postdivestiture year of 1986, the results are much less clear because of poor wage data.[38] In 1982

38. The 1986 *Statistics of Communications Common Carriers* data do not include the number of employees, making the construction of a *WAGE* variable difficult. When form M data on

GTE acquired U.S. Sprint, and in 1984 the BOCs were divested from AT&T. If local-exchange activities are used to subsidize forays into such competitive markets as long distance, one might expect the 1986 coefficient of a dummy variable for GTE to be significantly positive and greater than the estimate of the coefficient for a dummy for the BOCs. These expectations are confirmed in the net-plant equation, but not in the operating-cost equation. Such a result is consistent with the data in table 8 that show GTE's net plant per access line increasing in 1981–86 while the other groups in the sample generally show a decline.

Market Foreclosure

As mentioned above, Judge Greene was persuaded by the government's evidence that AT&T had exhibited preferential behavior toward Western Electric. The stability of the AT&T–Western Electric relationship can be seen in the share of Long Lines and BOC investment outlays accounted for by Western.

Before the divestiture, Western Electric published its sales data by customer class. Its total sales to AT&T divisions (other than to the unconsolidated operating companies) remained remarkably close to 67 percent of these divisions' plant and equipment outlays for the entire 1970–82 period (table 9, column 5). Western's share never fell below 63 percent nor rose above 70 percent during the entire period, even though technology was changing rapidly, exchange rates were vacillating, and new competition was developing from Canada, Japan, and Europe.

Immediately after the divestiture, sales by Western (now AT&T Technologies) dropped sharply even as AT&T's Long Lines (its interstate services now provided by AT&T Communications) and the seven regional holding companies (RHCs)[39] expanded their investment programs. Had AT&T Technologies' share of this market continued at Western's 67 percent, its sales to these buyers alone would have risen to $12.8 billion in 1986.[40] Instead, AT&T Technologies reported total product sales of

---

employees are used, the resulting wage data provide a very poor fit. Moreover, Continental data are clearly in error, since Continental reports the same number of employees in each of its operating companies.

39. Nynex Corp., Bell Atlantic Corp., BellSouth Corp., American Information Technologies Corp. (Ameritech), Southwestern Bell Corp., US West, Inc., and Pacific Telesis Group.

40. According to AT&T, part of the decline in AT&T Western/Technologies' sales after divestiture may have been due to a change in transfer pricing between AT&T Technologies and AT&T Communications. This change in pricing, however, could not explain much of the decline shown in table 9.

Table 9. *Western Electric Sales versus AT&T Expenditures on Plant and Equipment, 1970–86*
Millions of current dollars except as noted

	(1)	*(2)*	*(3)*	*(4)*	*(5)*
	Western Electric			*Western sales ÷ AT&T outlays (percent)*	*Western's share of AT&T outlays (percent)*
Year	*Total sales*	*Sales to AT&T divisions*	*AT&T plant & equipment outlays*	*(1) ÷ (2) × 100*	*(2) ÷ (3) × 100*
1970	5,856	4,843	7,159	81.8	67.6
1971	6,045	4,987	7,564	79.9	65.9
1972	6,551	5,438	8,306	78.9	65.5
1973	7,037	6,085	9,322	75.5	65.3
1974	7,382	6,601	10,074	73.3	65.5
1975	6,590	6,016	9,329	70.6	64.5
1976	6,931	6,477	9,847	70.4	65.8
1977	8,135	7,708	11,566	70.3	66.6
1978	9,522	8,801	13,670	69.7	64.4
1979	10,964	10,031	15,837	69.2	63.3
1980	12,032	11,175	17,301	69.5	64.6
1981	13,008	12,142	18,098	71.9	67.1
1982	12,580	11,706	16,798	74.9	69.7
1983	11,155	n.a.	14,127	79.0	n.a.
1984	10,189[a]	n.a.	16,902[b]	60.3	n.a.
1985	11,235[a]	n.a.	19,453[b]	57.8	n.a.
1986	10,178[a]	n.a.	19,152[b]	53.1	n.a.

Sources: Annual reports of the companies.
n.a. Not available.
a. Product sales by AT&T Technologies.
b. AT&T plus the seven Bell regional holding companies.

just $10.2 billion, including all nontelephone equipment sales that it was now allowed with the revocation of the 1956 consent decree. A simple regression analysis of Western's sales on AT&T or RHC investment expenditures for 1970–83 generates predictions of sales for AT&T Technologies that are more than $3.0 billion higher in 1984–86 than AT&T actually realized.

These data on AT&T product sales suggest that divestiture has freed the BOCs from their dependence on AT&T Technologies for central office and transmission equipment. Some of the decline in AT&T's product sales relative to equipment purchases by the BOCs and AT&T Communications may be attributed to lower CPE sales. But the decline in AT&T's CPE market shares was almost over by 1984 (see table 1), so that most of the loss must have been confined to central office and transmission equipment after the divestiture.

The Huber Report confirms these deductions. Between 1982 and 1986, AT&T's share of the U.S. switch market (in lines shipped) fell from 70 percent to 49 percent.[41] The opening provided to AT&T's competitors in the central office market led to substantial pressure on switch prices. Data supplied by Bell Communications Research (Bellcore), the research arm of the divested BOCs, show that the real price of large switches has fallen about 3 percent a year since 1982.[42] The Huber Report cites AT&T data showing that digital switch prices fell 11 percent a year in 1982–85.[43]

Thus it appears that divestiture opened up rather aggressive competition in the central office market between AT&T and Northern Telecom. AT&T now seems to be regaining much of its loss of the digital switch market, but its total central office sales suggest that it has lost substantial market share in other equipment. The divestiture has worked to free the BOCs to purchase equipment from other, perhaps lower cost, suppliers.

Current Policy Choices

The theory of the modified final judgment is reasonable if one looks only at the broad history of the telephone industry. But it fails to explain fully why the telephone industry evolved as it did, nor does it explain fully the pricing developments of the 1960s and 1970s. Nonetheless, it does appear that AT&T used its control of local-exchange carriers to try to frustrate entry into interexchange services and CPE, and that it favored Western Electric as a supplier.

AT&T has now been fragmented. There are eight companies where there was one. Do the structural separations mandated by the modified final judgment make sense three years later? Must the local-exchange carriers be isolated forever to ensure that there is an impartial gatekeeper to the nation's telephone network? In this section, I examine the likely sources of benefits and costs from the structural separations of the divested Bell operating companies from interexchange and information services (forward integration) and from manufacture (backward integration).

Forward Integration

The entry by the BOCs into interexchange and information services would raise two of the theoretical issues described above: cross subsidy

41. Huber, *Geodesic Network*, p. 14.7.

42. Data on large local switches (25,000 lines) supplied by Paul Brandon of Bell Communications Research, Livingston, New Jersey.

43. Huber, *Geodesic Network*, p. 14.11, note 50.

and denial of access. Were these carriers to offer such services, they might be able to deny competitors equal access to the local network and thus monopolize the downstream services. This monopoly would allow the BOCs to avoid the rigors of regulation even if they charge themselves the regulated access rate for connecting to the network. Moreover, they could effect such integration by cross-subsidizing the downstream activities from local-exchange or access services. In short, despite attempts to regulate access charges and local-exchange services, the BOCs might evade regulation and extract substantial monopoly rents.

For these evils of forward integration to occur, regulators would have to be unable

—to regulate interexchange service or information service tariffs (these markets, for example, might be "deregulated");

—to police cross subsidies; and

—to guarantee equal access or "open network" architecture to unintegrated competitors.

But even without forward integration by the carriers, the FCC would still be forced to regulate access rates for all providers of interstate message telecommunications service (MTS) interexchange or information services. If regulators find the problems of allocating common and joint costs or the welter of accounting forms too complicated to prevent cross subsidies, it is hard to see how they can set access charges efficiently. Simply identifying some costs as "non-traffic-sensitive" (NTS) and attempting to cover them by fixed consumer access charges is not enough. Who is to know if these costs are properly identified, or if they reflect efficient investment or allocation decisions by the carriers?

Nor is there reason to believe that regulation, an inherently political process, would allow efficient non-traffic-sensitive pricing in any event. The recent debate over the FCC's attempt to institute subscriber line charges does not augur well for efficient regulation of access charges. Integration by the local-exchange carriers could correct any errors in overpricing access. Were the carriers able to integrate forward into unregulated interexchange and information markets, they would have an incentive to undercut competitive prices where access charges are set too high. Of course, that would immediately trigger charges by competitors of monopolization in the downstream market.

To obtain improvements in economic welfare, it is more important to eliminate restrictions for markets that are concentrated than to ease them for downstream markets that are competitive. Eliminating restrictions on the BOCs in competitive markets may be less risky than easing them in concentrated markets, but also may confer fewer benefits. The obvious candidate for liberalization is the interexchange market, simply because it is now the largest of the markets downstream from the local-exchange or access carriers and remains relatively concentrated. The current inter-LATA switched message market is about $45 billion a year—a market that continues to be dominated by AT&T (see table 2). There are reasons to believe that AT&T's modest losses in market share will not continue. With the advent of equal access, the other common carriers no longer have an access cost advantage over AT&T.

As AT&T moves to lower many of its rates to meet its rate-of-return constraint, the other major common carriers are required to follow with rate reductions. Both MCI and U.S. Sprint have encountered financial difficulties as they attempt to build nationwide fiber-optic networks. These networks cannot easily be divested, nor can other firms quickly and easily enter this market as national carriers. In short, the interexchange market appears far from contestable except for brokers and resellers.[44]

Opening the interexchange market to the BOCs, however, could revive the bottleneck problem that the modified final judgment was designed to relieve if the equal access requirement proves unenforceable. Bypass options for gaining residential customers are not now available. Brock has estimated that with switched access charges at 6 cents a minute (in 1984 dollars), the elasticity of demand for the local-exchange carriers' switched circuits is no greater in absolute value than 0.26, suggesting that there are no ready substitutes for LEC access lines for terminating most calls.[45]

A possible solution to this problem is to allow the BOCs to offer interexchange services only to large customers or outside their local franchise areas.[46] These restrictions would lessen their ability to use their own access facilities (despite equal access) to gain an unfair advantage. On the other hand, if concentration in inter-LATA interexchange continues at its present level or even increases, it may eventually be worth the risk to allow the BOCs to offer interexchange services from all exchanges equipped with equal access facilities.

44. See the paper by Leonard Waverman in this volume for a discussion of this issue.

45. Gerald W. Brock, "Bypass of the Local Exchange: A Quantitative Assessment," Working Paper 12 (Washington, D.C.: U.S. Federal Communications Commission, Office of Plans and Policy, September 1984), table 5.

46. This was the original recommendation of the Justice Department in 1987. Department of Justice, "Report and Recommendations," pp. 59–77.

Why would the BOCs succeed where the other common carriers seem to be struggling or failing? There are two reasons, neither of which will please proponents of regulation: the existence of joint economies in local and interexchange services, and the ability of BOCs to use actual access costs in pricing the services, rather than regulatory-contrived access charges.

The potential joint economies in local and long-distance services have not been demonstrated empirically. If they derive from the opportunities to configure equipment to provide both services, the regulatory requirement for equal access might negate them. On the other hand, if they derive from common use of maintenance or administrative services, they probably admit far greater possibilities of cross subsidization. Truly separate subsidiaries could perhaps reduce the latter risk, but at a cost of eliminating the joint economies.

Using the proper price for access is likely to create similar political and regulatory problems. Even with equal access, there is no reason for integrated local carriers to use the regulated access charge in their investment and pricing decisions for interexchange services. The most important potential sources of misallocation in the current postdivestiture market derive from the possible exercise of monopoly power in the inter-LATA interexchange market by AT&T and from the regulatory and political pressures to overprice access charges and intra-LATA services. It is difficult for federal government policy to prevent state regulators from using high intra-LATA prices to subsidize local-exchange or network access services. Entry by the BOCs into inter-LATA services, however, would allow them to reduce rates to attract customers as long as their *actual* incremental access, transmission, and switching costs were covered by incremental revenues. In markets in which they own the local circuits, they would not necessarily use external access charges as their own measure of opportunity costs.

Given that the price elasticity of long-distance services is greater than the elasticity of demand for local service, the potential gains in economic welfare from allowing the BOCs to offer inter-LATA services *in their own franchise areas* could be substantial. In fact, these welfare gains are likely to be much greater than those obtainable by simply allowing the BOCs to offer services, perhaps as resellers, in markets in which they have no local franchises. Thus the dilemma is quite clear: the greatest benefits from liberalization are likely to occur precisely in those situations in which the greatest political and regulatory risks exist.

The welfare gains from allowing the BOCs to offer inter-LATA services do not depend on the demonstration of joint economies. If AT&T is able to dominate the inter-LATA interexchange market, or if regulators insist on overpricing access for inter-LATA carriers, the benefits from BOC competition in these markets is likely to dwarf the potential costs of BOC cross subsidization from local services. A simple calculation illustrates this fact.

Most studies of the demand for access to the telephone network conclude that the price elasticity of demand is between −0.1 and −0.2. On the other hand, the elasticity of "toll," or interexchange, demand is generally in the range of −0.3 to −2.7.[47] Assume that fully 10 percent of the BOCs' interexchange costs are fobbed off on unsuspecting local regulators. Even if all seven BOCs captured as much as 50 percent of the inter-LATA market, their total inter-LATA revenues would be no more than about $23 billion, or half of local-exchange and access revenues. Since half of their inter-LATA revenues would be access charges, overwhelmingly paid to other companies, the total cross subsidy from local service would amount to no more than 0.10 times 0.5 times 0.5, or 0.025, of local-exchange revenues.

Assume further that BOC entry into inter-LATA services reduces inter-LATA rates 5 percent and that lower inter-LATA rates are welfare enhancing. Then the welfare effects of entry *plus* cross subsidies for various demand elasticities are

Local service	(ϵ_d = −0.1)	(ϵ_d = −0.2)
$\Delta Q/Q$	−0.0025	−0.005
Welfare loss (millions)	$5.5 × 10^6	$11 × 10^6
Inter-LATA service	(ϵ_d = −0.5)	(ϵ_d = −1.0)
$\Delta Q/Q$	−0.025	−0.05
Welfare gain (millions)	$29 × 10^6	$58 × 10^6

The gains from lower inter-LATA rates overwhelm the losses in welfare from higher local rates even with a 10 percent "cross subsidy." Thus there seem to be large potential gains from allowing the BOCs to offer inter-LATA services.[48] These gains derive from the differences in demand elasticities, the likelihood that interexchange services will not become "contestable," or competitive, and the political pressures to overprice access. Equal access is not efficient access, and it is unlikely that poli-

47. Taylor, *Telecommunications Demand*, tables 3-1, 3-5.

48. See Bruce L. Egan and John T. Wenders, "The Implications of Economic Efficiency for US Telecommunications Policy," *Telecommunications Policy*, vol. 10 (March 1986), pp. 33–40. (I ignore the small effects of a 5 percent local rate increase on telephone subscription and the externalities of telephone service.)

ticians and regulators will allow access charges to be set efficiently until local-exchange services become a contestable market.

Backward Integration

The reason for barring the BOCs from equipment manufacture cannot be that they would use their procurement practices to monopolize switching, transmission, or customer premises equipment markets. There are seven regional holding companies: if all were fully integrated into equipment, there would still be at least seven competitive entities unless they were to engage in far-reaching joint ventures.

Instead, the theory must rely on the ability of the BOCs to evade regulation through overcharging their regulated operations for equipment. This concern may have been part of the 1949 antitrust suit settled by the 1956 decree, or the FCC's docket 19129, but it never coalesced as the major issue in the 1974 suit. There has been little dispositive evidence that AT&T used its unregulated equipment operations at Western Electric to evade common carrier regulation. It is too soon to know if prices for equipment other than digital switches have fallen dramatically since divestiture, but preliminary evidence in the Huber Report suggests that nondigital-switch prices have been falling more rapidly than in earlier years. Table 8 reported evidence that suggested that BOCs actually had lower plant costs per unit of service than the other telephone operating companies before competitive entry into interexchange services and CPE. After 1971, however, the net plant costs per main telephone rose sharply for BOCs, perhaps as a response to interexchange competition.

Most analysts agree that Western Electric was slow to innovate in digital switches for local-exchange offices. Even before divestiture, some of the BOCs were beginning to turn to Northern Telecom for their larger digital switches. Moreover, since divestiture, AT&T Technologies has lost a substantial share of the BOC market for telephone equipment. Northern Telecom has been the principal beneficiary, but Siemens A.G., NEC, Stromberg-Carlson, and Ericsson are looming as new competitors. The sharp rise in imported telecommunications equipment in 1983 is further evidence of this newly competitive situation.

With seven potential new players, many of the equipment markets could become even more competitive. It seems unlikely that many of the BOCs would venture into the extremely expensive development of large digital switches, but as long as there are seven such companies, the danger of concealing supracompetitive prices from regulators seems remote. If, however, the seven were to join forces in equipment manufacture and acquisition, the old danger of overpriced equipment could reappear.

Judge Greene's 1981 decision makes it clear that he wished to prevent the BOCs from repeating the practices of AT&T after the *Carterfone* decision—the use of the bottleneck to monopolize the sale or lease of customer premises equipment. Given the large increase in CPE competition even before the divestiture, and the considerable increase in CPE competition since then, this concern seems exaggerated. The CPE market is already so competitive that the injection of from one to seven new players would seem to matter very little. If the integrated AT&T was too weak to prevent competition from developing (table 1), one is hard pressed to see how one regional company could use its bottleneck facilities to monopolize this market.

Judge Greene's 1987 Opinion

In September of 1987, Judge Greene issued his opinion in the first triennial reexamination of the line-of-business restrictions as specified in the modified final judgment.[49] With few exceptions, he denied the regional holding companies' requests for liberalization of the restrictions because he was unpersuaded that their bottleneck monopolies had been eroded by technical change or market entry. The only changes approved by Greene were to remove all limits on noncommunications services and to allow the BOCs to offer the transmission services and protocol conversion for videotext services. In 1988 Greene ruled that the BOCs may also offer voice storage and retrieval services, but he continued to bar them from general participation in information services.

Greene's reasoning on keeping the line-of-business restraints was simple. The BOCs continue to control the essential local-access links to all but a handful of telecommunications users. With this control, the BOCs could repeat the perceived sins of AT&T in the predivestiture years if Greene allowed them to integrate into equipment manufacture, interexchange services, or information services. Changes in the structure of each of these markets since 1982 would quickly be reversed if the BOCs were to return to any of them. Greene did not examine the effects of changing relative prices of local and interexchange services. Nor did he give credence to the argument that seven integrated local-exchange companies could be more easily policed than one. Until Greene is persuaded that

49. *United States* v. *Western Electric*, 673 F. Supp. at 525.

there are economical alternatives to the local-subscriber loops of the BOCs for access to interexchange or information service carriers, he is unlikely to permit any substantial liberalization of the line-of-business restrictions.

Unfortunately, Greene is now hostage to the original reasoning in the modified final judgment—that the BOCs should be denied the right to participate in upstream or downstream markets until all dangers of vertical integration from a bottleneck position disappear. He is unwilling to balance prospective welfare gains from integration against possible losses from the abuse of the bottleneck. The result may be a considerable loss in economic welfare.

Summary

Regulatory problems rarely involve choices from among first-best solutions to resource allocation issues. The choices involved in delimiting the domain of BOC activities are among risky alternatives. To hem the BOCs into a corner called "local service" is to sacrifice the benefits of unleashing the competitive energies of seven large communications holding companies. On the other hand, to allow the BOCs free entry into such downstream services as information and inter-LATA services is to risk cross subsidization induced by regulation. Perhaps the best solution is to keep the line-of-business restraints until the states adopt alternatives to cost-based telephone rate regulation.

Deregulation in Japan

Tsuruhiko Nambu, Kazuyuki Suzuki, and Tetsushi Honda

THE JAPANESE telecommunications industry is undergoing a major transformation from a regulatory regime to a competitive structure. The Japanese style of telecommunications deregulation lies somewhere between the American and the British approaches. Nippon Telegraph and Telephone (NTT) is an integrated national network that provides local as well as long-distance services, as American Telephone and Telegraph did before the 1984 divestiture of its operating companies. Like AT&T but unlike British Telecom, NTT now faces competition from newcomers in both the long-distance and the local markets. Thus under the new regime, Japanese regulators face the difficult task of structuring a competitive market compatible with NTT's mandate to provide "universal" service.

In this paper, we tackle the Japanese regulatory problem by focusing on the government-business relationship in the telecommunications industry. The first section discusses the historical background of the privatization of the former NTT Public Corporation and the new regulatory system under the Japanese Ministry of Posts and Telecommunications (MPT). The important features of this regulation are analyzed in the context of the new competitive environment.

The next three sections concern the past and present organization of NTT. Its performance as a public monopoly, and how that performance compares with AT&T's, are central topics. The demand and cost functions are estimated to test the efficiency of the NTT tariff structure.

In the last section, we use a simple model of dominant firm–competitive fringe to analyze NTT's behavior under the burden of subsidizing universal service. We then look at several policy choices for shaping an industry structure that will be competitive while allowing every citizen access to telecommunications services.

The Institutional Framework

There were two kinds of regulated industries in Japan after the Second World War. One was the traditional type, in which private firms supplied goods and services under the supervision of regulatory agencies. Transportation, electricity and gas, and banking were and still are in this category. The other form was the public corporation, called *kosha*, which provided goods or services under the supervision of the Diet. These industries included telecommunications, the nationally integrated railroad system, and cigarettes. For these public corporations, governmental bodies administered the regulations, but the ultimate decisions on key matters were made by the Diet. In particular, all revisions of tariffs required the consent of special committees in the Diet.

From 1952 on, telecommunications services were provided by the Nippon Telegraph and Telephone Public Corporation as a public monopoly. Before the Second World War, both telecommunications and postal services had been supplied by the Japanese Ministry of Communications, or so-called Japanese PTT (Teishinsho), but after the war a public corporation was created to develop the telecommunications industry.[1] Many career people in the old PTT moved to the NTT Public Corporation, creating a climate of independence from the Ministry of Posts and Telecommunications (MPT). Until the mid-1970s continuous growth and capacity expansion characterized the telecommunications industry in Japan. During the 1970s, however, international and other external factors combined to induce a change in the industrial structure.

Internally, NTT reached its goals of satisfying the unfilled demand for access and of building up the direct distance dialing system (discussed in more detail below). It was thus forced to look for a new organizational objective. Externally, the demand for data communications began to grow (although almost ten years behind the U.S. growth), spurring complaints about the inefficiency of public monopolies and government regulation.[2]

The business world discovered promising opportunities in the telecommunications and information industries. Some well-known firms in the high-technology area launched trial balloons by building telecommuni-

1. The public monopoly, called Denden Kosha, was legally based on the Nippon Telegraph and Telephone Public Corporation Law of 1950.

2. For a general discussion about the need to reshape the industry, see Ken-ichi Imai, ''Some Proposals concerning Japan's Telecommunications Strategy,'' Discussion Paper 105 (Hitotsubashi University, Institute of Business Research, September 1982).

cations service facilities themselves.[3] The two government ministries concerned, the Ministry of International Trade and Industry (MITI) and the MPT, eager to increase their influence over the newly developing information industries, pushed for liberalizing the public monopoly. These elements created a favorable environment for reshaping the monopoly structure of the Japanese telecommunications industry.

The political enthusiasm for reforming public administration in Japan—in this case, paring the excessive costs of inefficient management in the public corporations—was another driving force for reexamining the performance of the public monopoly system. At first Japan National Railways was the prime target for privatization, but soon telecommunications services and the cigarette industry also came under scrutiny. The American deregulation movement provided further impetus for accelerating the privatization of these sectors.

Interministerial competition between the MPT and MITI, which regulates the electronics and computer industry, was also important in forcing a change in the organization of the telephone industry. MITI advocated liberalization of the telecommunications sector, and the MPT was forced to counter MITI's increased influence in this area. New legislation emerged as a result of an economic and political compromise between them.

The New Institutional Setting

In 1985 laws for restructuring the Japanese telecommunications industry, previously guided by the Public Telecommunications Law of 1953, were enacted by the Diet. One is the Telecommunications Business Law and the other is the Nippon Telegraph and Telephone Corporation Law. The business law required a radical reconstruction of the industry configuration, and the corporation law privatized the old public corporation, giving birth to a new NTT.

THE TELECOMMUNICATIONS BUSINESS LAW. This statute divided the telecommunications sector into two categories of firms: type I and type II. Type I carriers own the telecommunications facilities themselves; type II carriers do not own these facilities but rent them from type I carriers. (This dichotomy is unrelated to the conventional U.S. distinction

3. Daini Denden Kikaku Co. was created by Kyocera and others preparatory to entering the telecommunications business. Japan National Railways proposed to enter by utilizing its nationwide railroad network. Major banks, trading companies, and other big businesses expressed interest in joining new ventures in the telecommunications industry.

between basic and enhanced services.) Type I carriers are not necessarily nationwide enterprises; new entrants can be medium-sized or small firms. Three new entrants, called NCCs (new common carriers), have already begun to provide services between Tokyo and Osaka. They are Japan Telecom Company, Teleway Japan Corporation, and Daini-Denden Inc. In 1986 only private-line services were supplied by these new firms, with no access charge, but in 1987 they began to supply public telephone services. Another firm, the Tokyo Telecommunications Network (TTN), a subsidiary of the electric utility monopoly in the Kanto District (the Tokyo Denryoku Corporation) has begun to provide local services in the Kanto district. If one includes small firms, such as those that provide mobile services like paging devices, more than ten type I carriers were operating in 1987.

Type II carriers usually provide enhanced (or value-added) services, and they are further classified into "special" type II and others. The former now (1988) include seventeen carriers that supply nationwide service.

The MPT has become a true regulatory body of the telecommunications industry because all carriers are now free from the supervision of the Diet but subject to the business law.

The legal distinction between type I and type II is that the former must obtain permission from the MPT to enter the industry, but the latter has only to file notice of entry. For type I carriers, there are regulations of the kind often applied to the public utilities. One is the "demand and supply adjustment clause," by which the MPT can regulate the rate of entry to avoid purportedly ruinous competition. At the same time, exit of type I businesses requires approval of the minister of posts and telecommunications. The rate level and rate structure of telecommunications services are subject to MPT regulation based on traditional rate-of-return principles.[4]

THE NIPPON TELEGRAPH AND TELEPHONE CORPORATION LAW. NTT was privatized by the NTT Corporation Law and provides both type I and type II services. Though still in a dominant position in the national market, NTT faces competition from new entrants in the most profitable market, that between Tokyo and Osaka. In principle NTT is now a private company, but it has an extra burden: it must continue to supply "universal

4. The MPT has regulated with a rather heavy hand, much as the U.S. Federal Communications Commission used to do. According to the new tariff accounting rule, all rates of services must be based on the fully distributed cost. Fair rate of return is to be calculated based on the average rate of return of the major industrial sector.

and equitable'' service as if it were a public monopoly.[5] In other words, NTT is obliged to cross-subsidize the ''deficit'' services as a public utility if regulators set rates lower than costs for these services. This mandate obviously is inconsistent with the existence of competition in some markets and poses problems, discussed below, of efficiency and income redistribution.

NTT has faced another problem of corporate reorganization. Data communications services have been awarded to NTT, and NTT used to provide technical and enhanced services to customers. But some analysts feared that there might be unfair cross subsidization from the revenues of other NTT divisions to the data communications division. To eliminate the dominance of NTT in the type II area, this division was separated from NTT in 1988, and the NTT Data Communications System Corporation was created.

NTT does not have a manufacturing division, so it purchases its customer premises equipment from the outside. Because NTT is a huge purchaser, critics often claim that it abuses its dominant position to obtain favorable prices for equipment.[6] This assertion, however, is certainly open to question.

TYPE II BUSINESSES. Under the new regime, enhanced services are provided in a competitive market. The number of entrants into type II business through 1987 was more than 400.[7] ''Special'' type II carriers provide nationwide, large-scale network services and are subject to tighter regulation than regular type II carriers.

The MPT has estimated that the market for type II services is almost ¥520 billion ($3.4 billion).[8] For the moment, the largest market for enhanced services is in the distribution sector, particularly between wholesalers and retailers.

INTERNATIONAL TELECOMMUNICATIONS. International telecommunications services are provided by Kokusai Denshin Denwa Corporation (KDD) as a monopoly, but because of the 1985 Telecommunications

5. This charge was inserted in article 2 of the NTT Corporation Law during its final consideration by the Diet at the insistence of the Socialist party, although NTT was reluctant.

6. For example, NTT sells its own telephone sets, and it can exert significant purchasing power on manufacturers. NTT is also criticized for ''unfair'' bundling of telephones and dialing numbers when it sells telephone sets.

7. Council on Telecommunications, *Denki Tsushin Jigyoho Fusoku Dai 2 Jo ni Motozuki Kojiru beki Sochi, Hosaku no Arikata ni Tsuite (Toshin)* (Future Policy Recommendations on the Basis of Article 2 of the Telecommunications Business Law attachment [A Report]) (March 1988).

8. Ministry of Posts and Telecommunications, *Tsushin Hakusho, 1986* (White Paper on Telecommunications, 1986) (Tokyo: Ministry of Finance, 1986).

Business Law, it will soon face competition. At present the market for international telecommunications is less than one-twentieth the size of the domestic telecommunications market, but future prospects are bright and more carriers will probably appear.[9]

An Anatomy of NTT's Telecommunications Services

From its establishment in 1952 until the recent restructuring, the Nippon Telegraph and Telephone Public Corporation was the sole supplier of Japanese telecommunications services, except for its international division, which was transferred to KDD in 1953. The granting of a monopoly to NTT was based on the need to reconstruct the telecommunications network services, which were destroyed in the Second World War, as soon as possible to meet the rapid increase in demand. To accomplish this aim, policymakers decided they had to standardize the telecommunications network and avoid inefficient overinvestment by competitive firms. Besides, the task of financing the huge initial investment seemed beyond the capabilities of private firms at the time. Until the end of the 1970s, then, NTT's business had two principal goals: to fulfill pent-up demand for access to service, and to provide complete direct distance dialing service nationwide. And these two goals were part of a larger objective—to equalize access to telecommunications services.

When NTT was founded in 1952, it had 1.37 million subscribers, and unfulfilled demand for access amounted to 340,000 subscribers. The ratio of subscribers to the total population (the diffusion rate) was 1.8 percent. The ratio of residential subscribers to the number of households (the penetration ratio) was roughly 0.6 percent.[10]

NTT tried to meet this unfulfilled demand, but it grew further, to 2.91 million potential subscribers by 1970, because Japan's economy was expanding rapidly. The annual growth rate of subscribers (including unfulfilled demand) was about 14 percent from 1952 to 1970, whereas the real economic growth during the same period was 10 percent. Still, progress was made; the diffusion rate and the penetration ratio had reached 15.7 percent and 25.4 percent respectively by 1970.[11]

9. One long-term forecast has been made by the Ministry of International Trade and Industry, *2000 Nen no Joho Sangyo Bijon* (The Vision of the Information Industry in the Year 2000) (Tokyo, 1987).

10. These figures are for the beginning of 1952. See Takeshi Shoda, *Nippon Denshin Denwa Kosha: Johoka Shakai no Riidaa* (NTT Corporation: The Leader of the Information Society) (Tokyo: Asahi Sonorama, 1980), p. 117; and MPT Policy Research Group, *Denki Tsushin Gyosei, 1985* (Administration of Telecommunications, 1985) (Tokyo, 1985), p. 21.

11. Ibid.; and MPT, *Tsushin Hakusho, 1984*.

Figure 1. *Fulfillment of Demand for Access by NTT, Selected Fiscal Years, 1952–77*

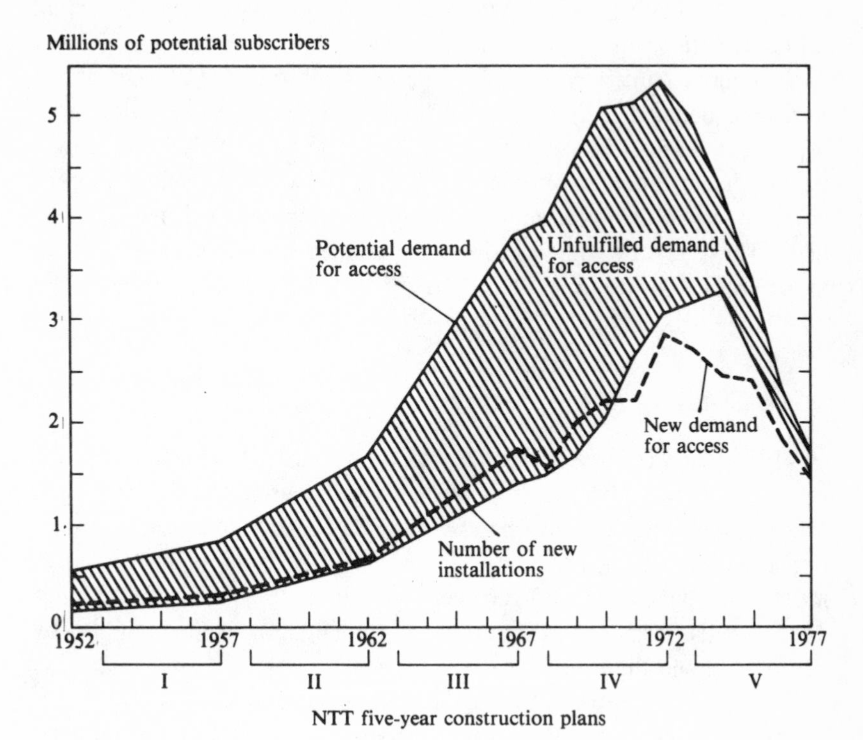

Source: Ministry of Posts and Telecommunications, *Tsushin Hakusho, 1984* (White Paper on Telecommunications, 1984) (Tokyo: Ministry of Finance, 1984), p. 42.

NTT required several five-year construction plans to meet its aims of no unfulfilled demand and nationwide service (see figure 1). Practically speaking, the first goal was reached by 1977 and the second by 1979. By 1984 subscribers numbered nearly 44 million, the diffusion rate was 37 percent, and the penetration ratio had reached 79 percent.[12]

The annual growth rate in the demand for calls (defined as the volume of three-minute calls) was 8.1 percent between 1977 and 1984, and it accelerated to 10.0 percent between 1980 and 1984. Broken down into figures for local versus long-distance service, the annual growth rates were 9.3 percent and 5.2 percent, respectively, between 1977 and 1984, and they accelerated to 11.8 and 6.8 percent between 1980 and 1984.[13] The

12. Authors' calculations based on data from MPT, *Tsushin Hakusho, 1986*, p. 333.
13. Authors' calculations based on unpublished data supplied by NTT.

more rapid growth of local-service demand compared with long-distance demand reflects the increased ratio of residential-user demand to total demand. Although residential telephones accounted for only 25 percent of total telephones in 1960, their share went up to 54 percent in 1972 and reached about 70 percent in 1984. The reasons for this sharp increase in household demand include the rise in incomes and the desire for more convenient services in the household sector. The relatively slow growth of long-distance demand might also be explained by the substitution of private-line and data communication services for switched message services. These private-line services corresponded to 5 percent of all telephone services in terms of sales in 1977 but reached about 8 percent in 1984.[14]

The Changing Rate Structure of NTT

NTT service is provided under a two-part tariff. For businesses, the basic fee or monthly charge is ¥2,600; for households, ¥1,800.[15] In principle the basic fee corresponds to the customer's cost (rental cost of the subscriber's equipment line connection). The price of local calls is not flat-rate but rather ¥10 for every three minutes. The price of long-distance calls is determined not only by the length of the call but also by the distance and the time of day.

The rate structure of NTT has been characterized by wide price differentials between local and long-distance service. The ratio of the most expensive long-distance service (in the daytime) to the price of local service was 72 to 1 until October 1980, far more than the ratio of the actual costs.[16] NTT has tried to correct this price gap by lowering the price of long-distance service beyond certain distances: 500 kilometers in 1981, and 320 kilometers in 1983 (see table 1). Despite these corrections, the ratio still stands at 40 to 1, while the ratios for the United States, West Germany, and France are around 15 to 1, and the ratio for the United Kingdom is only 4 to 1.[17]

Trends for the average prices of long-distance service are given in table 2. NTT lowered the price for middle-range (60–320 kilometers) service in 1984. The price of long-distance service was reduced about 40 percent

14. See Nippon Telegraph and Telephone, *Nippon Denshin Denwa Kosha Shashi: Keiei Keitai Henko Made no Hachi-nen no Ayumi* (History of NTT: Steps over the Past Eight Years to Modify the Structure of Operations) (Tokyo, 1986), p. 27.

15. Tariff data supplied by NTT.

16. See MPT Policy Research Group, *Denki Tsushin Gyosei, 1985*, pp. 60–61.

17. Quoted from unpublished materials prepared by NTT.

Table 1. *Price Differential between Local and Long-Distance Service in Japan, Selected Fiscal Years, 1953–83*[a]
Ratio of price of local service to most expensive long-distance service

Time of day	*1953*	*1969*	*1980*	*1981*	*1983*
Day	1:109	1:72	1:72	1:60	1:40
Night	1:60	1:45	1:45	1:36	1:24
After midnight	1:60	1:45	1:28	1:24	1:22

Source: Ministry of Posts and Telecommunications Policy Research Group, *Denki Tsushin Gyosei, 1985* (Administration of Communications, 1985) (Tokyo, 1985), pp. 60–61.
a. The Japanese fiscal year ends on March 31 of the named calendar year.

Table 2. *Two Measures of Trends in the Average Price of Long-Distance Service in Japan, Fiscal Years 1977–84*
1980 constant prices in yen

Year	*Measure I*[a]	*Measure II*[b]	*Year*	*Measure I*[a]	*Measure II*[b]
1977	163.8	154.0	1981	131.6	118.4
1978	157.7	152.0	1982	126.3	111.3
1979	152.2	147.1	1983	116.7	102.2
1980	141.0	134.0	1984	105.3	93.3

Source: Authors' calculations based on unpublished data supplied by NTT.
a. Calculated from the tariff (average price of service in the daytime, weighted by use of different distance intervals in 1980).
b. Average revenue per three-minute call deflated by the consumer price index.

between 1977 and 1984, but it is still far above the average cost. On the other hand, the basic fee and the price of local calls are generally considered to be far below the average cost.[18]

In 1984 NTT's total revenues reached ¥4,750 billion. Revenues were ¥2,500 billion from telephone calls, ¥990 billion from basic fees, and ¥380 billion from private-line service. Revenues from long-distance calls contributed more than 76 percent of the total revenue from telephone calls, although the volume of long-distance calls was only about 27 percent of the total volume of calls.[19]

These facts suggest the existence of a cross subsidy between local and long-distance services if the costs of the two types of calls are similar. There is also a cross subsidy between basic fees and long-distance service. Revenues from basic fees and local calls contributed only 33 percent of total revenue in 1984.

In essence, NTT's tariff system was designed by the authorities for the welfare of the residential user. We will discuss the problem of such cross subsidization later in connection with Ramsey pricing.

18. Based on discussions with NTT staff, who use fully distributed cost (FDC) allocation.
19. Ibid.

The new common carriers began to supply private-line services in 1986, and they launched into the general long-distance market in 1987. This market has been a main source of NTT's profit, but the new common carriers will charge considerably less than NTT's current prices.[20] Unless NTT adopts cost-based pricing, it will be difficult for it to continue its cross subsidy with the introduction of competition. The instability of NTT's revenue structure is shown by the fact that about 30 percent of NTT's revenues from telephone calls comes from only 3 percent of total subscribers.[21] In other words, NTT's revenues will be deeply affected by the shift of only a few large users to the NCCs. Some suggest that access charges be levied on the NCCs in order to maintain NTT's existing price structure. Considering that the aim of introducing competition into the telecommunications market is to enhance social welfare or allocative efficiency by driving the prices of both services to marginal cost, however, the adoption of an access charge does not seem to be a good solution.

The Performance of NTT

As table 3 shows, in fiscal 1985 NTT's operations were about half those of AT&T before its divestiture in terms of net sales, total assets, and number of employees (at the then-current exchange rate of ¥150 = U.S. $1.00). Note that the figures shown for the new AT&T are net of access charges. The ratio of NTT's operating income to total assets was 6.4 percent, and the ratio of income before taxes to net sales was 6.2 percent in fiscal 1985. NTT's low rate of return results not only from price regulation but also from its heavy dependence on interest-bearing liabilities, where annual interest paid is equivalent to income before taxes.

NTT's debt resulted from the accelerated increase in investment expenditure and related adjustment costs to fulfill pent-up demand in the 1970s and to incorporate further innovations in the 1980s. Analysts also note that NTT has been obliged to supply high-quality and complete service and to have ample reserves in the transmission system for national security purposes. The total amount of investment expenditure reached almost ¥11 trillion between 1968 and 1977 and was ¥12 trillion between 1978 and 1984.[22]

20. The NCCs' new tariff systems, which on average are about 20 percent lower than those of NTT, were approved by the Ministry of Posts and Telecommunications in June 1987.

21. Based on discussions with NTT staff.

22. Authors' calculations based on data from MPT, *Tsushin Hakusho*, annual issues; and NTT, *Nippon Denshin Denwa Kosha Sashi*, p. 27.

Table 3. *Comparison of Nippon Telegraph and Telephone with American Telephone and Telegraph, Fiscal Years 1983, 1985*[a]

Indicator	*NTT (after privatization) 1985*[b]	*AT&T (before divestiture) 1983*	*AT&T (after divestiture) 1985*
Scale			
Net sales (millions of dollars)	33,943	69,403	34,910[c]
Income before taxes (millions of dollars)	2,107	9,376	2,546
Total assets (millions of dollars)	72,844	149,530	40,463
Number of employees	308,789	837,790[d]	337,600
Rate of return			
Operating income to total assets (percent)	6.4	8.7	7.4
Net sales to total assets (percent)	0.47	0.47	0.86
Income before taxes to net sales (percent)	6.2	13.5	7.3
Stability			
Liabilities to equity (percent)	223	139	151
Productivity			
Net sales per employee (thousands of dollars)	109.9	77.7[d]	103.4
Number of subscribers per employee	137.7[e]	103.8[d]	n.a.

Sources: Japan Development Bank, "Financial Data of Industries" (September 1986), pp. F-2, F-3; Compustat data; and AT&T annual reports.

n.a. Not available.

a. NTT's fiscal year ends on March 31 of the named calendar year; AT&T's fiscal year ends on December 31 of the calendar year.

b. Calculated at ¥150 = US$1.00.

c. Net of access charges.

d. For 1982.

e. For 1984.

Although much of the role of NTT in the technological development of telecommunications services (including communications equipment) has been gradually transferred to equipment manufacturers belonging to the so-called NTT family (NEC Corporation, Fujitsu Ltd., Oki Electric Industry Company, Hitachi Ltd.), NTT still plays a leading role in this field, especially in basic research. NTT spent ¥91 billion on research and development in fiscal 1983, which represented 27 percent of total R&D expenditure (including spending by the government) in this sector (table 4). NTT's share of R&D expenditure in the telecommunications sector had exceeded 50 percent in the early 1970s. In the United States, the ratio of federal expenditures to total expenditures for R&D in the communications equipment industry was about 0.4 in 1980, according to a National Science Foundation survey.[23] NTT had, in a similar manner, borne much

23. National Science Foundation, *Research and Development in Industry, 1981*, NSF 83-325 (Washington, D.C.: Government Printing Office, 1981), tables B-3, B-12.

Table 4. *Japanese R&D Expenditure for Telecommunications, Fiscal Year 1983*

Group	*Millions of yen*	*Percent*
Government	4,330	1.3
Universities	6,646	2.0
NTT	90,800	27.2
KDD, NHK[a]	12,480	3.7
Communications equipment industry	222,000	66.0
Total	336,256	100.0

Source: Council on Telecommunications Research and Development Policies, *21 Seiki o Mezashita Denki Tsushin Gijutsu Kaihatsu Seisaku* (Research and Development Policies of Telecommunications into the 21st Century) (Tokyo: MPT, 1983), p. 86.

a. KDD, or the Kokusai Denshin Denwa Corporation, provides Japan's international telecommunications services. NHK, or Nippon Hoso Kyoku, is Japan's national television network.

of the burden of the technological development in telecommunications for the private sector in fulfilling its responsibility for the national interest. These expenditures contributed to the low rate of return of NTT. In fiscal 1985 NTT's ratio of debt to equity amounted to 2.23, while the figures for its U.S. counterparts were around 1.50.[24]

Finally, NTT's productivity measured in net sales per employee was $109,900 in fiscal 1985 (at ¥150 = US $1.00), and productivity growth after privatization was 8.7 percent in real terms during fiscal 1984–85. This growth exceeded its average annual productivity growth rate of 5.1 percent during fiscal 1980–84, just before the privatization. Labor productivity in AT&T in fiscal 1985 was $103,400, and it may be assumed that the Bell operating companies' productivity is comparable. Thus the level of productivity in terms of net sales per employee is comparable between Japan and the United States.

Estimation of the Demand Function for Long-Distance Services

We specify the demand function for long-distance service as follows:

$$Q = F\,(P_T/CPI,\, N).$$

The volume of long-distance service, Q (volume of three-minute calls), depends on the price of long-distance service (P_T) in relation to the consumer price index (CPI) and the number of subscribers (N).

As mentioned before, the situation of excess demand and problems with the quality of service (especially direct distance dialing) had been just about completely cleared up by 1977. We therefore estimate the

24. See Japan Development Bank, "Financial Data of Industries" (September 1986), p. F-3; Compustat data; and AT&T, *1985 Annual Report*, p. 19.

Table 5. *Estimation of the Demand Function for NTT Long-Distance Service*

	Independent variable		
Dependent variable	$LN(P_T/CPI)$	$LN(N)$	$\bar{R}^2$
Volume[a] of three-minute calls, $LN(Q)$	−0.605	0.249	0.98
	(3.53)	(0.60)	. . .
Number of messages, $LN(Q)$	−0.628	0.226	0.98
	(3.60)	(0.52)	. . .

Source: Unpublished time-series data from NTT for fiscal 1977–84. Numbers in parentheses are *t*-statistics. Values of the constant term are not reported here.

a. Volume of calls includes both residential and business calls.

demand function from 1977 to 1984. The price of long-distance service has changed several times during this period, so we estimate the cumulative effect of the change in prices of the demand for long-distance service. We also estimate the demand function using the number of messages, rather than the volume of three-minute calls, as the dependent variable.

Table 5 reports the results of estimating the demand function both ways; the results are quite similar. Of principal interest is the long-run price elasticity of −0.61 to −0.63, which is statistically significant. This value is analogous to the empirical results obtained in both the United States and United Kingdom.[25]

The estimation results for the demand function for local service are not reported here because we cannot obtain the data series for this service before 1977. Because the price of local calls has not been revised since 1976, we are estimating essentially the effect of changes in the consumer price level on local-service demand.

NTT tentatively estimates the short-run price elasticity of demand for local service to be −0.15, based on data for only a few months immediately before and after a price change in local service in November 1976. NTT estimates that local-service demand is price inelastic, a result consistent with similar findings in the United States and United Kingdom.[26]

The Cost Structure of NTT: Estimation of the Long-Run Marginal Cost of a Telephone Call

In this section, we estimate the marginal cost of a telephone call, which is the key measure of the economic efficiency of NTT's pricing system.

25. See James M. Griffin, "The Welfare Implications of Externalities and Price Elasticities for Telecommunications Pricing," *Review of Economics and Statistics*, vol. 64 (February 1982), pp. 59–66; and Lester D. Taylor, *Telecommunications Demand: A Survey and Critique* (Ballinger, 1980).

26. See Taylor, *Telecommunications Demand*, p. 125.

Although it is unclear whether NTT had sufficient incentives to minimize total cost in the past, we attempt to estimate the relation between cost and outputs using a simple multi-output cost function.

The multi-output cost function can be used to describe the total cost (TC) of producing local-service access (N), local-service usage (Q_1), and long-distance service usage (Q_2):

$$TC = F + sN + c_1Q_1 + c_2Q_2,$$

where F measures the total amount of fixed costs, and s, c_1, and c_2 are assumed constant marginal costs per subscriber, per local-service call, and per long-distance service call, respectively. We cannot estimate econometric cost functions that are more sophisticated, such as the translog cost function, because of data limitations. Moreover, it is also difficult to estimate the cost function expressed in the equation directly because of the high correlations among N, Q_1, and Q_2. To avoid this problem, we try to estimate the marginal cost of a telephone call using a method devised by Littlechild.[27]

We first distinguish those costs that are associated with subscribers being in the system (customer costs) from those costs that are associated with the making of telephone calls (call costs).[28] Customer cost (CM) depends on the number of subscribers (N). Call cost (CU) consists of two parts: a traffic-sensitive part and a non-traffic-sensitive part. The traffic-sensitive part is further divided into those costs that are required for both local calls and long-distance calls and those costs that are required for long-distance calls only. This distinction is needed because the long-distance call is transmitted using the first half of the local network ($\frac{1}{2}c_1$), then a toll network ($c_2 - c_1$), and finally the second half of the local network. The costs can be described in the following manner:

$$TC = CM + CU$$

$$CM = Fm + sN$$

$$CU = Fu + c_1(Q_1 + Q_2) + (c_2 - c_1)Q_2,$$

where

CM = customer cost;
CU = cost of telephone call;
Q_1 = volume of three-minute calls for local service;

27. S. C. Littlechild, *Elements of Telecommunications Economics* (London: Institution of Electrical Engineers, 1979), chap. 6.
28. Ibid., pp. 86–110.

Q_2 = volume of three-minute calls for long-distance service;
Fm = fixed part of a customer cost;
Fu = fixed part of a call cost;
s = non-traffic-sensitive cost per customer;
N = number of subscribers.

All the above costs are expressed in 1980 constant prices.

Traffic-sensitive costs and sN are deflated by the price index of capital goods, and fixed costs are deflated by a wage index because they are mainly labor costs. The data for Fm and Fu cannot be estimated separately because they are mainly common costs; thus we need some modification of the above equations. NTT has expanded its capacity in accordance with the increase in demand, and its costs include those investments for expansion. Taking this into account, we estimate the long-run marginal cost as follows. We distinguish the plant and equipment used exclusively for long-distance calls (toll office, toll switching equipment, major parts of the transmission system) as well as the customer plant and equipment (subscriber line and apparatus) among the total assets, and we calculate the weight of each type of plant and equipment among the total assets. Then we can compute $(c_2 - c_1)Q_2$ and sN using the total plant cost multiplied by the weights calculated above. Total plant cost is a long-run variable cost that consists of the annualized installation cost of equipment (including depreciation, maintenance cost, and property taxes), and the cost of capital (interest paid and so on). We compute $(c_2 - c_1)Q_2$ instead of $c_1(Q_1 + Q_2)$ because we can more easily identify the costs of the former expression.

The noncapital costs are not included in our long-run variable costs. Rohlfs incorporated noncapital costs (mainly labor costs), excluding the common costs, that are allocated to each service based on the embedded direct cost (EDC) study of AT&T into his estimation of marginal cost. But the weight of the increment of the noncapital cost factor (excluding common costs) in the total increment of costs is small, and thus its impact is relatively small in his estimation of the marginal cost.[29] There are two reasons why we exclude the noncapital costs of each service in our long-run variable costs. First, NTT lacks a report as detailed as the EDC study, so it is difficult to assign noncapital costs to each service. Second, the amount of those costs seems to be almost constant over time, as can be seen in the Rohlfs study's table IV-4. We deduct $(c_2 - c_1)Q_2$ and sN

29. Jeffrey Rohlfs, ''Economically-Efficient Bell-System Pricing,'' Bell Laboratories Economic Discussion Paper 138 (Murray Hill, N.J.: AT&T Bell Laboratories, January 1979).

Table 6. *Cost of NTT Local Access and Exchange Service Exclusive of Non-Traffic-Sensitive Customer Costs, Fiscal Years 1978–84*[a]

	Dependent variable $TC - sN - (c_2 - c_1)Q_2$			
Independent variable	*1978–82*		*1978–84*	
$Fm + Fu$ (millions of yen)	1,256,754	(14.16)	1,267,876	(39.59)
$Q_1 + Q_2$ (million)	6.9	(2.90)	6.6	(8.16)
$\bar{R}^2$	0.65	. . .	0.92	. . .

Source: Unpublished time-series data from NTT for fiscal 1978–84. Numbers in parentheses are *t*-statistics.

a. We can also estimate the marginal cost of long-distance service (exclusive of local costs), $c_2 - c_1$, to be 35.9 (or 25.9 in the later period) by deducting sN, and 6.9 $(Q_1 + Q_2)$ (or 6.6 $(Q_1 + Q_2)$), from TC.

from TC and estimate c_1, the cost of providing local access and exchange service (excluding non-traffic-sensitive customer cost), from the following equation:

$$TC - sN - (c_2 - c_1)Q_2 = Fm + Fu + c_1(Q_1 + Q_2).$$

We estimate the above equation for the two alternative sample periods in order to test the robustness of the parameters.

The empirical results are shown in table 6. From the above investigation we arrive at the following findings:

—The marginal cost of local service ranges between ¥6.6 and ¥6.9 for a three-minute call, and the marginal cost of long-distance service ranges between ¥32.5 and ¥42.8 (three-minute call). The latter estimates are equal to $c_2 - c_1$ plus the estimates of the marginal costs of local calls in table 6.

—The marginal cost of local service is almost constant over the different sample periods. On the other hand, the marginal cost of a long-distance call drops by ¥10.3, almost 25 percent, if we include 1983 and 1984 in the sample period. Although a lag does exist between increases in capacity (investment in plant) and increases in demand, the drop in long-distance costs reflects the rapid technical progress in recent years in the transmission system for long-distance calls. Between the late 1950s and the late 1970s, for both analog coaxial cables and digital transmission systems with metallic coaxial cable, the transmission cost per line declined drastically with the development of new systems. Even more important, since the beginning of the 1980s, the introduction of fiber-optic digital transmission systems has caused the transmission cost per line to fall radically.[30]

—The levels of the marginal costs of both local and long-distance service in Japan are quite similar to those in the United States, if we

30. NTT, *Nippon Denshin Denwa Kosha Shashi*, pp. 598–99.

Table 7. *Estimates of the Long-Run Marginal Cost of Three-Minute Telephone Calls, Japan and the United States, Various Years*

Country and item	*Cost*
Japan[a]	
Local call (cents)[b]	4.4–4.6
Local call (yen)	6.6–6.9
Long-distance call (cents)[b]	21.7–28.5
Long-distance call (yen)	32.5–42.8
United States	
Local call (cents)	
Littlechild-Rousseau estimate	3.7
Rohlfs estimate	4.4
PNR & Associates estimate	5.1
Long-distance call (cents)	
Perl estimates:	
Intra-LATA	20.1
Inter-LATA	24.0
Interstate	27.0
AT&T Telecommunications Policy Model estimates:	
Message telephone service—Day	39.0
Message telephone service—Non-day	18.0

Sources: Japan averages from NTT unpublished data; Littlechild-Rousseau, Rohlfs, and PNR estimates from Walter G. Bolter and others, *Telecommunications Policy for the 1980s: The Transition to Competition* (Prentice-Hall, 1984), p.423; Perl estimates from Lewis J. Perl, "Social Welfare and Distributional Consequences of Cost-Based Telephone Pricing," paper presented at Thirteenth Annual Telecommunications Policy Research Conference, Airlie, Va., April 1985, p. 13; AT&T estimates from the AT&T Telecommunications Policy Model (Telpol); and Stephen J. Brown and David S. Sibley, *The Theory of Public Utility Pricing* (Cambridge University Press, 1986), p. 242.

a. Average of 1978–84.

b. Calculated at the rate of ¥150 = US$1.00.

evaluate them at the then-current exchange rate of ¥150 to the dollar (table 7). The marginal cost of local service in Japan ranges from 4 to 5 cents for a three-minute call, and in the United States it also ranges from 4 to 5 cents. The marginal cost of long-distance service in Japan ranges from 22 to 29 cents (three-minute call), and in the United States it ranges from 20 cents in intra-LATA to 27 cents in interstate. According to a study by Perl, the marginal cost of long-distance service increases as the call distance increases,[31] but recently other analysts have concluded that the transmission cost per line becomes insensitive to the call distance once it exceeds some fixed level. And indeed the transmission cost per line in NTT's new fiber-optic digital transmission system is said to be invariant beyond 100 kilometers. The resemblance of the marginal cost structures for Japan and the United States, despite the obvious difference of call

31. Lewis J. Perl, "Social Welfare and Distributional Consequences of Cost-Based Telephone Pricing," paper presented at the Thirteenth Annual Telecommunications Policy Research Conference, Airlie, Va., April 1985.

distances within each country, implies that the technology adopted by both Japan and the United States is relatively insensitive to distance.

—The real effective price of NTT's local service in 1984 was ¥16.5 (real revenues divided by the volume of three-minute calls), which was 140 to 150 percent above its marginal cost. This discrepancy from marginal cost is in sharp contrast to that of AT&T reported by Rohlfs;[32] in the United States local service was priced 50 percent below marginal cost. On the other hand, the real price of NTT's long-distance service in 1984 was ¥93.3, which was two to three times the marginal cost; this discrepancy from marginal cost is similar to the result for the United States found by Rohlfs. The last result shows that in both Japan and the United States the price of long-distance service has been set quite high to cross-subsidize other services.

Efficiency of the Existing Tariff System

We have shown that, during the last decade, there has been a large price gap between local and long-distance service. In this section, we estimate the extent of the distortion in prices that developed when NTT was a public monopoly.

In Japan, unlike the United States, there has not been much analysis of the efficiency of the pricing system for telephone service. This has been due mainly to the fact that the data have not been publicly available. NTT's pricing system is a two-part tariff consisting of fixed and variable charges. The amount of the fixed charge is slightly different for each locality. Unfortunately, we have not been able to obtain sufficient data about either the fixed charges or the demand for connections, so we must limit ourselves to the analysis of the problem of the gap between long-distance and local variable charges.

Compared with the case in which the prices of long-distance and local-call services are calculated by the Ramsey principle, the loss of total surplus under current pricing is rather small, from 1 to 8 percent of the revenue from the variable charges. However, this loss is the inefficiency resulting from variable charges only. The fixed charges are estimated to be far less than the costs of connection, so NTT's two-part tariff probably causes substantial distortions on the whole.

MARGINAL COST PRICING. According to NTT, in 1984 local service incurred a deficit of ¥400 billion. The directory assistance service ran a

32. Rohlfs, "Economically-Efficient Bell-System Pricing."

Table 8. *Estimates of Potential Welfare Gains from a Shift to Marginal Cost Pricing by NTT*
Billions of yen

	Assuming constant-elasticity demand curves		*Assuming linear demand curves*	
Potential change in:	*Case 1*	*Case 2*	*Case 3*	*Case 4*
Consumer surplus				
Local calls	342	356	335	345
Toll calls	1,473	1,473	1,332	1,332
Subtotal	1,815	1,829	1,667	1,677
Producer surplus				
Local calls	−316	−316	−316	−316
Toll calls	−1,113	−1,113	−1,113	−1,113
Subtotal	−1,429	−1,429	−1,429	−1,429
Total surplus	386	400	238	248

Source: Authors' calculations.

deficit of ¥330 billion.[33] The local-service deficit arises from the fact that, in NTT's calculations, the customer cost is included in the local-service cost. These customer costs are far greater than the fixed charges, thus the current two-part tariff is far from optimal. The combined deficits of the directory assistance service and the local-service businesses are covered by the high price of long-distance calls. Eliminating the inefficiency will probably require an overhaul of this sort of pricing system. In particular, the fixed charges should be raised.

Unfortunately, we lack the necessary data to analyze the inefficiency of the entire pricing system. Without the necessary data, we cannot estimate what would be an efficient level of subsidization from phone service to other lines of business. Our analysis is limited to only one part of the entire distortion of the price system: the distortion that arises between the variable charges of the local and the long-distance rates.

To compute the welfare loss of the variable-charge pricing system, we postulate two types of demand functions: constant-elasticity demand functions and linear demand functions. Table 8 shows the welfare gains that occur if NTT shifts from it current pricing system to marginal cost pricing, assuming the marginal costs are ¥6.6 and ¥32.5 for local and long-distance calls, respectively. The potential increases in consumer surplus, producer surplus, and total surplus are given for four cases. The columns

33. "Shinai Tsuwa Kuiki o Hiroge, Ippun-kan Ju-en ni Neage, Chokyori Nesage: NTT ga Ryokin Kaitei-an" (NTT Proposes Rate Reform . . .), *Asahi Shinbun*, evening ed. (Tokyo), August 19, 1986.

for cases 1 and 2 show the changes in surpluses assuming constant-elasticity demand curves, while those for cases 3 and 4 assume linear demand curves. Because the data to estimate the long-run elasticity of local-call demand directly are not available, we have made a "guesstimate" that it ranges between 0.2 and 0.3. (This guesstimate is based on the estimate of very short-run elasticity, 0.15, which was made by NTT in 1976.)

The constant-elasticity demand functions for each market take the form

$$Q_i = A_i P_i^{-E_i},$$

where A_i's are constants and the subscript i denotes local use and long-distance calls. The values of E_i for local use and long-distance calls in case 1 are 0.200 and 0.605 respectively, and in case 2, 0.300 and 0.605 respectively.

Linear demand functions are

$$Q_i = A_i - B_i P_i,$$

where A_i's and B_i's are constants. The coefficients of local-use demand functions are computed using assumed values of price elasticities of 0.2 for case 3 and 0.3 for case 4, and the actual prices and quantities in 1984.

The computed potential welfare gains vary widely depending on the shape of the demand curves and the elasticity. The change in the total surplus would be somewhere between ¥238 billion and ¥400 billion, which represent 9 percent and 16 percent of the variable-charges revenue from telephone calls in 1984 (¥2,500 billion).

RAMSEY PRICES. Another benchmark of the inefficiency of the variable call rate is the magnitude of its deviation from the Ramsey price. Further, the comparative extent to which the producer surplus from call service (¥1,429 billion) places a burden on the consumers of long-distance call service, as opposed to the consumers of local-call service, is represented by the estimated Ramsey number.

If the charge is efficient in the Ramsey sense, the Ramsey number as defined below should be the same for both local and long-distance call services:

$$k_i = \frac{P_i - C_i}{P_i} E_i.$$

Under the current system, the Ramsey number for local-call service is 0.12 for $E_i = 0.2$ and 0.18 for $E_i = 0.3$. However, the Ramsey number for long-distance call service is 0.39, which indicates that consumers of long-distance call service bear a relatively heavy burden.

To calculate the Ramsey price, we must set up a revenue constraint. For the reasons already mentioned, however, we cannot identify how much of the producer surplus under the current pricing system is actually necessary for justifiable policy goals, such as efficiency and redistribution of income. Assuming that the producer surplus in 1984 was entirely necessary, we have adopted the constraint that it must be at least ¥1,429 billion (the producer surplus in 1984; see table 8).

The other requirement for efficiency in the Ramsey sense is that the Ramsey numbers for local and long-distance calls be equalized. Denoting the equalized Ramsey number as $\tilde{k}$, the Ramsey prices are

$$\tilde{P}_i = \frac{E_i C_i}{E_i - \tilde{k}} .$$

Given these conditions, we can compute $\tilde{k}$, $\tilde{p}_1$, $\tilde{p}_2$ for the four cases mentioned above. The results are shown in table 9.

The Ramsey price of a local call lies between ¥21.9 and ¥47.7, which is 33 to 190 percent more than the current effective price, depending on the specification of the demand function. As for long-distance calls, the

Table 9. *Estimates of Potential Welfare Gains from a Shift to Ramsey Pricing by NTT*
Billions of yen unless otherwise specified

Item	*Case 1*	*Case 2*	*Case 3*	*Case 4*
Ramsey number	0.172	0.252	0.258	0.308
Ramsey price				
Local call (yen)	47.7	41.7	25.6	21.9
Local call (cents)[a]	31.8	27.8	17.1	14.6
Toll call (yen)	45.5	55.7	76.5	83.2
Toll call (cents)[a]	30.3	37.1	51.0	55.5
Potential change in:				
Consumer surplus				
Local calls	−881	−688	−275	−164
Toll calls	1,068	797	324	191
Subtotal	187	109	49	27
Total surplus	187	109	49	27

Source: Authors' calculations.
a. Calculated at the rate of ¥150 = US$1.00.

Ramsey price would be between ¥45.5 and ¥83.2, or 10 to 52 percent lower than the current price.

Table 9 reports also the potential welfare gain of shifting to a Ramsey pricing system. If NTT adopts Ramsey prices, the resulting increase in the total surplus will amount to between ¥27 billion and ¥187 billion, or from 1.1 to 7.4 percent of the revenue from variable charges in 1984.

The Emergence of Competition and Universal Service

The Japanese telecommunications market of type I carriers has a structural particularity in the sense that competition is minimal despite the new liberal legislation. NTT inherited its public obligations to provide "universal and equitable services" as before but also faces competition from newcomers who have entered the most profitable long-distance telephone market, that between Tokyo and Osaka. Unlike AT&T, NTT continues to provide an integrated national network under the Telecommunications Business Law. It must, therefore, find the funds to finance the deficit services, funds that used to be raised from the Tokyo-Osaka telephone market.[34] The new common carriers, of course, never miss the opportunity to exploit this market, and their emergence naturally reduces NTT's traditional profits. In this section, we analyze the effect of new entrants upon NTT by making use of the dominant firm–competitive fringe market model.

NTT is a dominant firm that can set the prices in the local and long-distance markets, though its prices are regulated by the Ministry of Posts and Telecommunications. NTT is different from the usual profit-maximizing dominant firm, but its dominance prevails because competitors presume NTT's prices as given and file their prices with the MPT. The NCCs are pure profit maximizers who supply services as if in a competitive market. The NCCs' tariffs, approved by the ministry, are usually about 20 percent lower than NTT's in order to take account of the difference in reputation and also to enhance the new carriers' growth in a new market.

The Model

We can derive the market demand and profit functions of NTT in the presence of new competition; analyze the effects of key variables like

34. Local telephone and telegraph services are historically in the red, and NTT always faces political opposition in proposing tariff restructuring.

economies of scale, market size, and price elasticity; and look at the problem of cross subsidization. To simplify the analysis, we have made the following assumptions:

1. The services of NTT and the NCCs are perfect substitutes; that is, NTT's services are not differentiated.

2. NTT is in a position to determine its price (P_N), since the MPT cannot reject the proposed price without due reason.

3. P_N cannot fall below a reasonable level because of rate-of-return regulation.

4. The maximum supply of NCCs (Xc) is under the MPT's supervision.

5. NTT acts as a dominant firm, and the NCCs are in the competitive fringe.

6. The marginal cost is identical between NTT and the NCCs.

Several comments must be made about these assumptions. Some analysts argue that, in the short run, NTT's services are differentiated and that differentiation is a huge market barrier for the NCCs. But it is also true that NTT has no technological advantage over the NCCs. Sooner or later, customers will realize this, and thus in the long run the NCCs can provide perfectly substitutable services (assumption 1).

If assumption 1 is valid and the NCCs can undercut NTT's price, they can capture the whole market.[35] To prevent that, the new regime gives the MPT authority to regulate the threshold. The difference between NTT and the NCCs is represented in assumption 5, that NCCs are followers in the market. Of course, competitive fringe firms can grow and challenge the dominant price leader. In the regulatory framework, however, we disregard such a possibility for the moment. The assumption of identical marginal costs is not essential, and the following analysis is valid even if we take account of the differences of marginal costs among competitors.

To simplify the analysis, we added several other assumptions:

7. The linear market demand curve is

$$P = a - bX \ (a > 0, b > 0),$$

where p = price of the long distance call; X = service amount of telephone calls.

35. The NCCs have a problem, however: their calls must be identified when they enter NTT's local switches. But not all of the switches have an automated identification function. Accordingly, the business area for a newcomer is restricted to the switches that can provide identification data.

8. The average cost of NTT service is

$$AC = k/X + q \ (k > 0, q > 0).$$

So the identical marginal cost of NTT and the NCCs is given by q.

The supply curve of the NCCs is q, assuming that the maximum output cannot exceed Xc. The residual demand curve of NTT is given by deducting the NCCs' supply from the market demand.

The profit function of NTT is defined as

$$\pi = \frac{(P_N - q)[(1 - s)a + sq - P_N]}{b} - k,$$

where s is the market share of the NCCs, and

$$s = X_c / \overline{X},$$

where X is the maximum output that obtains when services are supplied at the marginal cost (q). The level of π depends on P_N. NTT will choose a lower price to realize the same level of profit, although the profit function takes the form of a parabola.

Comparative Statics

We have the comparative statics of π with respect to parameters as follows:

$$\frac{\partial \pi}{\partial b} < 0, \frac{\partial \pi}{\partial k} < 0, \frac{\partial \pi}{\partial s} < 0, \frac{\partial \pi}{\partial a} > 0, \frac{\partial \pi}{\partial q} < 0.$$

NTT's profits vary (1) inversely with the slope of the demand curve, (2) directly with the size of the market, (3) inversely with the market share of the NCCs, and (4) inversely with average and marginal cost.

In our linear model, NTT's profit after the entry of the NCCs is dependent on their market share.

We can calculate the amount of profits left for NTT for each market share of the NCCs. There exists a certain minimal level of profits that allows NTT to cross-subsidize the deficit services. If we know this amount, we can derive the critical market share of the NCCs compatible with NTT's cross subsidization. Let us denote the necessary profits and corresponding market share $\hat{s}$, expressed as follows:

$$\hat{s} = \frac{(P_N - q)(a - P_N) - b(\hat{\pi} + k)}{(a - q)(P_N - q)}.$$

In the model, $\hat{s}$ depends on parameters P_N, a, k, q. We have the following comparative statics:

$$\frac{\partial \hat{s}}{\partial P_N} > 0, \frac{\partial \hat{s}}{\partial a} > 0, \frac{\partial \hat{s}}{\partial k} < 0, \frac{\partial \hat{s}}{\partial q} < 0.$$

The level of P_N has a positive effect on $\hat{s}$. As the price P_N rises, the allowable NCCs' share can be larger. The NCCs' share can also be larger when the market size is larger and average and marginal costs are lower.

NCCs' Profits and Access Charge

The profit of the NCCs (π_c) is expressed as follows when their fixed costs (k^*) are given. For simplicity, we assume k^* is identical among new entrants.

$$\pi_c = P_c X_c - k^* - qX_c = (P_c - q)X_c - k^*.$$

Comparing k^* with k, NTT's fixed cost, it is clear that k is greater than k^*, since NTT must maintain reserve capacity while the NCCs avoid this responsibility. Though P_c is lower than P_N, and X_c is a small share of the market, this is the reason that the NCCs can earn a handsome profit, unless P_c and X_c are too small.

In the dominant firm–competitive fringe model, the latter can outweigh the former if opportunities exist for fringe firms to accumulate their profits and grow rapidly. In our model the NCCs can invest their profits to obtain a certain market share, at which NTT can no longer raise enough funds for cross subsidization. To avoid this difficulty, NTT is asking for an access charge to be levied upon newcomers for the purpose of collecting contributions toward universal services among type I carriers. This tax can take several forms. For example, it can be a lump-sum tax that does not respond to the supply of NCCs. In that case the NCCs' profits are reduced, but the supply schedule is unaffected. Or the access charge can be proportional to the amount of supply, thereby changing the shape of the NCCs' supply schedule. A third possibility is a combination of a minimal fixed and proportional charge that is similar to a two-part tariff system. In the first case, the reduction of profits may limit the investment by the NCCs and affect their maximal output. In the other cases, the access charge affects the NCCs' supply of services as well as their future investment.

The Available Policy Choices

In a newly competitive market, several policy choices are available to maintain universal and equitable service as before. In the model above, we derive the critical market share of the NCCs with the necessary profits of NTT. The first policy choice is to levy access charges on the NCCs according to the excess burden on NTT to fulfill its public obligations. The system of access charges can take various forms. When the access charge is a lump-sum tax, it does not affect the NCCs' supply curve but increases the revenue of NTT. In this case, NTT's necessary profit is reduced by the amount of the access charges, and the allowable market share of the NCCs increases. The NCCs' growth in the market can be compatible with NTT's cross subsidization in the presence of an access charge. It is, however, clear that this tax should not be excessive or it will discourage further investment by the NCCs.

The second choice may be a tax regime to affect the marginal cost of NCCs and hence change their maximum output. It may lead directly to the regulation of the market share of the NCCs. In this case, the regulation can be excessive and run counter to the spirit of opening the telecommunications market to new competition.

Under the new legislation, the level of the access charge is to be determined through negotiation among the related carriers. Because NTT can negotiate with each NCC separately, the outcome of such negotiations cannot be predicted. At the same time, access charges are a sensitive policy measure, like a tariff in international trade.

A direct subsidy payment to NTT is another possibility. It is an established theorem that a subsidy is superior to a tariff from the viewpoint of resource allocation. But in Japan a subsidy to NTT in any form is politically infeasible, since the privatization policy is designed to reduce the government's burden and make public corporations independent.

Another policy choice is to revise the local–long-distance tariff structure, raising the tariff on local calls in line with the reduction of the long-distance call tariff. In the face of competition, this policy is a natural alternative to the access charge plan, but it also is likely to encounter political difficulties because of the existence of stakeholders in the telecommunications industry. For one thing, customers are often against any increase of the local-call tariff, and such a policy may cause a political crush in the National Diet similar to that seen in the United States when the Federal Communications Commission proposed the original access charge plan. For another, the NCCs are against the lowering of long-distance telephone rates because it may be disastrous to their price-cost margins. The MPT might take their side, since it wishes to protect the NCCs until they are better established. In contrast with the U.S. situation, the revision of tariff structures has not been immediately followed in Japan by the emergence of new entrants. Japan's situation is also different from Britain's, where British Telecom is in a freer position to shift competitive pricing. In Japan the MPT may provide the greatest barrier to competition in the telecommunications market, probably because it believes in "regulated competition," which we consider a myth. As long as a "subsidy" is out of the question in the Japanese political environment, either a lump-sum access charge or a revision of the tariff structure will be needed if universal and equitable service is to be attained. In the short run, an access charge might be a feasible candidate, but in the long run it may harm efficient resource allocation and delay a more radical reshaping.

Problems under Liberalization

The Economics of Telecommunications Standards

Stanley M. Besen and Garth Saloner

NOT LONG AGO technical standards in the U.S. telephone industry were determined for the most part internally by the American Telephone and Telegraph Company. To be sure, AT&T had to coordinate with foreign telecommunications entities, independent telephone companies, and the U.S. Department of Defense, but the degree of coordination was relatively minor, and AT&T had substantial latitude in determining the standards that were employed. Three forces have now caused this situation to change dramatically.

First, because of the entry of large numbers of competing suppliers of equipment and services into the U.S. telecommunications industry, standard setting has ceased to be the technical concern of a single firm and has become a factor with important implications for competition. As a result, the processes by which standards are set are now subject to detailed scrutiny by both the regulatory authorities and the courts. In a sense, telecommunications standards have become too important to leave their determination solely to the telephone companies.

Second, the divestiture of the Bell operating companies (BOCs) by AT&T has, by fragmenting the telephone industry, reduced the ability of AT&T to determine standards as it had in the past. Horwitt has noted that "the market has changed since predivestiture days, when Ma Bell set telecommunications standards and other carriers and equipment vendors had no choice but to follow. Now, AT&T is just one more vendor—albeit a formidable one—lobbying for industrywide adoption of the technologies and protocols it wants to use."[1] To an increasing degree, AT&T must

The authors wish to acknowledge helpful comments by John A. Arcate, Stephanie M. Boyles, Donald A. Dunn, Joseph Farrell, Hendrik A. Goosen, Charles L. Jackson, Leland L. Johnson, Ian M. Lifschus, Walther Richter, Leonard Strickland, and Clifford M. Winston on earlier versions of this paper.

1. Elisabeth Horwitt, "Protocols Don't Stand Alone," *Computerworld*, vol. 20 (October 20, 1986), p. 27.

accommodate to the choices made by others rather than dictate the standards to which others must conform.

Third, a result of the growing internationalization of telecommunications technology and services is broader scope for international standard-setting bodies. The autonomy previously enjoyed by the United States in setting standards has consequently been reduced, and the needed degree of coordination with suppliers in other countries has increased. According to Pool: "Until now in the telecommunications field there have generally been two sets of standards, the CCITT [International Telegraph and Telephone Consultative Committee] standards of the International Telecommunications Union [ITU] followed in most of the world and the Bell System standards which prevailed in America (about half the world market). In the future . . . CCITT standards will become more influential in this country, and AT&T will have an incentive to reduce its deviations from them."[2] The major effort under way at the CCITT to establish standards for integrated services digital networks (ISDNs), in which the United States is only one of a large number of players, is an important indication of this change.

Determinants of the Standard-Setting Process

There is no standard way in which standards are developed. In some cases standards are *mandated* by government agencies using administrative processes. In others, *voluntary* standards are established cooperatively, with information being exchanged, technologies being altered, or side payments—payments made to a party to obtain its cooperation—being made to achieve a consensus. Finally, standard setting may be left to the market, where de facto standards emerge noncooperatively.

Two factors that affect the nature and outcome of the standard setting process are especially important. The first concerns the private incentives that each of the interested parties—developers, manufacturers, buyers—has to promote the universal adoption of any standard. Such incentives might be low because, even when all parties benefit from the existence of a standard, the private costs of participating in the adoption process may overwhelm the benefits of participating. This is especially likely in the establishment of systems of weights and measures and of standards relating to the use of common terminologies.

2. Ithiel de Sola Pool, "Competition and Universal Service: Can We Get There from Here?" in Harry M. Shooshan, ed., *Disconnecting Bell: The Impact of the AT&T Divestiture* (Pergamon Press, 1984), p. 119.

The incentive to promote standards may also be low when standardization eliminates a competitive advantage that is more important than the benefits of having a standard. For example, Brock has reported that IBM was unwilling to accept the COBOL-60 specifications for its business language because it wished to avoid the competition it would face if there were a common business language.[3] More recently, Horwitt has reported that American computer vendors such as IBM and telecommunications carriers such as Telenet have been reluctant to adopt the CCITT X.400 electronic mail standard.[4] Although adoption of this standard would permit communication among subscribers to different electronic mail systems, it would also permit subscribers to move easily from one vendor to another.[5]

At the opposite extreme are cases in which the expected gains to all parties from promoting the universal adoption of a standard exceed the costs of doing so. For example, the early, highly fragmented automobile industry was plagued by problems of incompatibility.[6] All manufacturers stood to gain greatly if standards were established, and a high degree of participation was required if standardization was to be achieved. As a result, many incurred the costs of participation.[7] Later, with consolidation of the industry, the benefits of standardization have become less important.

The second factor affecting the way standards are set is the extent to which the interested parties have different views about the standard that should be chosen. Differences in preferences are especially unlikely when there are no important differences among technologies, so that what is important is only that a standard be chosen, not what the standard is. Time keeping and the use of calendars may be examples where no individual cares which system is chosen, so long as there is some widely accepted method. Moreover, even when there are differences among technologies,

3. Gerald Brock, "Competition, Standards and Self-Regulation in the Computer Industry," in Richard E. Caves and Marc J. Roberts, eds., *Regulating the Product: Quality and Variety* (Ballinger, 1975), pp. 75–96.

4. Elisabeth Horwitt, "X.400 Flies in Europe, Lags in U.S.," *Computerworld*, vol. 21 (November 2, 1987), p. 49.

5. In contrast, in Europe the strong demand for X.400 products has apparently forced U.S. vendors to support the standard in order to participate in the electronic mail market.

6. David Hemenway, *Industrywide Voluntary Product Standards* (Ballinger, 1975).

7. If everyone benefits from having a standard, but the benefits are unequally distributed, those who obtain the largest benefits may be willing to incur the costs of setting standards while those with smaller benefits "free ride." This outcome, in which a public good is provided by those users who receive the largest benefits, has been referred to as "the exploitation of the great by the small." For a discussion, see Mancur Olson, Jr., *The Logic of Collective Action: Public Goods and the Theory of Groups* (Harvard University Press, 1965). Olson discussed, among other examples, the case of international alliances in which large countries often pay a disproportionate share of the costs.

so that the parties are not indifferent toward them, the same technology may still be everyone's preferred standard.

Agents frequently differ, however, in the standards they prefer. For example, manufacturers of VHS and Beta videocassette recorders would have different preferences as to which technology was adopted if standardization were attempted. Similarly, computer manufacturers who have designed their machines to work with specific operating systems would prefer different systems as the industry standard. Still another example is that some users of videotext prefer the North American Presentation Level Protocol Syntax (NAPLPS), with its sophisticated graphics capability, whereas others are content with the less expensive, text-only American Standard Code for Information Interchange (ASCII) standard.[8]

When preferences differ and side payments are impossible, each party will promote as the standard the technology that maximizes its own private benefits, not the technology that maximizes benefits for society at large. In these cases standard setting can no longer be viewed solely as a search for the technically best standard, or even as a process for establishing one of several "equivalent" technologies as the standard. Instead, standard setting is a form of competition in which firms seek to gain advantages over their rivals.

We can now identify four cases that differ in whether the interest in promoting any universal standard is large or small and in whether preferences are similar or diverse. The case in which there is a large interest in promoting a universal standard and preferences are similar is what can be called the *pure coordination case*. Either there are several possible standards toward which everyone is indifferent, or the same technology is preferred by all, and the *per capita* rewards to participation in standard setting are large enough to induce everyone to participate. The standardization process is simply a matter of agreeing on which alternative to use. The agreement, once it is reached, is self-enforcing because no party has an incentive to deviate unilaterally. In the language of game theory, there are either multiple equilibria with identical payoffs or a unique equilibrium

8. Stanley M. Besen and Leland L. Johnson, *Compatibility Standards, Competition, and Innovation in the Broadcasting Industry,* R-3453-NSF (Santa Monica, Calif.: Rand Corp., November 1986), pp. 80–84. Even if agents have no strong preferences when a technology is first introduced, they may develop them over time after the technology has been adopted. For example, when Sweden decided to switch from the left- to the right-hand side of the road in the late 1960s, the change was overwhelmingly voted down in a national referendum. See Charles P. Kindleberger, "Standards as Public, Collective, and Private Goods," *Kyklos*, vol. 36, no. 3 (1983), p. 389. Similarly, owners of railroads with incompatible gauges would each have a preference for the gauge used by their rolling stock.

that is Pareto superior; that is, everyone prefers that equilibrium to any other. The standardization process serves to select an, or the, equilibrium.

Much standardization approximates the pure coordination case. Although there may be some differences in preferences, these differences are small relative to the gains from achieving standardization. Standard setting is likely to be seen as an activity in which experts seek the best technical solution or, at least, choose one standard from several that are equally good. In short, standard setting is a game in which everyone obtains a positive payoff; moreover, it is a game in which the choice that maximizes the payoff to any party maximizes the payoffs to all. This view dominates descriptions of the standard-setting process advanced by standard-setting organizations.

Even when preferences do not differ, however, standardization achieved through private voluntary agreement may not occur because the gains to any party may be so small relative to the cost of participation that free riding on the part of everyone results in no standard at all. In what might be called the *pure public goods case*, the *per capita* gain from standardization is too small for anyone to find participation in the process worthwhile. Although everyone desires that standardization be achieved, and differences in preferences are small, no agent has a sufficiently large interest to develop the standard. This outcome is especially likely in industries that are highly fragmented, or when the beneficiary of standardization is the public. Here, if standardization is achieved it is likely to require government intervention, as in the establishment of standards for weights and measures, time, and language. Alternatively, several incompatible technologies may exist at the same time. Paradoxically, when standardization cannot create a competitive advantage, so that achieving a consensus should be easy, the incentive to free ride is greatest.

A third case involves large differences in preferences and little incentive to promote the adoption of a universal standard. This does not mean that there are no benefits from standardization, only that the distribution of benefits is sensitive to the standard that is chosen. In the *pure private goods case*, if there is no dominant firm, standardization cannot be expected to be achieved voluntarily. Private parties would not promote the creation of a formal standard-setting body and, if such a body were established, the objectives of participation would be to promote a favored candidate as the standard or to prevent the adoption of another. Unless side payments are possible, the most likely result is stalemate. Participants in standards meetings may attempt to stall the proceedings, for example, by continually introducing new proposals and providing other participants insufficient time to analyze them. The outcome will be either simultaneous use of incompatible technologies, the selection of a de facto standard through the market, or the failure of technological development because of the absence of a standard.

Although in principle government intervention can break a stalemate, such intervention may itself be the object of controversy, so that the government may be reluctant to intervene. The stalemate might also be broken if there is a dominant firm. If the dominant firm is opposed to universal standardization, however, it will be a *reluctant leader* and may attempt to prevent its rivals from producing compatible products.[9]

A firm with a large market share may be reluctant to promote its technology as an industry standard if it fears that the demand for the products of its rivals will increase, at its expense, if its rivals can offer compatible products. For example, it was recently reported that Ashton-Tate has attempted to prevent the adoption of its dBase language as an industry standard; the firm's chairman, Ed Esber, is quoted as stating: "the Dbase standard belongs to Ashton-Tate and Ashton-Tate intends to vigorously protect it. It's proprietary technology."[10]

Another possible example of reluctant leadership occurs when a firm is dominant because it controls access to an input that its rivals need to market either complete systems or individual components. Under certain circumstances, such a firm may prefer that its rivals be unable to offer components that are compatible with its "essential" input. The argument that IBM attempted to make it difficult for competing manufacturers of peripheral equipment to offer products that were compatible with IBM's mainframes was an important element of the government's case in the 1969 Sherman Act antitrust suit against the company. A similar argument was made in the government's 1974 suit that led to the divestiture of the Bell operating companies by AT&T, in which the essential input was access to the local distribution facilities of the operating companies.

In the fourth case, there are large differences in preferences, and each of the interested parties has a large interest in promoting the universal adoption of a standard. In this *conflict case*, a dominant firm may, if it desires, attempt to establish a de facto standard. Here, the dominant firm

9. For a discussion of allegedly anticompetitive standards practices in the computer, photography, and telecommunications industries, see Yale M. Braunstein and Lawrence J. White, "Setting Technical Compatibility Standards: An Economic Analysis," *Antitrust Bulletin*, vol. 30 (Summer 1985), pp. 337–55.

10. Alan J. Ryan, "Esber: Dbase Language Not for Public Domain," *Computerworld*, vol. 21 (October 19, 1987), p. 133.

will be a *cheerful leader*, and other firms may be forced to adopt the technology that it prefers. This is apparently what occurred in the emergence of the IBM personal computer as the industry standard.

In the absence of a dominant firm, the interested parties will all participate eagerly in the standardization process. The process can be expected to involve side payments and formation of coalitions. For example, Horwitt has reported that several computer software and hardware vendors recently agreed "to surrender market dominance based on proprietary products in favor of a standardized, public-domain Unix environment. . . . One major thrust behind the standards is vendors' realization that a fragmented Unix cannot effectively compete in the mid-range system arena against emerging proprietary products from the likes of Digital Equipment Corporation and IBM."[11] The vendors were reported as "willing to cooperate with their competitors—or even adopt a competing product—in order to hasten commercial availability of the multivendor programming and networking products that their customers demand."[12] Similarly, all major European equipment manufacturers, together with Digital Equipment and Sperry, have formed the X/Open Group to promote a standardized version of the Unix operating system. Their objective is to permit the portability of applications software among computers made by different manufacturers in order to "preempt any attempt by IBM to establish de facto minicomputer standards, as it has for mainframes and personal computers."[13]

Firms can also be expected to promote their own products in the market during the standardization process to make more credible their threat to go it alone. They may also attempt to use the government to increase their leverage either in the market or in cooperative standard setting.[14] There will be considerable pressure, however, for a standard to be adopted.

The four-way classification of the standards process described above is summarized in figure 1.[15]

11. Elisabeth Horwitt, "Vendors Pull Together to Boost Unix Standards," *Computerworld*, vol. 21 (January 26, 1987), p. 6.

12. Ibid.

13. Robert T. Gallagher, "Europeans Are Counting on Unix to Fight IBM," *Electronics*, vol. 59 (July 10, 1986), p. 121.

14. The case of AM stereo may be illustrative. After the FCC decided not to adopt a standard but to leave standard setting to "the market," some of the other contenders succeeded in having the FCC revoke its "type acceptance" of Harris Corporation's exciter-generator, without which Harris's equipment could not be used. This forced Harris to withdraw temporarily from competition and forced stations using its system to cease operating in stereo, an example of the use of governmental processes to gain a competitive advantage. Later, Harris dropped out of the competition, and stations using its technology switched to using Motorola's, an example of coalition formation. See Besen and Johnson, *Compatibility Standards*, for a fuller account.

15. Note in figure 1 that all firms in an industry may not be in the same cell. The examples of dBase and Unix discussed above are apparently cases in which the dominant firm preferred

Figure 1. *Determinants of the Standards Process*

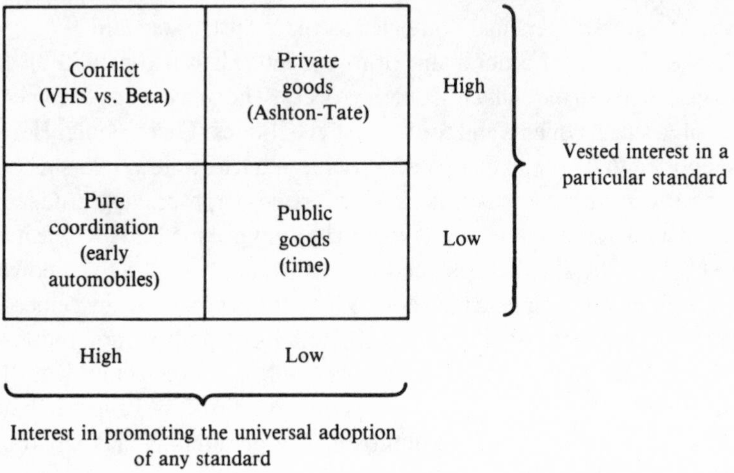

	High	Low	
High	Conflict (VHS vs. Beta)	Private goods (Ashton-Tate)	Vested interest in a particular standard
Low	Pure coordination (early automobiles)	Public goods (time)	

Interest in promoting the universal adoption of any standard

As noted above, when standard-setting bodies describe their activities, they typically characterize them as involving pure coordination. In these descriptions the participants are willing to expend considerable resources to achieve compatibility, and any conflicts about what the standard should be reflect differences in technical judgments. Although standardization may not come easily in these cases, standard-setting bodies will in general be able to achieve the needed degree of coordination. At the same time, the conventional description of standard setting fails to encompass a large and important number of cases in which differences about what the industry standard should be are not primarily technical—the conflict case—or in which some of the parties actually prefer incompatibility—the private goods case. Much of the remainder of the analysis examines situations in which the interests of the parties are not necessarily congruent because such cases raise the most interesting and difficult standardization issues from the standpoint of public policy. We do not mean to suggest that the pure coordination case is unimportant, however; indeed, we provide a detailed analysis of the possible role for cooperative standard setting in this case.

that no standard be adopted, because it thereby would retain a competitive advantage, and in which smaller firms preferred that a standard be chosen, because a standard would enhance their ability to compete.

Whether consensus will be achieved in private cooperative standard setting depends on several factors including (1) the importance of the benefits of standardization, (2) whether a small number of participants can prevent an effective standard from emerging,[16] (3) the extent to which the interests of the participants diverge, and (4) whether side payments are possible.

The prospect of achieving consensus increases with the benefits from the network externalities that standardization produces. At one extreme, if consumers are reluctant to purchase a good from any vendor because they fear that they may be stranded with the wrong technology, for example, with the wrong videocassette recorder or computer, all vendors have a strong interest in agreeing on a standard. In such cases, firms may be willing to agree to conform to a standard that is not the one they prefer if the alternative is to have no sales at all. In contrast, the greater the ability of a firm to have sales even where there are no compatible products, the more reluctant it will be to conform to a standard other than the one it prefers.

If the success of a standard depends on obtaining agreement from all participants, standardization is less likely than when a smaller majority is required. When unanimity is required, any participant can hold out, refusing to support a standard unless it obtains a large share of the resulting benefits. This can involve either insistence that its preferred technology be chosen as the standard or demand for payment in some other form. Because all participants can behave in this manner, consensus is unlikely. Thus standard-setting bodies typically require less than unanimous consent for a standard to be adopted.[17]

Clearly, the more divergent are the interests of the participants, the less likely it is that a consensus will emerge. Where preferences are similar, standardization involves only learning that this is the case.[18] Once everyone knows that everyone else prefers the same technology, each can proceed to adopt the technology in complete confidence of conformance. Here,

16. Such behavior can arise either when all participants want a standard but differ strongly as to what that standard should be, or when some participants do not want any standard at all to emerge. In the latter case, those firms that do not want a standard will not participate in the process, as apparently occurred in the cases of the COBOL and dBase standards noted earlier.

17. When less than unanimous agreement is required, a small number of firms may agree to support a standard, leaving other firms to decide whether to conform. Recently, several computer and hardware manufacturers, not including IBM, discussed the creation of a standard for extending the bus (high-speed data path) for the IBM PC-AT from 16 to 32 bits. See editorial, "Inside the IBM PCs," *Byte* (1986 Extra Edition), pp. 6, 8. Rules relating to the adoption of voluntary standards by committees are discussed in the next section.

18. Joseph Farrell and Garth Saloner, "Standardization, Compatibility, and Innovation," *Rand Journal of Economics*, vol. 16 (Spring 1985), pp. 70–83.

information sharing can promote the adoption of a standard that otherwise would not emerge. In contrast, if preferences diverge, not only will such confidence be lacking, but each participant will tend to exaggerate the differences in order to have its technology chosen. Thus, each participant may contend that it will not follow the lead of another even if it actually would. The result is to reduce the likelihood that anyone will attempt to start a "bandwagon."

Finally, the ability to make side payments may overcome resistance to agreement on a standard. Especially when the difficulty in reaching agreement stems from large divergences in preferences, the reluctance to conform may be overcome if those who gain most from the standard that is adopted share those gains with others. The sharing of gains need not involve cash transfers but could, for example, require that the "winners" license their technologies on favorable terms to the "losers."[19]

Cooperative Standard Setting in Practice

The foregoing suggests that there are many instances in which cooperative standard setting is viable and productive. An important response to the need for coordination of product design has been the evolution of a strikingly large and complex standard-setting community charged with the responsibility and authority to negotiate and adopt standards for its various industries. In addition, liaisons and affiliations among standard-setting bodies have been formed across industrial and national boundaries as the need has arisen. The result is a standards community comprising hundreds of committees and involving more than a hundred thousand people. It is particularly remarkable that, for the most part, this community has emerged at the initiative of industry participants and without governmental intervention or direction. Indeed, governmental agencies often take their guidance from the industry bodies and formally adopt as mandatory standards the voluntary standards that these bodies produce.

Voluntary Standard Setting

At some stage, usually fairly early in the development of a new product, manufacturers and purchasers often realize that economies can be reaped by adopting voluntary standards for some of the product's components or

19. An alternative is the adoption of "compromise standards" that borrow aspects of the technologies preferred by the different participants in a way that leaves no participant with an advantage. This approach may be used because arranging for side payments is often difficult.

features. Through a subcommittee of an existing trade association or standard-setting organization, comments are obtained from all interested parties by a lengthy and formal procedure.[20] Acceptable standards emerge under the "consensus principle," which usually connotes "the largest possible agreement . . . among all interests concerned with the use of standards."[21]

NATIONAL STANDARDS. A central clearinghouse is used to keep track of, and to disseminate information about, standards. In the United States, this function is provided by the American National Standards Institute (ANSI), a private organization with more than 220 trade associations and professional and technical societies, and more than 1,000 corporations as members.[22] ANSI approves a standard when it finds that its criteria for due process have been met and that a consensus among the interested parties exists. Some 8,500 American National Standards have been approved in this manner.[23]

In the United States the decisions of standard-setting bodies, and their operating procedures, have been subject to antitrust scrutiny. At least three organizations have been held to have violated the antitrust laws when they refused to certify that a new technology conformed to an industry standard.[24] As a result, the principle has been established that antitrust liability

20. Charles D. Sullivan, *Standards and Standardization: Basic Principles and Applications* (New York: Marcel Dekker, 1983).

21. Lal Chand Verman, *Standardization: A New Discipline* (Hamden, Conn.: Archon Books, 1973), p. 12. The consensus principle is explicitly not taken to imply unanimity. See T. R. B. Sanders, *The Aims and Principles of Standardization* (Geneva, Switz.: International Organization for Standardization, 1972). Certainly it does not imply a simple unweighted majority of industry participants. Hemenway has noted, for example, that "a number of negative votes of groups that are only distantly concerned with the subject matter may be discounted in the face of affirmative votes of parties that are vitally affected by the standard." See Hemenway, *Industrywide Voluntary Product Standards*, p. 89.

22. Robert B. Toth, ed., "Standards Activities of Organizations in the United States," National Bureau of Standards, special publication 681 (Washington, D.C.: Government Printing Office, 1984). Originally organized in 1918 as the American Engineering Standards Committee, comprising four engineering societies—mining, civil, chemical, and mechanical—its name was changed to the American Standards Association in 1928. At that time, its membership was opened to trade associations and government bureaus. Finally, from 1966 to 1969 it was reorganized under the name of ANSI, and the focus of its role shifted from standards creation to a broader coordinating role. See Sullivan, *Standards and Standardization*, p. 33; and Hemenway, *Industrywide Voluntary Product Standards*, p. 88.

23. See Toth, ed., "Standards Activities," p. 72.

24. See *American Society of Mechanical Engineers, Inc.*, v. *Hydrolevel Corp.*, 456 U.S. 556 (1982); *Radiant Burners, Inc.*, v. *People's Gas Light & Coke Co.*, 364 U.S. 656 (1961); and *Indian Head, Inc.*, v. *Allied Tube & Conduit Corp.*, 817 F. 2d 938 (2d Cir. 1987), *aff'd* No. 87-157, slip opinion (U.S. Sup. Ct. June 13, 1988). See also Federal Trade Commission, Bureau of Consumer Protection, *Standards and Certification*, final staff report (Washington, D.C.: FTC, April 1983), for an extended analysis of the potential for anticompetitive behavior in the development of standards.

may be incurred by private voluntary standard-setting organizations if their actions are anticompetitive, and these organizations must now expect that their activities may be subject to challenge.[25] In one case, a trade association actually declined to adopt an industry standard because it feared that it could avoid antitrust liability only by adopting costly procedures to ensure that its actions would be perceived as "fair."[26]

INTERNATIONAL STANDARDS. The need for standards transcends national boundaries. The same forces that produced the formation of national standards bodies have also led to the creation of organizations for international standardization. In 1946 delegates from sixty-four countries established the International Organization for Standardization (ISO).[27] In 1947 the International Electrotechnical Commission (IEC), formed some forty-three years earlier, became affiliated with the ISO as its electrical division, considerably expanding the ISO's scope. There are two striking features of the ISO: its extent and the rate of growth of its output. Of the roughly 7,500 international standards that had been written by early 1985, some 5,000 had been developed, promulgated, or coordinated by the ISO, compared with the mere 37 ISO recommendations that had been approved by the ISO's tenth anniversary in 1957 and the 2,000 standards written by 1972.[28]

As with ANSI, the ISO is a nongovernmental, voluntary institution. It has seventy-two "full members" and seventeen "correspondent members." The full members are national standards associations, such as ANSI, that have voting rights in the technical committees of the ISO as

25. Collective activity to influence *government* standard setting in general has been immune from liability under the antitrust laws. The Noerr-Pennington Doctrine adopted by the courts provides substantial antitrust immunity to firms acting collectively to influence legislative or regulatory behavior. See Daniel R. Fischel, "Antitrust Liability for Attempts to Influence Government Action: The Basis and Limits of the *Noerr-Pennington* Doctrine," *University of Chicago Law Review*, vol. 45 (Fall 1977), pp. 80–122; and James D. Hurwitz, "Abuse of Governmental Processes, the First Amendment, and the Boundaries of *Noerr*," *Georgetown Law Journal*, vol. 74 (October 1985), pp. 65–126.

26. See the discussion of the behavior of the National Association of Broadcasters in deciding whether to adopt an AM stereo standard in Besen and Johnson, *Compatibility Standards*.

27. The ISO was preceded by the International Federation of the National Standards Association (ISA), formed in 1926 by about twenty of the world's leading national standards associations. The ISA disbanded in 1942 because of the war. See Sanders, *Aims and Principles of Standardization*, p. 64. In 1981 the ISO changed its name to the International Organization for Standardization but retained the abbreviation ISO. See Anthony M. Rutkowski, *Integrated Services Digital Networks* (Dedham, Mass.: Artech House, 1985), p. 20.

28. For these statistics and additional information on the structure and procedures of the ISO, see Edward Lohse, "The Role of the ISO in Telecommunications and Information Systems Standardization," *IEEE Communications Magazine*, vol. 23 (January 1985), pp. 18–24; and Sanders, *Aims and Principles of Standardization*, pp. 12, 64–68.

well as in its Council and General Assembly.[29] The correspondent members are governmental institutions from countries that do not have national standards bodies. The writing of the standards is carried out by the 164 technical committees and about 2,000 of their subcommittees and working groups. It is estimated that the number of individual participants has grown from some 50,000 in 1972 to more than 100,000 today. Some 400 international organizations, including the CCITT (discussed below), have formal liaison with the ISO.

The same process for achieving consensus that characterizes national standard setting is present in the international arena. Although the consensus principle is held as an ideal for the standards process at the international level as well, formally a Draft International Standard (DIS) must be approved by 75 percent of the ISO's full members who have elected to participate in the relevant technical committee, with two or more negative votes receiving special consideration. Once a DIS has been approved by a technical committee, it must be adopted by the ISO Council as an International Standard.

It is significant that the number of ANSI standards exceeds the number of International Standards. Because international standardization is a fairly new phenomenon, standardization is often achieved at the national level before it is taken up internationally. The principal function of the ISO in its early years was to coordinate existing national standards.

Standard Setting in the Telecommunications and Computer Industries

Telephone services have traditionally been provided by government-run (or, in the United States, government-regulated) monopolies. In Europe, these are the PTTs (post, telegraph, and telephone administrations); in the United States, until recently this position was held by AT&T. So long as these organizations had complete control over the design and use of the network, standardization within countries involved only a single firm. International standardization, requiring coordination among many firms, however, involved consultation and agreement among national gov-

29. The ISO accepts as a member the national body that is "most representative of standardization in its country." Most of these (more than 70 percent) are governmental institutions or organizations incorporated by public law. See Rutkowski, *Integrated Services Digital Networks*, p. 21.

ernments. It is not surprising, therefore, that there is a treaty-based organization to deal with standardization issues.

The International Telegraph Union was formed by an agreement of twenty countries in 1865. In 1932 it merged with the organization created by the International Radiotelegraph Convention and was renamed the International Telecommunication Union (ITU).[30] The main goal of the ITU, which currently has 162 members, is to promote cooperation and development in telecommunications. The branches of the ITU most concerned with issues of standardization are the CCITT and the International Radio Consultative Committee (CCIR). The latter is concerned with matters specifically related to radio propagation and facilities, while the former deals with all other telecommunications issues.

The results of CCITT and CCIR deliberations are usually adopted as *recommendations*. While these are not legally binding, countries find it in their interests to adhere to them in order to permit interworking of national systems. Although rarely done, the ITU can adopt CCIR and CCITT recommendations as treaty agreements (known as *regulations*). While these have been restricted mainly to issues relating to radio, the 1988 World Administrative Telegraph and Telephone Conference will consider regulations affecting "all existing and foreseen new telecommunications services."[31]

Because the CCITT is a part of a treaty organization, the United States is represented there by a delegation from the Department of State. Two public advisory committees, the U.S. Organization for the International Telegraph and Telecommunications Consultative Committee (USCCITT) and the U.S. Organization for the International Radio Consultative Committee (USCCIR), provide advice to the State Department on matters of policy and in preparation of positions for meetings of the CCITT.[32] The State Department is also able to provide accreditation to organizations and companies that allows them to participate directly in CCITT and CCIR activities. U.S. representation has historically been made in this way

30. For details of the history of the ITU, see W. H. Bellchambers and others, "The International Telecommunication Union and Development of Worldwide Telecommunications," *IEEE Communications Magazine*, vol. 22 (May 1984), pp. 72–83.

31. Resolution 10 of the Plenipotentiary Conference of the ITU (Nairobi, 1982), cited in Rutkowski, *Integrated Services Digital Networks*, p. 261.

32. "Membership [in the USCCITT] is extended to all parties interested in telecommunications standards, including users, providers, manufacturers, national standards organizations, and Government Agencies." Dorothy M. Cerni, "The United States Organization for the CCITT," *IEEE Communications Magazine*, vol. 23 (January 1985), p. 38.

through companies involved in the provision of telegraph and telecommunications services.[33]

Several U.S. domestic voluntary standards organizations are also involved in the telecommunications standardization process. One of the most important of these is Committee T1 sponsored by the Exchange Carriers Standards Association (ECSA), which was organized after the divestiture of the Bell operating companies by AT&T to deal with standardization issues previously handled internally by AT&T.[34] This committee, whose members include exchange carriers, interexchange carriers, and manufacturers, develops interface standards for U.S. networks. Although the private sector plays a large role in the development of U.S. telecommunications standards, it does so subject to the substantial authority of the FCC to regulate domestic and international communications under the Communications Act of 1934.[35]

Standardization decisions lie at the core of the establishment of telecommunications networks, although translators—devices that permit communication between otherwise incompatible components—can often substitute for interface standards. The same was not always true of computer hardware technology. Especially in the days when the mainframe reigned supreme, computers were mainly used as stand-alone processors. Standardization issues generally revolved around the ability of manufacturers of peripheral equipment to connect their products to the central processing units of other manufacturers. Because there were only a few mainframe manufacturers, and they provided integrated systems and thus were not dependent on the equipment of peripheral manufacturers, the mainframe makers had little incentive to ensure that interfaces were standardized. Users of computer languages, in contrast, had obvious incentives to achieve standardization and used the typical voluntary committee structure.

Several factors have combined to increase the desirability of intercomputer communication. These include the desire to make corporate and external data available to a wide range of company employees; the need to share information generated in a decentralized way as a consequence of the emergence and rapid acceptance of the microcomputer; and the increased use of computer technology in the service economy (for example,

33. Rutkowski, *Integrated Services Digital Networks*, p. 25.

34. Ibid.; see also Ian M. Lifschus, "Standard Committee T1-Telecommunications," *IEEE Communications Magazine*, vol. 23 (January 1985), pp. 34–37.

35. Title I of the act provides the FCC with general jurisdiction over communications services, title II with specific jurisdiction over common carrier telecommunications services, and title III with jurisdiction over the use of radio stations.

banking, airline and theater reservations) and the desire to access these and other potential services (for example, educational and library services, grocery ordering, mail) from the home.

The first important successes in standardizing data communications were not achieved until the mid-1970s. One of the most important early standards was CCITT Recommendation X.25, which established interface specifications between data terminal equipment and public data networks.[36] These early standards were imperative for meeting immediate requirements—they were not components of a grand design that would ensure compatibility of different protocols and system architectures.[37]

The initiative for developing an overarching framework for information transfer between any two end-systems was taken by the ISO. The ISO initiative is perceived as a bold and farsighted attempt to avoid the haphazard evolution of incompatible protocols. In contrast to many standards proceedings, this initiative anticipated future needs rather than merely waiting to react to them.

The result of this initiative was the Open Systems Interconnection (OSI) reference model. This model provides a framework for structuring communication among separate end-users. The term *open* conveys the ability of any end-user to connect with any other. The forum in which such communication takes place is called the "OSI environment." An end-user is best thought of as a particular applications process.[38] For example, an end-user could be a production-line control program or a person operating a manual keyboard terminal.

The communication between application processes requires that several functions be performed, which the OSI reference model structures into seven layers.[39] Broadly speaking, the upper three layers provide support for the particular application being used; that is, the services that allow the application process to access the open system and to interpret the information being transferred to the application process. The lower three

36. These protocols are essential for packet-switched networks. In such a network, data to be transmitted from one user to another are arranged in "packets." Besides the data, each packet includes such information as the users' addresses. Protocols establish, among other things, call origination and acceptance formats, error checking, and speed and flow parameters. See Antony Rybczynski, "X.25 Interface and End-to-End Virtual Circuit Service Characteristics," in William Stallings, ed., *Computer Communications: Architectures, Protocols, and Standards* (Silver Spring, Md.: IEEE Computer Society Press, 1985), pp. 153–63.

37. Harold C. Folts, "A Tutorial on the Open Systems Interconnection Reference Model," in Stallings, ed., *Computer Communications*, pp. 7–26.

38. Ibid.

39. Ibid.; see also Andrew S. Tanenbaum, "Network Protocols," in Stallings, ed., *Computer Communications*, pp. 27–63.

layers concern the transmission of the data themselves from one applications process to another. The middle layer (the "transport" layer) links the support layers to the information transmission layers.

At the same time opportunities from intercomputer communication were blossoming, there was an important change in the technology of telecommunications networks. Voice communication requires both a transmission and a switching technology. The transmission technology carries the voice signal through the network, and the switching technology is responsible for its routing. The traditional, analog technology amplifies the voice signal in such a way that it can be transmitted. Each time the signal is switched, the signal must be interpreted and then transformed again, and this process causes an accumulation of "noise."

The alternative, digital technology immediately creates a digital representation of the voice signal. This digitized signal can then be switched repeatedly without decoding and redigitizing. Because the signals are in digital form, the switching is performed by computer. As the cost of computer technology has fallen, so has the cost of digital technology. Accordingly, telecommunications networks are rapidly being transformed from analog to digital transmission and switching. The entire telecommunications network will eventually be digital, forming an integrated digital network (IDN).

Once the telecommunications network transmits digital information, this network itself can be used for the kind of intercomputer communication discussed above. A single, expanded digital network that will be used for voice, data, facsimile, and video transmission is an integrated services digital network (ISDN). Because of the obvious connection between the work of the ISO on the Open Systems Interconnection model and the interests of the CCITT in telecommunications, these two bodies are working together closely in developing standards for the ISDN.[40]

Noncooperative Standard Setting

An alternative to setting voluntary standards through committees is for standards to evolve through the adoption decisions of market participants. To evaluate the usefulness of the committee system, or the desirability of imposing mandatory standards, one must understand how well the market sets de facto standards.

40. Technical Committee 97, headquartered at ANSI in New York, is the ISO subcommittee responsible for ISDN standards. See Rutkowski, *Integrated Services Digital Networks*, p. 17.

The market's performance may be evaluated in several ways. These include whether the market selects the appropriate standard; whether inferior standards are abandoned when new, superior, technologies become available; whether the proper trade-off between variety and standardization is made; and whether translators are appropriately developed. These are important economic issues, and until quite recently, they have been virtually ignored by economists. This section briefly reviews the burgeoning theoretical literature that attempts to correct this failing.

The distinctive feature of the models discussed here is that the standardization creates a demand-side economy of scale. In particular, where there are benefits to compatibility, users of a particular technology reap benefits when others adopt the same technology. Thus one person's adoption decision confers a positive externality on other adopters. Because individual decisionmakers ignore these externalities in making their decisions, one cannot in general expect the outcomes to be efficient. Indeed, as discussed below, these and many other inefficiencies can arise.

Two other issues about how standardization affects market structure and firm behavior are also important. The first is whether, in the presence of benefits from compatibility, firms can take strategic actions to put their rivals at a disadvantage. When a firm has the ownership rights to a given technology (such a firm is often called a "sponsor" of the technology in this literature), the adoption of the technology as a standard will confer some monopoly power on the owner. Thus each firm may be expected to take measures to encourage the adoption of its technology as the standard and to protect and extend its monopoly power once it has been achieved.

The second issue is of particular importance in such markets as telecommunications, where customers use a primary product (such as the telephone network) in conjunction with secondary products (such as customer premises equipment and enhanced telecommunications services). For such markets the question arises whether firms with a dominant position in the primary market can employ control of interface standards to extend their dominance profitably to the secondary market. (These two issues are discussed under "Standards and Competition" below.)

Inertia and Momentum in the Adoption of a New Standard

The benefits from standardization may make users of a standardized technology reluctant to switch to a new, and perhaps better, technology because of fear that others, bound together by the benefits of compatibility, will not abandon the old standard. If this is the case, it may be difficult for a new standard to be adopted. As a result, de facto standardization may retard innovation.

The first theoretical model of this phenomenon was developed by Rohlfs, who considered what happens when a given number of agents are simultaneously considering adopting a new technology.[41] Suppose that all potential users of a technology would adopt if each knew that the others would do so as well, but that no agent would adopt if it thought that it would be the only adopter. Rohlfs pointed out that there usually are multiple equilibria in this situation. One is for everyone to adopt the new technology; another is for no one to adopt it. Similarly, if some subsets of potential users are in favor of adoption, but others are not, still other equilibria are possible.

Consider four potential adopters. Suppose that agents 1 and 2 will adopt if the other does, but that agents 3 and 4 will adopt only if the other does and if 1 and 2 also adopt. Even if all four agents are better off adopting, it is conceivable that inertia will lead to an equilibrium in which only 1 and 2 adopt, if that outcome is somehow "focal."[42]

A second problem is that it may not be an equilibrium for all four agents to adopt, yet that may be the most socially desirable outcome. This occurs, for example, when agents 3 and 4 are moderately reluctant to adopt the technology, but their adoption would make agents 1 and 2 much better off. Because 3 and 4 ignore the benefits that they confer on 1 and 2 in making their decisions, too little adoption may occur.[43] Indeed, 1 and 2 may not adopt the new technology if they are unsure that 3 and 4 will do so.[44]

Farrell and Saloner have demonstrated that some of these potential inertia problems disappear if one allows for sequential rather than simultaneous decisionmaking and complete information.[45] In that setting, they

41. The model is actually cast in terms of agents choosing whether to join a telecommunications network, but the analogy to the choice of a standard is complete. See Jeffrey Rohlfs, "A Theory of Interdependent Demand for a Communications Service," *Bell Journal of Economics*, vol. 5 (Spring 1974), pp. 16–37.

42. How agents 1 and 2 manage to coordinate their behavior is, of course, important. The point of the example, however, is that agents 3 and 4 may fail to achieve coordination even if agents 1 and 2 can do so. Note that this is an example of pure coordination if all four agents are better off adopting. If this is the case, a standard-setting body would succeed in promoting adoption of the new technology.

43. See Philip H. Dybvig and Chester S. Spatt, "Adoption Externalities as Public Goods," *Journal of Public Economics*, vol. 20 (March 1983), pp. 231–47. The authors have demonstrated that, in some cases, simple subsidy schemes may alleviate both problems.

44. Note that this is an example of conflict and cannot be resolved by replacing noncooperative standard setting with a standard-setting body. Agents 3 and 4 will not switch even if 1 and 2 agree to do so.

45. See Farrell and Saloner, "Standardization, Compatibility, and Innovation."

showed that when all agents prefer joint adoption of the new technology to the status quo, adoption is the unique equilibrium.[46] Moreover, if the agents do not all prefer joint adoption of the new technology, the only equilibrium involves the largest set of possible adopters. Of those that do not adopt, there is no subset that desires to switch. This result suggests that the intuition about the possible innovation-retarding effects of standardization does not extend to a model in which the timing of the adoption decision is endogenous and information is complete.

Although this model provides a useful benchmark, it suffers from a lack of realism. First, the assumption that all potential adopters are perfectly informed about one another's preferences is not innocuous. Second, the model has a timeless quality to it—there are no transient costs of incompatibility, adoption is not time consuming, and all potential adopters of the technology are extant at the time the adoption is first contemplated. In reality, some potential adopters will make their decision only some time in the distant future.

Richer models have been developed to incorporate each of these features. The conclusion that emerges uniformly from these studies is that the outcome of the adoption process may be inefficient. The inefficiency is not, however, only that a socially efficient standard may not be adopted. It is also possible that a new standard may be adopted too readily; that is, when, from a social point of view, it should not be.

For example, Farrell and Saloner have considered what happens when two potential adopters are imperfectly informed about each other's preferences.[47] They found that the outcome resembles a "bandwagon": if one potential adopter is keen on the adoption of the new technology, that agent will adopt early in the hope of inducing the other to follow. If a potential adopter is only moderately enthusiastic, the agent will favor a wait-and-see strategy, adopting only if another is more eager and gets the bandwagon rolling.[48]

Wait-and-see behavior can have the effect of stalling the bandwagon even when both potential adopters hope that adoption will occur. Thus

46. The proof uses the following backward-induction argument. Suppose that there are n potential adopters and $n - 1$ have already adopted the technology. In that case, the nth adopter will as well. Therefore, consider the $n - 1$st adopter when $n - 2$ have already adopted. That potential adopter knows that if he adopts, the final adopter will also, and so he, too, adopts. The same logic can be applied all the way back to the first adopter. This explains why a standard-setting body can succeed in achieving universal adoption only when all four agents prefer the new technology.

47. Farrell and Saloner, "Standardization, Compatibility, and Innovation."

48. Steven R. Postrel, "Three Essays in Industrial Organization" (Ph.D. dissertation, Massachusetts Institute of Technology, 1988), has extended these results to the n-agent case.

there may be too little standardization. The converse is also possible, however. Suppose that two firms are currently both using an existing technology when a new technology becomes available, and that only one firm favors switching to the new technology. That firm may adopt the new technology, leaving the other with the choice between remaining the lone user of the old technology or switching as well. If the benefits to compatibility are large, the latter firm may find switching is its best alternative. However, the firm that opposes the switch may be hurt more than the firm that favors the switch benefits, so that firms in the aggregate are worse off than if both had remained with the old technology. Thus although bandwagons may overcome the need to employ cooperative standard setting to achieve efficient adoptions, they may also promote inefficient ones.

Not only has it been shown that incomplete information can lead to either "excess inertia" or "excess momentum" in the adoption of a new technology, but in another paper Farrell and Saloner have provided two models in which this can occur even with complete information.[49] The first model examines the case in which only new adopters consider a new technology, but the installed base of users of an old technology does not find switching profitable. Excess inertia can arise if the first potential adopters to consider the new technology are not prepared to give up the transient benefits from being compatible with the installed base of the old technology. They then adopt the old technology, swelling the ranks of the installed base and making the old technology even more attractive. In that case the new technology may not get off the ground, even if it would be much preferred by most new users if it became established. The failure of the market in this case is that the first potential users to consider the new technology confer a benefit on later adopters that the first potential users do not take into account in making their adoption decisions. Cooperative standard setting will not be able to overcome this problem because, by assumption, early potential adopters highly value the benefits of compatibility with the installed base.

Excess momentum can occur when the new technology is adopted, but the harm imposed on users of the old technology, who are thereby stranded, exceeds the benefit to new adopters from the new technology. This result is important because it suggests that simple public policies intended to

49. See Joseph Farrell and Garth Saloner, "Installed Base and Compatibility: Innovation, Product Preannouncements, and Predation," *American Economic Review*, vol. 76 (December 1986), pp. 940–55.

encourage the adoption of new technologies can exacerbate an existing bias in the market.[50]

The second model examines what happens when adoption takes time, and all potential adopters of a new technology are users of an old one. The first adopter of the new technology will lose any compatibility benefits it currently enjoys until others also adopt the new technology. At the same time, any user who does not switch to the new technology may find itself temporarily stranded with the old technology if other users switch before it does. If the first of these effects is strong, excess inertia may arise, with no potential adopter willing to take the plunge, and the result will be that all remain with the old technology. If the latter effect is strong, excess momentum may arise, with each user rushing to be the first to adopt the new technology out of fear of being temporarily stranded.

In the models above, potential adopters choose between the status quo and a single new technology. Arthur has shown that the "wrong" technology may be chosen even when a sequence of first-time potential adopters are choosing between two new technologies.[51] As in the Farrell and Saloner model discussed above, early adopters are pivotal.[52] If most favor one of the technologies and adopt it, it becomes relatively less expensive for later adopters who, in turn, may find it uneconomical to adopt the other technology. If the majority of later adopters would have preferred the other technology, however, society may have been better served by its adoption. In that case, the chance predisposition of early adopters to

50. Such policies have also been shown to have the unexpected effect of slowing the adoption of new technology; see Nathan Rosenberg, "On Technological Expectations," *Economic Journal*, vol. 86 (September 1976), pp. 523–35; and Norman Ireland and Paul L. Stoneman, "Technological Diffusion, Expectations, and Welfare," paper presented to the Technological Innovation Project (TIP) Workshop, Stanford University, February 1984. An adopter of a new technology knows that these policies provide an incentive to new innovation, increasing the chance that the new technology will itself soon be obsolete. For a discussion of these and other implications of public policy intended to hasten technology adoption, see Paul A. David, "Technology Diffusion, Public Policy, and Industrial Competitiveness," in Ralph Landau and Nathan Rosenberg, eds., *The Positive Sum Strategy: Harnessing Technology for Economic Growth* (Washington, D.C.: National Academy Press, 1986); and Paul A. David and Paul L. Stoneman, "Adoption Subsidies vs. Information Provision as Instruments of Technology Policy," TIP Working Paper 6 (Stanford University, Center for Economic Policy Research, April 1985).

51. See Brian W. Arthur, "Competing Technologies and Lock-in by Historical Small Events: The Dynamics of Allocation under Increasing Returns," Discussion Paper 43 (Stanford University, Center for Economic Policy Research, January 1985). In the simplest version of his model, the demand-side externalities arise from "learning by using": each time a potential adopter selects one of the technologies, the costs to later users of the same technology are reduced. The model can, however, easily be extended to the case of compatibility. For a discussion of this point, see Paul A. David, "Some New Standards for the Economics of Standardization in the Information Age," in Partha Dasgupta and Paul Stoneman, eds., *Economic Policy and Technological Performance* (Cambridge University Press, 1987), pp. 206–39.

52. See Farrell and Saloner, "Installed Base and Compatibility."

the socially inferior technology, and the fact that they serve their own rather than society's interests, results in the less preferred technology being chosen as the standard.[53]

Communication, Cooperation, and Contracts

Where an inefficient standard emerges—for example, when a new standard is adopted despite the great harm inflicted on the installed base, the failure of the market to select the "right" standard could be avoided if all potential adopters coordinated their activity and made appropriate side payments. If such contracts and side payments could overcome inefficiencies, it is important to know why they do not naturally arise within a market setting.

There are several possible reasons. The most important is that many of the agents whose adoption decisions are relevant are not active market participants at the time the new technology becomes available but arrive much later. In principle, one could imagine a scheme whereby a fund is made available by current users to provide subsidies to later adopters as they arrive. Each current member of the installed base, however, would have an incentive to free ride on the contributions of the others, or, if a method of taxes and subsidies were used, to understate their true aversion to stranding. (The free rider problem would arise even if the model were "timeless.") Moreover, if, as in Arthur's model, there is uncertainty about the preferences of future adopters, even a central authority would often err in its choice of a standard.[54]

An additional difficulty arises if there is asymmetric information about adopters' preferences. Farrell and Saloner have explored the implications of communication in their asymmetric-information bandwagon model and found that communication is a mixed blessing.[55] When potential adopters are unanimous in their desire to adopt the new technology, communication

53. The same phenomenon is analyzed from a different perspective in Robin Cowan, "Backing the Wrong Horse: Sequential Technology Choice under Increasing Returns," unpublished paper (Stanford University, December 1986). As in Arthur's model there is learning by using. In addition, however, potential adopters are unsure which technology is better. Each trial of a technology provides some information about its desirability. Thus, as in the above models, there is a connection between the welfare of late adopters and the decisions of early ones. Since early adopters ignore the value of the information they provide to late ones, from a social point of view there may be too little exploration of the value of alternative technologies.

54. Such a central authority has been called a "blind giant" by Paul A. David, "Narrow Windows, Blind Giants, and Angry Orphans: The Dynamics of Systems Rivalries and Dilemmas of Technology Policy," TIP Working Paper 10 (Stanford University, Center for Economic Policy Research, March 1986).

55. Farrell and Saloner, "Standardization, Compatibility, and Innovation."

helps coordination and eliminates excess inertia. If adopters have differing preferences, however, communication can actually make matters worse. Potential adopters who are only slightly averse to the adoption of the new technology will exaggerate their degree of aversion, making it even less likely that a bandwagon will get started. This suggests that there are circumstances in which inertia may actually be *increased* if there is an attempt to set voluntary standards through industry committees.

Another portion of the literature addresses the trade-off between standardization and variety. Farrell and Saloner have shown that when the degree of standardization is left to market forces, too little variety may be provided if the existence of a historically favored technology prevents an otherwise viable alternative from getting off the ground.[56] Matutes and Regibeau have addressed the case in which products are combined in "systems" and show that standardizing the product interface can *increase* the variety of systems by promoting "mix-and-match" purchases. However, it can also lead to higher prices.[57]

The compatibility of components may also have implications for technology adoption. Berg has compared a regime in which there are two competing technologies with one in which there is only one technology.[58] In the former case, one of the technologies may eventually become the de facto standard. The adopters of the abandoned technology may then find that compatible components are no longer provided. The realization of this possibility will tend to dampen the demand for both technologies, leading to slower technology adoption. Farrell and Gallini have shown that a monopolistic supplier of the primary good may encourage competition in the component market in order to mitigate this problem.[59]

The preceding discussion assumes that if all the potential adopters are present at the time the adoption decision is made, and if there is complete information, participants will negotiate an efficient outcome. Even if these conditions are satisfied, however, if the adopters have different preferences, bargaining may delay the adoption of a standard. Farrell and Saloner have compared a setting in which a coordinated outcome can be negotiated

56. Joseph Farrell and Garth Saloner, "Standardization and Variety," *Economics Letters*, vol. 20, no. 1 (1986), pp. 71–74.

57. Carmen Matutes and Pierre Regibeau, "Mix and Match: Product Compatibility without Network Externalities," *Rand Journal of Economics*, vol. 19 (Summer 1988), pp. 221–34.

58. Sanford V. Berg, "Technological Externalities and a Theory of Technical Compatibility Standards," unpublished paper (University of Florida, Department of Economics, February 1985).

59. Joseph Farrell and Nancy T. Gallini, "Second-Sourcing as a Commitment: Monopoly Incentives to Attract Competition," Working Paper 8618 (University of California, Berkeley, Department of Economics, December 1986).

(which represents the committee system) with one in which adoption takes place unilaterally in the market.[60] In their model, the committee is more likely to achieve coordination than the market. In addition, although the committee is slower, it outperforms the market mechanism, even allowing for the value of speed.

The Development of Translator Devices or "Gateway" Technologies

In the above analyses, potential adopters face the choice between two inherently and unalterably incompatible technologies. In practice, however, technical compatibility is not required for two components of a system to communicate. Where components have not been designed to be compatible, devices variously known as translators, adapters, converters, or gateways can often be used to permit the components to interact.[61] If translation were always costless and technically perfect, so that messages sent in either direction and then returned were identical to those that were originally transmitted, standardization would be unnecessary, but this is often not the case. Nonetheless, there is a thriving business in the sale of devices that permit communication in the absence of compatibility.[62] The existence of translators has several implications for standardization, most of which have not been addressed in the theoretical literature.

First, in some circumstances, the use of translators may be more efficient than the development of standards. Standard setting is costly, and if only a few users wish to combine incompatible components, it may be less costly for them to employ translators than to attempt to achieve standardization. If the principal uses of the incompatible components are to serve users with different needs, moreover, important benefits may be lost if standardization is required.

60. Joseph Farrell and Garth Saloner, "Coordination through Committees and Markets," *Rand Journal of Economics*, vol. 19 (Summer 1988), pp. 235–52.

61. For a brief discussion of translators as a substitute for standards, see Braunstein and White, "Setting Technical Compatibility Standards."

62. Some examples of translation devices are Word For Word, a "software document converter that converts files and documents from one PC-compatible word processing system to another" (advertisement in *Byte* [1986 Extra Edition], p. 229); a series of products offered by Flagstaff Engineering that "can connect . . . incompatible computer systems using diskette, tape, communications, or printed media" (advertisement in *Byte*, vol. 11 [September 1986], p. 320); and PC<>488, which "allows [the] IBM PC/XT/AT or compatible to control IEEE-488 instruments" (advertisement in *Byte*, vol. 11 [November 1986], p. 155).

Second, translators are likely to be important during the period in which incompatible technologies are vying to become the industry standard and consumers wish to have access to a larger "network" than any single technology can provide. Translators permit the choice of a standard to be deferred until more information about the various technologies becomes available. This does not mean, of course, that either the market or standard-setting bodies will necessarily select the efficient standard after the period of experimentation, but better choices may be possible if there are more data about the competing technologies.

Third, the existence of translators may promote the development of specialized uses for particular technologies and thus narrow the range of uses of each. David and Bunn argue, for example, that the development of the rotary converter for "translating" AC to DC electrical current delayed the development of high-voltage DC transmission.[63]

Finally, the presence of translators may reduce the incentives to achieve standardization. So long as incompatible components can be combined into a system, consumers are likely to be less willing to demand that manufacturers standardize, and manufacturers are likely to be less willing to incur the costs of doing so.

Nonetheless, one can easily overstate the extent to which translators can and will substitute for standards. There are likely to be cases in which translation is technically inefficient, in which the costs of achieving translation are high, or both.[64] Several large communications users have emphasized to us the value of their having standardized communications networks and have argued strongly that, for them, translators are a poor substitute. They are thus likely to be an important force in promoting standardization.

Standards and Competition

For the most part, the models discussed in the preceding section do not explicitly consider the prices for the different technologies that the potential adopters face. This is consistent with markets in which the various technologies are competitively supplied so that adopters face competitive

63. Paul A. David and Julie Ann Bunn, "The Battle of the Systems and the Evolutionary Dynamics of Network Technology Rivalries," TIP Working Paper 12 (Stanford University, Center for Economic Policy Research, January 1987).

64. Katz and Shapiro have shown that firms providing incompatible technologies will in general not have the correct incentives to provide converters. See Michael L. Katz and Carl Shapiro, "Network Externalities, Competition, and Compatibility," *American Economic Review*, vol. 75 (June 1985), pp. 424–40.

prices. This feature of the models is important because if the technologies were offered instead by firms with some market power, the firms might have an incentive to behave strategically. In this section we examine strategic actions of three kinds. First, we analyze the effect of strategic pricing on the market's choice of technology. Second, we look at the effect of truthful advance announcements that firms will introduce a new product. Finally, we study the contention that leading or dominant firms, or firms with control over bottleneck facilities, might use their positions to choose or change standards to disadvantage their rivals.[65]

Strategic Pricing and Product Preannouncements

Katz and Shapiro have examined the implications of strategic pricing in a two-period model in which there is competition between two technologies.[66] The most interesting case they have considered is one in which each technology is offered by a single firm and one technology has lower costs in the second period but higher costs in the first.[67] They found that the sponsor of the technology that will be cheaper in the future has a strategic advantage. This is a somewhat surprising result and its flavor is exactly the reverse of that in the models considered earlier, in which there was a tendency for adopters to choose the technology that is more attractive at the time that they adopt.

The intuition behind the result is the following. Where each technology is provided by a single sponsor, that firm has an incentive to price low early on, even below its cost, to achieve a large installed base and thus become the industry standard. Potential adopters know, however, that later on (in the second period) the firm will no longer have an incentive to use promotional pricing and will charge a higher price. Potential adopters therefore expect the firm that will have the lower future costs also to have the lower future prices. If both firms charge the same first-period price, potential adopters will prefer the technology that will have lower future costs. Put differently, the firm that has higher first-period costs can overcome that disadvantage by promotional pricing. The firm that has

65. For an example of this view, see Walter Adams and James W. Brock, "Integrated Monopoly and Market Power: System Selling, Compatibility Standards, and Market Control," *Quarterly Review of Economics and Business*, vol. 22 (Winter 1982), pp. 29–42.

66. Michael L. Katz and Carl Shapiro, "Technology Adoption in the Presence of Network Externalities," *Journal of Political Economy*, vol. 94 (August 1986), pp. 822–41.

67. Katz and Shapiro also have studied the case in which both technologies are competitively supplied. Their results were similar to those of Farrell and Saloner, "Installed Base and Compatibility," discussed in the preceding section.

higher second-period costs, however, cannot do the same because consumers will rationally expect the firm to exploit its dominant position at that stage.

Strategic behavior yields lower prices for consumers. It does not, however, guarantee that the technology with the lower overall cost is adopted. At the same time, a ban on promotional pricing might prevent the adoption of the technology with the lower cost.

Similar problems arise in the the model developed by Farrell and Saloner.[68] Recall that in that model there is an installed base of users of an old technology when a new technology becomes available. As a polar case, they considered what happens when the new technology is supplied by a competitive industry, whereas the old technology is supplied by a monopolist. They showed that in some circumstances the monopolist will be able to prevent the new technology from being adopted by offering a discount to potential adopters. (The same advantage exists when a monopolist is the supplier of a new technology that is incompatible with one offered by a competitive industry.) The discount need not be offered to all adopters. Instead, there may be some critical level of the installed base at which the old technology will become invulnerable because the compatibility benefits from joining the installed base are so large. Once that point is reached, the monopolist need no longer offer a special inducement. There is thus a window of opportunity for the new technology that the monopolist may be able to close through strategic pricing. Moreover, this tactic to prevent entry may be successful even when the new technology would have been superior from a social point of view.[69]

The Farrell and Saloner model can also be used to demonstrate that a simple announcement that a product will be available in the future (a "product preannouncement") can make the difference between the adoption and nonadoption of a technology.[70] Suppose that the old technology is competitively supplied, but that the new technology is supplied by a monopolist. By the time the monopolist is ready to introduce its product, the installed base of the old technology may make entry impossible. By preannouncing the introduction of a new product, the monopolist may be able to induce some potential adopters to wait for the product's arrival. If that occurs, the new product will begin with an installed base of its

68. Farrell and Saloner, "Installed Base and Compatibility."

69. Katz and Shapiro found the same result in their two-period model; see Michael L. Katz and Carl Shapiro, "Product Compatibility Choice in a Market with Technological Progress," *Oxford Economic Papers*, vol. 38 (November 1986), supplement, pp. 146–65.

70. Farrell and Saloner, "Installed Base and Compatibility."

own, making it the more attractive technology to later adopters. As in the case of strategic pricing, the preannouncement can result in the adoption of the socially less preferred technology, in this case because the preannouncement leads to the stranding of users of the old technology.

Standards and Bottleneck Facilities

For the most part, the theory of noncooperative standard setting discussed thus far has focused on the market for a "primary" good such as computers, for which compatibility is sought or avoided because of its effect on demand in the primary market. In the analyses cited, the effects that compatibility, pricing, and preannouncement decisions in the primary market have on the market for the secondary good has not been analyzed in detail because it is implicitly assumed that producers of the primary good do not participate in the secondary market.

The situation in the telecommunications market is somewhat different. Here, one set of firms, the local telephone companies, is assumed to control the market for basic telephone transmission capacity, the primary market.[71] At the same time these firms are, or would like to be, participants in the secondary markets for customer premises equipment (CPE) and enhanced telecommunications services. The questions that face regulators are whether control of the primary market can be extended, through the use of standards or in other ways, to the secondary markets, and whether the local telephone companies will have the incentive to attempt to "leverage" their market power in this manner.[72]

The use of standards to increase profits in either the system (primary) market or in the market for a complementary good has been analyzed in

71. Whether this presumption is true is not addressed in this paper, although the conclusions would be affected if there were effective competition in the transmission market. Similar issues arise in countries where a single entity controls the entire telecommunications system and competes with outside suppliers. This explains the large role given to the achievement of common standards by the Commission of the European Communities. The commission is concerned with "the promotion of Europe-wide open standards, in order to give equal opportunity to all market participants." See Commission of the European Communities, "Towards a Dynamic European Economy," Green Paper on the Development of the Common Market for Telecommunications Services and Equipment (Brussels, June 1987), p. 5.

72. This is akin to the issues raised in the various antitrust cases involving IBM, in which it was alleged, among other things, that IBM manipulated its interconnection standards to extend its putative monopoly in the market for mainframe computers to the market for peripheral equipment. We raise the example of the IBM cases only because they present analogies for policy questions in the telephone industry. For a vigorous defense of IBM's actions, see Franklin M. Fisher, John J. McGowan, and Joen E. Greenwood, *Folded, Spindled, and Mutilated: Economic Analysis and U.S.* v. *IBM* (MIT Press, 1983).

detail by Ordover and Willig.[73] They consider a firm that is either the only supplier of one component of the system, the "primary" component, or that has a cost advantage in producing that component; it should be clear that the component is called primary not because it is any more necessary than any other component but because of the advantage that the firm has in producing it. Other components of the system can be produced by rivals at the same cost.

It is well known that a firm with monopoly over one component will often be able to obtain maximum profit without regard to the presence of rivals in the competitive market *so long as there are no constraints on the price, or prices, that it can charge*. Consider the case in which all consumers place the same value on a system, a system consists of one primary and one secondary component, all firms have the same costs of producing the secondary components, and only one firm can produce the primary component. Suppose that the cost of producing the primary component is 10, the cost of producing the secondary component by any firm is 5, and the value that each consumer places on a system, or its constituent components, is 25. If there are no constraints on the prices that the producer of the primary component can charge, it can set the price of a system at 25, the price of the primary component at 20 (= 25 − 5), and the price of the secondary component at 5. This firm obtains a profit of 10 (= 25 − 10 − 5) on each system that it sells directly to consumers. Even where a consumer purchases only the primary component from the firm, however, it still obtains a profit of 10 (= 20 − 10). The firm is thus indifferent to whether consumers purchase the entire system or only the primary component from it because its profits are the same in either case. If rival firms can produce the secondary component more efficiently—say at a cost of 4—the profits of the firm are actually increased if it leaves the

73. Janusz A. Ordover and Robert D. Willig, "An Economic Definition of Predation: Pricing and Product Innovation," *Yale Law Journal*, vol. 91 (November 1981), pp. 8–53. See also Janusz A. Ordover, A. O. Sykes, and Robert D. Willig, "Nonprice Anticompetitive Behavior by Dominant Firms toward the Producers of Complementary Products," in Franklin M. Fisher, ed., *Antitrust and Regulation: Essays in Memory of John J. McGowan* (MIT Press, 1985), pp. 115–30. Ordover and Willig actually describe a large number of ways in which firms might attempt to exercise such leverage. These include refusing to sell the primary good to a rival; selling only complete systems and not their components; selling both systems and components but setting high prices for components if purchased separately; "underpricing" components that compete with those sold by rivals; and "overpricing" components that are needed by rivals to provide complete systems. Thus standards are only one of a number of tools that a firm can use strategically to disadvantage its rivals and to increase its profits. It should also be observed that these are all variants of the "raising rivals' costs" strategies analyzed in detail in Thomas G. Krattenmaker and Steven C. Salop, "Anticompetitive Exclusion: Raising Rivals' Costs to Achieve Power over Price," *Yale Law Journal*, vol. 96 (December 1986), pp. 209–93.

market for the secondary component to them. It can charge a price of 21 (= 25 − 4) for the primary component and obtain a profit of 11 (= 21 − 10), which is larger than the profit of 10 it obtains from selling an entire system.

It may pay to eliminate a rival, however, if there are limits on the prices that can be charged for the primary component. Thus, in the previous example, if the producer of the primary component can charge at most 12 for it—say, because of regulation—then so long as it can charge any price above its cost on the secondary component it will wish to eliminate its rivals and dominate the secondary market as well. If it can, for example, charge 6 for the secondary component, its profits are 3 (= 12 + 6 − 10 − 5) if it can sell both components, or an entire system, but it can earn only 2 (= 12 − 10) if it is limited to selling only the primary component. If the firm can charge 13 or more for the secondary component, it can earn the entire monopoly profit even with the restriction on the price it can charge for the primary component. If, however, there are rivals in the provision of the secondary component, and if the firm must make the primary component available at a price of 12, its profits are limited to 2 (= 12 + 5 − 10 − 5). This reduction occurs because consumers will buy the secondary component from the firm's rivals if it attempts to charge a price in excess of 5. Thus firms have an incentive to eliminate rivals. One way in which a firm can do so is to make its primary component incompatible with the secondary component manufactured by its competitors.[74]

The producer of the primary component might also wish to eliminate its rivals if consumers place different values on systems and if these differences are proportional to the consumers' use of the secondary component. Suppose, for example, that there are two consumers, one who places a value of 25 on a system consisting of one primary and one secondary component and another who places a value of 40 on a system consisting of one primary component and two secondary components. The firms' costs are the same as in the previous example.

If there is no competition in the secondary market, the producer of the primary component can offer it at a price of 10 and the secondary component at a price of 15 and capture the entire consumers' surplus. Its

74. The ability to use standards in such an anticompetitive manner is severely limited if efficient, low-cost translators are available. For example, a firm that seeks a competitive advantage by designing interfaces that cannot directly accommodate the products of its rivals will find the strategy unsuccessful if users can easily connect incompatible devices through the use of translators.

profits in this case are 30 (= 40 + 25 − 20 − 15). If there are rival suppliers of the secondary component who can produce at a cost of 5, so that the firm must obtain its profits entirely on the primary component, it will sell the primary components for 20 and earn profits of only 20 (= 20 + 20 − 10 − 10).[75] Eliminating a rival is desirable because it permits price discrimination that would not otherwise be possible.[76] Once again, a possible strategy for eliminating rivals is to design the primary component so that it is incompatible with the components produced by competing firms.

The two elements necessary for the use of the strategies analyzed by Ordover and Willig both appear to be present in the telephone industry. First, there are regulatory constraints on the prices that can be charged for the primary product, access to the transmission network. These constraints take the form of limits both on the overall rate of return that the firm can earn and on the prices of individual services. Second, the primary product may be a bottleneck or "essential facility" that is needed if the suppliers of enhanced services or CPE are to be able to sell their wares. (To the extent that suppliers of enhanced services or CPE can "bypass" the local transmission facilities of a telephone company, the ability of the telephone company to use standards anticompetitively is reduced.)

At the same time, one of the assumptions in the examples presented by Ordover and Willig must be questioned. In their examples, the firm that controls the primary market does not, as a result, have a cost advantage in producing the secondary goods. In such cases, no loss in efficiency results from a ban on the participation of suppliers of the primary good in the secondary markets. Similarly, there is no loss from requiring suppliers of the primary good to participate in these markets through separate subsidiaries, so that instances of anticompetitive behavior can be more easily detected.

However, in addressing the effects of the limitations placed on AT&T by its *Computer II* decision, which required that customer premises equipment and enhanced services be offered through separate subsidiaries, the FCC noted that "the inability to realize . . . scope economies was one cost of structural separation for AT&T's provision of CPE; and we believe the elimination of such costs could well result in efficiencies for AT&T's

75. The firm's profits are the same if it sells only one primary component at 30. The analysis assumes that the firm cannot offer only complete systems at discriminatory prices.

76. This is analogous to the argument that firms will vertically integrate forward in order to permit them to practice price discrimination. See J. R. Gould, "Price Discrimination and Vertical Control: A Note," *Journal of Political Economy*, vol. 85 (October 1977), pp. 1063–71.

provision of enhanced services, to the extent that such services could be integrated into or collocated with AT&T's basic network facilities.''[77] In examining the effects of similar restraints on the BOCs, the commission observed ''that structural separation imposes direct costs on the BOCs from the duplication of facilities and personnel, the limitations on joint marketing, and the inability to take advantage of scope economies similar to those we noted for AT&T.''[78] If the economies of scope noted by the FCC are important, a blanket ban on BOC participation in the customer premises equipment and enhanced services markets, although it might prevent anticompetitive behavior, might also prevent efficient supply.[79]

We conclude that the conditions are present under which standards might be used to place at disadvantage the competitors of those who control access to the telecommunications transmission system. To prevent these and other forms of anticompetitive behavior, the FCC and the courts have either prohibited the telephone companies from providing certain services or have required that these services be provided through fully separated subsidiaries. If telephone companies have lower costs than these competitors, however, either a blanket prohibition or a separate subsidiary requirement might be economically inefficient. As a result, the FCC has begun to pursue an alternative approach whereby the restrictions on the telephone companies are eliminated and, at the same time, a regulatory framework to make the anticompetitive use of standards more difficult is established.

Telecommunications Standards, Telephone Regulation, and the FCC

Until the 1960s, standardization was not a major issue of telecommunications policy because there were no competing providers of equipment, or of communications services, who might be adversely affected by the standards that were chosen by AT&T. (Consumer welfare, of course, could depend on those choices.) Beginning with the FCC decision in the

77. Federal Communications Commission, ''Report and Order in the Matters of Amendment of Section 64.702 of the Commission's Rules and Regulations (Third Computer Inquiry); and Policy and Rules Concerning Rates for Competitive Common Carrier Services and Facilities Authorizations Thereof; and Communications Protocols under Section 64.702 of the Commission's Rules and Regulations,'' CC Docket No. 85-229, adopted May 15, 1986, released June 16, 1986, 60 RR 2d 603 (1986), para. 80.

78. Ibid., para. 90.

79. For a forceful statement of the proposition that substantial efficiency losses will result if the BOCs are confined to providing basic service, see Almarin Phillips, ''Humpty Dumpty Had a Great Fall,'' *Public Utilities Fortnightly*, vol. 118 (October 2, 1986), pp. 19–24.

Carterfone case, which introduced competition into the supply of equipment to telephone customers, standards have become an increasingly important policy concern.[80] In adopting its equipment registration program, in which it sought to eliminate technical barriers to the entry of independent equipment suppliers, the commission required, with one minor exception, that "all terminal equipment be connected to the telephone network through standard plugs and jacks."[81] In its *Computer II* decision, in which it sought to promote competition in the market for equipment and enhanced services, the FCC required that technical information that independent suppliers might need to compete had to be provided to them on the same terms as to the subsidiaries of the telephone companies.[82] In this regard, the commission singled out "information relating to network design and technical standards, including interface specifications [and] information affecting changes which are being contemplated to the telecommunications network that would affect either intercarrier connection or the manner in which CPE is connected to the interstate network."[83]

The decisions by the FCC to require standardized interconnection for terminal equipment and to provide technical information to independent suppliers were part of an effort designed to make it possible for independent equipment vendors to compete effectively in the supply of this equipment. Although the commission did not itself participate in the process of establishing interconnection standards, leaving their determination to the industry, its policy has been enormously successful, at least when judged by the wide variety of equipment that is now available and by the sharp declines in the market shares of the telephone companies.[84]

80. See *Use of the Carterfone Device*, 13 F.C.C. 2d 420, *reconsideration denied*, 14 F.C.C. 2d 571 (1968).

81. 56 F.C.C. 2d 593, 611 (1975).

82. *Second Computer Inquiry*, 77 F.C.C. 2d 384 (1980).

83. Ibid., para. 246. Although the requirement that competitors be provided with information limited the ability of AT&T to use standards to disadvantage its rivals, AT&T might still prefer different standards from those desired by the competitors.

84. In the early 1980s, AT&T's share of the customer premises equipment market had declined to somewhat more than 60 percent, and by 1986 its share of total lines shipped had fallen further, to about 36 percent for handsets, 25 percent for key systems, and 20 percent for PBXs (private branch exchanges). See Majority Staff of the Subcommittee on Telecommunications Consumer Protection and Finance of the House Committee on Energy and Commerce, *Telecommunications in Transition: The Status of Competition in the Telecommunications Industry*, 97 Cong. 1 sess. (GPO, 1981); and Peter W. Huber, *The Geodesic Network: 1987 Report on Competition in the Telephone Industry* (U.S. Department of Justice, Antitrust Division, January 1987). In the United Kingdom, where entry of independent suppliers of terminal equipment did not begin until much later than in the United States and where a somewhat different equipment registration program exists, non-British Telecom suppliers have captured half of the addition of the installed base of telephones since 1980 and about 10 percent of the key system market. See

Under the modified final judgment that settled the government's antitrust suit against AT&T, the Bell operating companies are "prohibited from discriminating between AT&T and other companies in their procurement activities, the establishment of technical standards, the dissemination of technical information . . . and their network planning."[85] Moreover, the judgment "requires AT&T to provide [the] Operating Companies with, *inter alia*, sufficient technical information to permit them to perform their exchange telecommunications and exchange access functions. . . . The Operating Companies, in turn, are prohibited from discriminating in the 'establishment and dissemination of technical information and procurement and interconnection standards.' "[86]

Finally, in its *Computer III* decision, the commission indicated it would waive its requirement that the Bell operating companies provide enhanced services only through separate subsidiaries if competitors were provided with Comparably Efficient Interconnection (CEI) and offered an open network architecture (ONA) plan acceptable to the commission.[87] CEI is intended to give competing suppliers access to the telephone transmission system on the same basis as the subsidiaries of the telephone company that are providing the same services. ONA means that the components of the telephone system are to be made available to competing suppliers on an "unbundled" basis so that they can be combined with the services of these suppliers in any manner desired. The nature and identities of these components—the basic service elements (BSEs)—in ONA are likely to be contentious issues because they will affect the potential for competition. Competing suppliers will undoubtedly wish to have highly disaggregated components with which they can easily interconnect. The telephone companies are likely to argue for a higher level of aggregation.

Both the interfaces with the basic service elements and the number and nature of these elements are standards issues. The first involves an obvious standards concern because the design of these interfaces will determine whether a competing supplier can employ a particular element in offering its services. Less obvious is why the second is a standards issue. If components can be obtained on a bundled basis only, the interface between them is completely inaccessible to the competing supplier. But the eco-

---

J. H. Solomon, "Telecommunications Evolution in the UK," *Telecommunications Policy*, vol. 10 (September 1986), pp. 186–92.

85. *United States* v. *American Telephone & Telegraph Co.*, 552 F. Supp. 131, 142 (D.D.C. 1982).

86. Ibid., pp. 176–77.

87. 60 RR 2d 603 (1986).

nomic effect of an inaccessible interface is exactly the same as if it were accessible but incompatible with the supplier's equipment. Providing components only on a bundled basis is the limiting case of incompatibility.

Two broad lessons can be drawn from this history. First, the range of services that independent suppliers can offer to telecommunications customers has increased markedly over the past three decades as the restrictions previously imposed by AT&T have been eliminated by regulation. Indeed, the initial effect of many regulatory interventions was either to deny AT&T, and later the BOCs, the ability to provide certain services or to restrict the way in which the services could be offered.

Second, the elimination of the restrictions on what services the telephone companies can provide is being conditioned on the imposition of behavioral constraints designed to help independent suppliers compete. These constraints include requirements that information about changes in network design be promptly provided to competing vendors, that these vendors be provided with interconnection to the telephone system that is "comparable" to that provided when a telephone company itself offers a service, and that the components of the network be available on an unbundled basis so that customers can acquire from the telephone companies only those portions of network services that they desire.

The Determination of Telecommunications Standards

In early sections of this study we examined why suppliers may seek to standardize their services to increase the value of their offerings to consumers, as well as the difficulties involved in achieving such standardization. In the preceding two sections we discussed how standards can be used as a competitive weapon. This section examines two cases of standard setting in telecommunications, the integrated services digital network (ISDN) and open network architecture (ONA), to illustrate both phenomena.

ISDN Standardization

A worldwide effort involving thousands of people is currently under way to develop standards for an ISDN.[88] This effort is intended to define

88. Rutkowski, *Integrated Services Digital Networks*, pp. 5–21, 33–175, contains an extensive description of this process.

the architecture of the ISDN and to promote common ISDN standards in different countries. The countries involved in attempting to set ISDN standards through the CCITT are interested in achieving compatibility among their various national telecommunications networks to achieve the demand-side economies of scale discussed earlier. Yet even when compatibility is highly valued, it may not be easily achieved.

The principal reasons are that, even if all parties value compatibility, countries may not agree on what the single standard should be and that some countries may prefer a degree of *incompatibility* to shelter their domestic telecommunications suppliers from foreign competition. As a result, achieving agreement on common standards is likely to be a slow process, and differences among national systems may persist. There even is some danger that the slowness of the process may encourage the development of incompatible systems by those unwilling to wait for international consensus.

In attempting to achieve standardization among national ISDNs, the CCITT has not confined its activities to the specification of a single dimension of each interface through which information can move. Instead, it has pursued a strategy of attempting to achieve compatibility at a variety of "layers," ranging from the physical interconnections that will be permitted to the forms in which data will be recognized, that are patterned on those in the Open Systems Interconnection (OSI) reference model. Because communication must be effected at all layers at each interface, the specification of standards is quite complex.

Not only are the various interface specifications being spelled out, but so is the architecture of the ISDN. This means that the standards will encompass where the interfaces will be and whether they will be accessible to users or independent suppliers. Clearly, the more alike the various national systems are, the simpler and less costly will be the required interfaces among them. But the fact that the architecture of ISDN will be specified by CCITT may create problems in those countries, such as the United States (and, increasingly, the United Kingdom and Japan), where there are a large number of competing suppliers of telecommunications services.

The concern is that the design of the ISDN—in particular, restrictions on user access—can be used to limit the competition faced by the operators of the transmission network. As a result, there may be significant conflicts between users and suppliers. Rutkowski has made the point succinctly: "Users generally have an interest in maximizing their service options, while providers (particularly telecommunication network providers) have an interest in limiting those options to maximize their operating efficiencies and minimize losses to competitive providers.''[89]

From the perspective of establishing standards, the most significant aspect of the development of an ISDN is the increase in the number of interfaces at which access to the telecommunications network can occur and the ways in which such access can take place.[90] Whereas before the *Carterfone* decision, ''access'' was available only at an AT&T-supplied terminal, now subscribers, or providers of enhanced services, can obtain access to the system at several points by using different kinds of equipment. The ISDN is likely to further increase this number. A significant degree of standardization of interfaces and terminals must be accomplished, however, for this to occur.

Consider a message that must access—that is, pass through—a particular node in the telecommunications network if it is to reach its intended destination. To obtain access, several components are required to establish a ''path.'' The first such component is the subscriber's terminal equipment, either a device with a standard ISDN interface (such as a digital telephone) or one that requires an adapter to access a digital network. Second, there is network equipment required to perform switching and concentration functions (for example, a digital exchange). The third kind of component is network termination equipment, which lies between the transmission system and the subscriber's premises and enables connection of the subscriber's premises to the local telephone loop. Certain types of equipment permit the second and third components to be combined. Finally, there is the link between the local loop and the network itself.

The subscriber can use these components in various ways and, depending on the regulatory regime, may choose to obtain many or few of them from the telephone company. In the United States, for example, a subscriber might employ a terminal requiring an adapter, as well as the adapter and both kinds of termination equipment, from the telephone company. Alternatively, he might obtain the adapter from an independent vendor and the termination equipment from the telephone company. Or he might also purchase the ''switch'' from an independent vendor and only the last link from the telephone company. Or he may acquire all of the components from independent vendors. Similarly, a subscriber may

89. Ibid., p. 46.

90. The introduction of open network architecture in the United States will have a similar effect. A recent article argues that the effect of ONA is likely to be an increase in the number of interfaces by ''an order of magnitude.'' See ''Part 68 Is Not Compatible with ONA,'' *Telecommunications*, North American ed., vol. 21 (January 1987), p. 8.

use a terminal that does not require an adapter but may purchase any or all of the remaining components from independent vendors. Of course, this wide range of options is available only where competitive suppliers exist. In many countries, all components must be acquired from the telephone company.

The ISDN model currently under consideration does not contemplate an interface at which a subscriber, or an independent service provider, can obtain access to the system without employing the telephone company's local loop.[91] This is consistent with the views of most post, telephone, and telegraph administrations in Europe and, probably, with those of the BOCs, which would like to require use of this loop. It is not, however, consistent with the views of independent suppliers who wish to increase the number of points at which they can obtain access so that they can use as much or as little of services supplied by the telephone company as they desire. Thus even if there were no controversy about the designs of the interfaces that were actually offered, there might still be a dispute over how many were offered and where they were located. In the context of earlier discussion, denying access would be equivalent to providing an interface that was totally incompatible with the equipment of one's rivals.

U.S. policy is likely to vary from international ISDN standards if the latter do not permit access to the network without use of the local loop. For example, U.S. vendors can expect to obtain access throughout the telephone company network, and there have even been discussions of whether Comparably Efficient Interconnection requires that the equipment of these vendors be located at telephone company central offices.[92] One continuing policy concern is thus likely to be which interfaces are available to independent suppliers and on what terms.

One way to assuage this concern is for the telephone companies to provide, as they are currently required to do, unbundled private-line service—that is, pure transmission capacity—along with the ISDN, as apparently contemplated by the CCITT (and a likely element of U.S. telecommunications policy). Thus an ISDN would not completely replace the existing telecommunications system, and some elements of the old system would remain. As a result, independent suppliers would have substantial freedom to construct their own networks using private lines provided by the telephone company and other components of their own

91. In the language of CCITT, this is not a "reference point." See Rutkowski, *Integrated Services Digital Networks*, pp. 145–46.

92. Note that denying access to independent vendors at the central office may be equivalent to the strategy, discussed by Ordover and Willig, of not making certain components of a system available to rivals. See Ordover and Willig, "Economic Definition of Predation," pp. 32–36.

choosing. These systems would use none of the "intelligence" in the telephone company's ISDN but would be able to provide many, or all, of the same services. As a result, even if all of the elements of an ISDN were not available on an unbundled basis, enough other resources could be available to make feasible the provision of competitive offerings. This would also protect competing vendors from the possible manipulation of the design of interfaces for strategic purposes. Thus, although a requirement that private-line service continue to be provided does not appear to be a standards issue, it may be a partial substitute for complete agreement on standards.[93]

Yet another way to prevent carriers from using standards in an anticompetitive manner is to limit their ability to provide certain types of services, or to limit the way in which they may do so.[94] As we have already noted, however, drawing the line between the provision of basic (transmission) and other services is becoming increasingly difficult. It will become more difficult with the introduction of an ISDN, in which the network itself will contain a substantial amount of intelligence. Economies may be lost, moreover, if such restrictions are imposed. In any event, existing restrictions are being relaxed, so that competition between exchange carriers and independent service suppliers is likely to increase. The result is that these suppliers will remain concerned about where they can obtain access and what the nature and terms of that access will be. Regardless of how ISDN standardization issues are resolved by the CCITT, these issues are unlikely to go away anytime soon.

Open Network Architecture

The Bell operating companies are currently involved in developing open network architecture plans that, if accepted by the FCC, will relieve the companies from some of the restrictions they face in offering enhanced telecommunications services. The commission will require that ONA plans offer users the opportunity to purchase unbundled basic service elements so that users can configure their own telecommunications networks using

93. The requirement for private-line service can also be thought of as providing an alternative interface. It should also be noted here that the pricing of private lines, as well as of competing telephone company offerings, will affect the nature of competition. As Ordover and Willig note, "underpricing" components that compete with those sold by rivals and "overpricing" components that are needed by rivals may be part of a competitive strategy. Thus even if private lines are available, they will not be an attractive alternative to an ISDN if they are very costly. See ibid.

94. This is, of course, the approach taken in the FCC's *Computer II* decision and in the modified final judgment.

as much or as little of the BOC networks as they desire. Two standards issues are raised by these developments. First, what will the basic service elements be? Second, will the BOCs offer standardized ONA plans?

The choice of basic service elements is a standards issue because offering parts of the network on a bundled basis only is economically equivalent to making it impossible for users to connect to the interface between them. As in the case of the ISDN, network users will desire that the elements of the network be offered in small "bundles" so that they can purchase from the BOCs only the portions they want. Other elements will be purchased from other vendors or provided directly by the user, who may be an independent service provider. In contrast, the BOCs will presumably want to offer more aggregated bundles so as to limit the choices of users. Alternatively, the BOCs can make many small elements available but, by failing to standardize their interfaces, may force users to buy more elements from the BOCs than they would desire. For these reasons, considerable controversy is likely to accompany announcement of the ONA plans.

Whatever the ONA plans contain, there will also be the issue of whether they are the same for all BOCs. Although some large telecommunications users have expressed concern that lack of uniformity will increase their costs,[95] there is now no formal mechanism to coordinate the standards that will be used in the various regions of the United States (but see the discussion of the Information Industry Liaison Committee below).

The economic theory of standards suggests three possible reasons why standardized ONA plans may fail to emerge: large differences in preferences among the BOCs, difficulties in coordination among the BOCs even if preferences are similar, and the desire of some or all BOCs to achieve a competitive advantage through the adoption of different ONA plans.[96] If the BOCs do not adopt uniform plans, the outcome may be the absence of a national standard, with slow development of new technologies because users who operate in different regions find employing incompatible tech-

95. M. Betts, " 'Open' Nets Trigger Fears," *Computerworld*, vol. 20 (November 3, 1986), pp. 1, 15, describes a large user who "is worried that the lack of standard protocols will increase cost for equipment, staff expertise and software and complicate operating procedures as well as hamper the diagnosis and resolution of network problems" and quotes the counsel for the International Communications Association, a group of large users, that "we have an overall concern that we may end up with seven separate, incompatible, ONA plans."

96. A particular technology may be favored because it reduces the competition that a BOC faces from suppliers of equipment that compete with equipment offered by the BOC, or because the technology reduces the ability of other suppliers to offer equipment providing services that could otherwise be offered through the network. The stated goal of the proposed European policy to achieve common telecommunications standards among countries is to promote competition. See Commission of the European Communities, "Towards a Dynamic European Economy."

nologies too costly; the simultaneous use of incompatible technologies in different regions, despite the higher costs and lower benefits for users; or the emergence of one technology as the standard through a bandwagon in which those BOCs using other technologies are forced to switch to the standard.

A uniform national standard may fail to develop rapidly if users, uncertain about whether a national standard will emerge and what that standard will be, adopt a wait-and-see posture. If the fear of being stranded with the wrong technology leads to such behavior by a large number of users, "excess inertia" may result.[97] Excess inertia is especially likely if the BOCs have different preferences, but it can also occur if such differences do not exist. Coordinated standard setting might overcome this inertia.

A second possibility is the rapid adoption of incompatible technologies in different regions. This result is likely if there are many customers whose communications are confined to a single region, so that incompatibility is unimportant to them, or if the benefits of using the new technologies exceed translator costs for users who communicate between regions. Note that, although the new technologies develop rapidly in this case, the cost of incompatibility to users—in terms of translator costs or services not used because their benefits are less than the cost of translation—may still be substantial, and the outcome may be less efficient than if there were a common standard.

Third, one technology may emerge as the national standard. This can occur if a bandwagon that is started by early adopters produces changes in the offerings of those BOCs using other technologies. Once again, however, it is important to observe that the winning technology is not necessarily the one that is most economically efficient.

Finally, of course, the BOCs may adopt standardized ONA plans. As we noted early in this study, three conditions seem especially important for such adoption to occur. First, there can be no important differences in the preferences of the various BOCs, which is possible as long as none of the companies has made significant investments in a particular technology. Second, the growth of the market must be highly dependent on the existence of a common standard because users place a great value on compatibility. Finally, the competitive advantages from incompatibility

97. Besen and Johnson, *Compatibility Standards*, pp. 32–60, conjecture that the absence of an AM stereo standard may be responsible for the slow rate of diffusion of that technology by radio stations and listeners. Inertia can also result if the benefits to users are reduced because incompatibility raises their costs.

must be small. If these conditions are met, standards may emerge through agreements among the BOCs. Recently, the Exchange Carriers Standards Association announced formation of the Information Industry Liaison Committee to "act as an ongoing national forum for the discussion and voluntary resolution of ONA issues."[98] Although the committee is not a formal standard-setting body, its presence may still promote agreement on common standards.

We do not mean to suggest that absence of a formal mechanism to achieve national uniformity will necessarily produce inefficient outcomes, or that the existence of such a mechanism will always overcome these inefficiencies. The main lesson of the theory discussed above, however, is that there is no guarantee that uncoordinated standard setting by the BOCs will achieve an efficient outcome and that there are many instances in which it will not. It may be difficult, moreover, to tell even after the fact whether the outcome is an efficient one. The emergence of a common standard and its rapid diffusion are still consistent with the choice of the "wrong" technology.

Conclusion

Two basic lessons can be drawn from the economic theory of standard setting. The first is that, even when everyone benefits from standardization, there is no guarantee that standardization will be achieved or, if it is, that the "right" standard will be chosen. The second is that standards may be used as tools of competitive strategy, with firms either seeking incompatibility or promoting their preferred technology as the standard to gain an advantage over their rivals. Both problems are present whether de facto standardization occurs through the market or voluntary standards are chosen cooperatively.

Not surprisingly, both lessons can be applied in telecommunications. The fragmentation of the industry—among regions in the United States, internationally, or among user groups—may create coordination problems. The central role played by telecommunications carriers may create competitive ones. The examples of ISDN and ONA are only two instances of the growing importance of standards issues in this industry.

Much may be learned about the best way to set standards by observing the performances of the differing approaches to telecommunications stan-

98. "ECSA Sponsoring Information Industry Liaison Committee on 'Open' Network Architectures," *Telecommunications Reports*, vol. 53 (October 19, 1987), p. 15.

dardization being pursued in the United States and Western Europe. In the United States, standardization dictated by AT&T has been replaced by a system in which all participants have substantial autonomy, and voluntary standard setting has become increasingly important. In Europe, by contrast, a system in which each country has had substantial freedom to establish its own standards appears to be changing to one in which countries forgo some independence to obtain the benefits of more rapid technical change.

Telecommunications Policy and U.S. National Security

Ashton B. Carter

THIS PAPER analyzes the ways in which changes in the information industries—deregulation, divestiture of the telephone system, internationalization of the information market, and new technologies—might affect U.S. national security, and the ways in which the U.S. government might therefore seek to influence the evolution of the commercial information sector. National security is taken here in its narrowest sense, that is, military security: the ability of the armed forces, intelligence agencies, and diplomatic agencies to prepare for and to conduct war.

Obviously national security in this narrow sense can come into conflict with economic security. Indeed, such conflicts are a theme of this chapter. Insofar as national security requires the U.S. telecommunications or computer industries to do things they would not do for purely commercial reasons, such as hardening telephone switching equipment or satellites to electromagnetic pulse effects, national defense becomes an added overhead on the U.S. industry. Industry cannot accept this overhead as easily as it has in the past. For one thing, telecommunications companies can no longer absorb these costs into a guaranteed rate base. For another, foreign companies without a large defense overhead, like Japanese companies, might gain competitive advantages over a burdened U.S. industry. To see that its defense needs are met adequately by an industry increasingly absorbed with commercial competition, the U.S. national security establishment will need to isolate the needs that are genuinely vital and technically practical, communicate them convincingly to industry, and pay for more of them directly than it has had to in the past. A major observation in this chapter is that up to now the government has failed to take any of these steps.

On the other hand, there is no inherent reason that defense needs must conflict with commercial needs. The two can be complementary. Defense research and development (R&D) and procurement can invigorate commercial industry. And the defense community is becoming increasingly aware that information systems, even more than weapon systems, hold the key to security—especially where nuclear weapons are concerned. In developing its information systems—called command, control, communications, and intelligence (C^3I) in the jargon—the Defense Department depends for its quality and cost on a strong *commercial* industry as never before in the postwar period. Seeking out and reinforcing points of common military and commercial interest must surely be a key goal of national policy.

Yet military security must always be the overriding imperative. It must therefore be invoked selectively, and never as a political convenience to cover policies that really have other aims. To do otherwise debases the national security coinage. True military needs will get lost if policymakers forget how to sort them out from other policy goals. Using the national security imperative loosely also risks providing its powerful political support to policies that would not stand on their merits if they were analyzed according to their true goals. At this time, the Defense Department follows the commercial field in information technology, yet it has the instincts of a technology leader, forged in past decades. Meanwhile the nation as a whole is firmly committed to military strength yet groping for a politically acceptable industrial policy to secure commercial preeminence. This is a recipe for a marriage. A second observation of this chapter is that the "military imperative" is often a much weaker partner analytically than it is politically, where information industry policy and military policy join.

Military Need for the Commercial Telecommunications Networks

The Department of Defense and other agencies depend on the commercial networks for certain national security telecommunications. These networks are changing under the impact of divestiture, deregulation, and new technology. The changes are both physical and administrative. The federal government has naturally been concerned that the evolving networks meet its needs. Older companies like American Telephone and Telegraph (AT&T), the Bell operating companies, and Western Union have a tradition of cooperating with the government, but newer entrants into the telecommunications marketplace do not. And competition turns all away from public service.

In response to flux in the networks, the federal government created a category of national security emergency preparedness (NSEP) telecommunications policy, lodged mainly in the Department of Defense. NSEP concerns itself chiefly with military contingencies, but it includes natural disasters and sabotage as well. NSEP policymaking is not prominent within the Department of Defense. It falls under the rubric of command, control, communications, and intelligence, which has always been a fragmented and weak party in DOD councils (though there are heartening signs that its strength is gathering). The budget available for NSEP is still meager.

It is often noted that 95 percent of the Defense Department's circuits in the United States are provided by the commercial common carriers,[1] but that figure is not a good measure of national security reliance on the commercial networks. A much smaller fraction of that traffic is critical; the rest is the normal "housekeeping" traffic of a large organization. In a crisis, the volume of critical traffic increases, and the networks may suffer physical destruction. On the other hand, the military provides its own survivable backup systems in case the terrestrial networks are destroyed. The picture therefore changes as the state of military emergency deepens.

One can perceive in the government's NSEP activities four distinct goals. The first and most mundane is to help the Defense Department, as a large user of telecommunications, to manage its affairs in the changing marketplace, since it can no longer rely on the public spirit or monopoly power of AT&T. The second—and the one invoked most often by the department to justify its actions—is to ensure continuity of national military command and support for retaliatory military operations in the aftermath of a massive nuclear attack involving thousands of nuclear detonations on the United States. A third is to rebuild the telecommunications networks after a massive nuclear war as a precondition for rebuilding civil society. A fourth is to support military operations in the midst of and following a limited attack involving conventional weapons, sabotage, or at most tens of nuclear detonations.

These goals are not clearly distinguished in current NSEP programs. Yet they call for quite different technical measures. In general, NSEP planners tend to emphasize military operations conducted by surviving members of the presidential chain of succession after a massive nuclear attack. If the networks could be made able to handle this task, they would a fortiori certainly be able to handle most needs of civil reconstruction and limited war. But it is impractical to harden the networks to this degree. It also is unnecessary, since the nuclear retaliatory forces that would

1. See, for example, John F. Judge and Bruce Gumble, "The Defense Communications Agency: Coping with Changing Times," *Defense Electronics*, vol. 17 (May 1985), p. 232.

respond to massive nuclear attack have narrowband communications systems dedicated to this task that are independent of the commercial networks. Yet NSEP planners know that if such hardening were feasible, it would be their most compelling goal. The other goals—civil recovery from large-scale attack and military operations in limited war—enjoy far less support in Congress, in regulatory bodies such as the Federal Communications Commission (FCC), in industry, and above all in the rest of the Defense Department. What is more, civil recovery and limited war would make demands on the telecommunications system that are quite different from one another and from the needs of nuclear deterrence. Government NSEP initiatives therefore need to be more focused, and more technically persuasive, if they are to deserve support.

Because the common carriers can no longer absorb the costs of meeting DOD requirements into a guaranteed rate base, the government will need to find other mechanisms for meeting its emergency needs. The most obvious choice is direct funding. A precedent exists in the Civil Reserve Air Fleet program, where the department pays commercial airlines to modify the floors and doors of large aircraft so they can supplement military airlift in wartime. At this time, however, NSEP funding is still quite small. Carefully crafted sets of minimal measures for meeting the goals of civil recovery and limited war, which are the most realistic of the four NSEP goals, might be affordable. Direct regulation by the FCC is another option and will be used at least to require carriers to continue restoring critical DOD circuits to service as the highest priority if the network is disrupted. Other, indirect policy levers for the federal government include tax breaks, antitrust protection (necessary if different carriers are to cooperate in NSEP planning), and barter (for example, granting rights-of-way along interstate highways in return for certain routings of transmission facilities).

Early Defense Department Concerns over AT&T's Divestiture

On February 21, 1981, soon after taking over the Defense Department, Secretary Caspar Weinberger wrote to Attorney General William French Smith to object on grounds of national security to the Department of Justice's antitrust case against AT&T. These appeals, while strongly worded, were not specific about the source of the department's concern:

> The purpose of this letter is to express the deep concern which the Department of Defense feels over the reports of the proposed settlement

of the government's old antitrust suit against the American Telephone and Telegraph Company. . . .

Our concern is based upon the fact that a *great deal* of the current capability for communications command and control of our *strategic weapons* depends upon the continued existence of the *only* communications network in the United States capable of providing the services required for the strategic weapons system. . . .

The Department of Defense recommends very strongly that the Department of Justice not require or accept any divestiture that would have the effect of interfering with or disrupting *any part* of the existing communications facilities or network of the American Telephone and Telegraph Company that are essential to defense command and control.[2]

In testimony before the Senate Armed Services Committee in March, Weinberger repeated the same concerns:

The American Telephone & Telegraph network is the most important communication net we have to service our strategic systems in this country. Because of the discussions I have had concerning the effect of the Department of Justice suit that would break up part of that network, I have written to the Attorney General and urged very strongly that the suit be dismissed, recognizing all of the problems that might cause and because of the fact it seems to me essential that we keep together this one communications network we now have, and have to rely on.[3]

Frank C. Carlucci, at the time deputy secretary of defense, in a letter of April 8, 1981, to Assistant Attorney General William Baxter, head of the Antitrust Division, cited "severe problems [that would] confront the Department of Defense if this network is broken up."[4]

What were the "severe problems" foreseen by Defense? They could not literally have been the retaliatory command and control of "strategic weapons," if by these are meant strategic nuclear missiles in silos and submarines, bombers, and cruise missile carriers, and their wartime com-

2. *Departments of Justice and Defense and Antitrust Litigation*, Hearing before the Subcommittee on Government Information and Individual Rights of the House Committee on Government Operations, 97 Cong. 1 sess. (Washington, D.C., 1982), p. 66. (Emphasis added.)

3. *Fiscal Year 1981 Department of Defense Supplemental Authorization*, Hearing before the Senate Committee on Armed Services, 97 Cong. 1 sess. (Washington, D.C., 1981), p. 21. See also George H. Bolling, *AT&T—Aftermath of Antitrust: Preserving Positive Command and Control* (Washington, D.C.: National Defense University, 1983), pp. 51–52.

4. *Departments of Justice and Defense*, Hearing, p. 67.

mand centers and warning sensors. Strategic forces do not rely on the commercial telecommunications network to receive their "go codes" from national command authorities: the commercial network's dependence on a relatively small number of easily targeted switching and control nodes precludes dependence on it for the assured retaliatory capability needed for nuclear deterrence. The Department of Defense maintains several custom-designed mobile radio and satellite systems ranging over the entire radio spectrum to direct strategic nuclear forces.[5]

It is true that the Defense Department relies on commercial common carriers to build and operate many of its telecommunications systems for use in circumstances short of retaliation to nuclear attack. The systems range from the department's Automatic Voice Network (AUTOVON), Automatic Secure Voice Network (AUTOSEVOCOM), and Automatic Digital Network (AUTODIN) for ordinary peacetime traffic to specialized hot line networks like the Strategic Air Command's Primary Alerting System (SAC PAS) and the North American Aerospace Defense Command's (NORAD's) Alerting System. Some of these networks are leased from common carriers, and some are government-owned but operated and maintained by common carriers. These DOD networks are called "strategic" to distinguish them from the "tactical" communications systems used by forces in combat, and from theater networks in Europe, Korea, and elsewhere. This meaning of strategic, and not any connection with strategic nuclear weapons, presumably lies behind Weinberger's usage, which suggests a stronger military imperative. Though the department's reliance on common carriers is extensive, what reason is there to believe that the divested industry could not serve it as well as AT&T did in the past?

Until divestiture, the department had relied on AT&T to provide one-stop shopping for complete, end-to-end provision of telecommunications services.[6] The Defense Communications System therefore had not needed to maintain teams of technically proficient engineers and network managers to plan its networks, to guarantee priority for its communications in emergencies, and to adapt to rapid changes in needs (for example, during presidential trips). In this, the Defense Department was like many

5. For technical descriptions of the nuclear command and control system, see Bruce G. Blair, *Strategic Command and Control: Redefining the Nuclear Threat* (Brookings, 1985), pp. 182–240; and Ashton B. Carter, "Communications Technologies and Vulnerabilities," in Ashton B. Carter, John D. Steinbruner, and Charles A. Zraket, eds., *Managing Nuclear Operations* (Brookings, 1987), pp. 217–81.

6. Where independent (non-Bell) telephone companies provided exchange services, AT&T often served as intermediary with the Defense Department.

large businesses, which found themselves after divestiture suddenly in need of telecommunications managers who could sort through the welter of new companies, products, and services offered by the divested, deregulated industry. The department's problems were perhaps even greater than those of other telecommunications users because of the scale of its use (hundreds of millions of dollars of long-haul business alone every year), its need to avoid even short disruptions in key networks like SAC PAS, and perhaps most important, its need to follow government procurement regulations while dealing with a growing family of service providers.

In retrospect one can see that it was mainly fear about its ability to manage the day-to-day tasks that AT&T had always managed for it, and not a well-founded conviction that a divested telephone system would be somehow inherently incapable of performing crucial military services, that spurred Defense's challenge to the antitrust suit against AT&T. The department stated as much in a 1981 report prepared by the Defense Communications Agency and introduced into the trial by AT&T's lawyers:

> DOD can unequivocally state that divestiture . . . would cause substantial harm to national defense and security and emergency preparedness telecommunications [because it] would substantially reduce, or eliminate entirely the incentives . . . to engage in that prior joint [that is, among different service providers as well as between providers and government] planning and preparation [necessary] to conduct centralized network management. . . . No contractual arrangements can be made to overcome the resulting severe degradation of timely response capability which currently exists.
>
> From our perspective, the most critical element . . . has been the ability of the Government to rely upon the Bell System, as a regulated communications monopoly, to provide the required planning, design, standards, operations and maintenance, reconstitution, and overall network management necessary to assure a high quality, interoperable, redundant, credible, and rapid telecommunications response to all types of emergencies or disasters. . . .
>
> The Defense Department totally disagrees that divestiture would have no adverse effect on the Nation's ability to rely upon the nationwide telecommunications network. Instead, we believe that it would have a serious short-term effect, and a lethal long-term effect, since *effective* network planning would eventually become virtually non-existent.[7]

7. Bolling, *AT&T—Aftermath of Antitrust*, pp. 53–54. (Emphasis in original.)

Judge Harold Greene seems to have recognized that the department's concerns were largely administrative, and also that it would be incapable, at least temporarily, of piecing together the services previously provided in toto by AT&T. The 1982 decree (often referred to as the "modified final judgment," or MFJ) said, "The BOCs [Bell operating companies] shall provide, through a centralized organization, a single point of contact for coordination of BOCs to meet the requirements of national security and emergency preparedness."[8] The judge's opinion of August 11, 1982, addressed the specifics of the single point of contact, requiring BOCs to support it with people and funds, giving it regulatory protection, and guaranteeing its authority within the BOCs. Though the single point of contact referred only to the BOCs, similar mechanisms were ultimately set up by other telecommunications companies. Some government communications systems were exempted from many of the provisions of both the modified final judgment and Computer II,[9] including AUTOSEVOCOM, SAC PAS, the NORAD Alerting System, several Federal Emergency Management Agency (FEMA) systems, the Federal Aviation Administration (FAA) National Airspace System, and the White House Communications Agency.

Meanwhile, the government prepared to create a coordinating mechanism for the telecommunications industry and a liaison office within the government. An executive order of September 13, 1982, created the presidential-level National Security Telecommunications Advisory Committee (NSTAC), comprising the chief executive officers of the large telecommunications companies: American Telephone and Telegraph, Bell Communications Research (Bellcore), Fairchild Industries, Ford Aerospace, International Business Machines (IBM), International Telephone and Telegraph (ITT), Microwave Communications, Inc. (MCI), Martin Marietta, Motorola, Northern Telecom, General Electric, Hughes Aircraft, U.S. Telephone Association, Western Union, and so on, for a total of twenty-seven companies.[10] One of the first issues NSTAC took up was creating a national coordinating mechanism. At the same time, the government promulgated its own policy in several National Security Decision Directives (NSDDs) and executive orders.[11] The upshot of this activity was to

8. *United States* v. *American Telephone and Telegraph Co.*, 552 F. Supp. 131, 227 (D.D.C. 1982), *aff'd sub nom. Maryland* v. *United States*, 460 U.S. 1001 (1983).

9. *Second Computer Inquiry*, 77 F.C.C. 2d 384 (1980) (referred to as "Computer II").

10. Executive Order 12382, "President's National Security Telecommunications Advisory Committee," September 13, 1982.

11. The key directives are NSDD-47, "Emergency Mobilization Preparedness," July 22, 1982; NSDD-97, "National Security Telecommunications Policy," 1983 (NSDD-97 replaced

create a joint industry-government National Coordinating Center (NCC) and to designate the National Communications System (NCS) as the government's own agent for national security telecommunications.

The National Communications System was created in 1963 after the Cuban Missile Crisis had revealed to President John F. Kennedy that the communications networks of different government agencies were often unable to communicate with one another. The manager of the NCS is also the head of the Defense Communications Agency. The NCS has nominal authority to coordinate, though not really to control, the telecommunications resources of Defense, State, Treasury, Energy, CIA, FEMA, FAA, and a host of other government entities. Executive Order 12472 of April 3, 1984, gives the NCS the charter to make and carry out national security telecommunications policy. It is the NCS, working with the advice of NSTAC, that has nominal authority for much of national security telecommunications, and its activities are the focus of much of the following discussion.

The National Coordinating Center was installed in the Defense Communications Agency headquarters building near the Pentagon. The NCC houses a representative of each of twelve major telecommunications companies.[12] The NCC representatives are supposed to coordinate their companies' efforts with those of other companies and with the government's efforts in emergencies. If these relatively low-level representatives cannot get their companies to cooperate to the satisfaction of NCS, there is a well-defined process by which the issue can "escalate" to higher levels within both the government and the companies.

When asked recently about the "fundamental aspects" in which the business of government communications has changed in recent years, J. Randolph MacPherson, chief regulatory counsel on telecommunications for the Defense Department and the NCS, replied that the government must now do its own planning, define its own needs, make its own technical decisions, and work more closely with regulators and contract lawyers.[13] The communications manager of any large company has faced

the Carter administration's famous Presidential Directive 53); NSDD-145, "National Policy on Telecommunications and Automated Information Systems Security," September 17, 1984; and Executive Order 12472, "Assignment of National Security and Emergency Preparedness Telecommunications Functions," April 3, 1984.

12. American Satellite Company, AT&T, Bell Communications Research, Inc. (Bellcore), the Communications Satellite Corp. (COMSAT), General Telephone and Electronics (GTE), ITT, MCI, Pacific Telecommunications, Inc., RCA, TRT Communications Corporation, U.S. Telephone Association, and Western Union.

13. Judge and Gumble, "Defense Communications Agency," p. 239.

the same concerns since AT&T's divestiture. They are entirely administrative. It therefore seems that the national security community's early concerns over divestiture failed to distinguish clearly enough its fiscal and administrative convenience from security imperatives, transient disruption from basic threats, and peacetime operations from critical wartime needs. The breakup of AT&T has not materially disadvantaged national security, at least on the grounds most strongly adduced by the Defense Department during the trial. Presumably transient disruptions will recede as the government develops its own internal technical competence and learns to manage its own telecommunications as well as AT&T used to do it.

Current Preoccupations of National Security Telecommunications Policy

Aside from the task of learning to manage a large number of telecommunications services, what are the goals of national security telecommunications policy, and how might they affect the commercial sector? In particular, are there ways the emerging networks need to be configured to serve national security needs?

The chief preoccupation of those charged with policymaking for national security telecommunications has been the survivability of the network under massive nuclear attack. In the past, the Bell System took steps to reduce the vulnerability of the network, like locating key facilities outside expected target zones and hardening them to electromagnetic pulse. These costs were passed on to ratepayers, instead of appearing directly in the Defense budget. As regulators tightened AT&T's revenues, the Bell System did less and less of this "free" hardening of the system. In a completely competitive environment the carriers can be expected to do little at all. The question now is whether the government is going to demand survivability features of the deregulated networks, and, if so, who is going to pay for them. Since there is predictable disagreement among government agencies and between government and industry about how to allocate the burden of altering the networks for noncompetitive purposes, the government needs to focus on efforts that are important and feasible at reasonable cost.

On the surface it would seem attractive to make use of the telephone system and other commercial networks for critical postattack military functions. With a physical plant valued at more than $200 billion, 19,000 local exchanges connecting 120 million lines to one another, and 725,000 employees, the exchange component alone has assets that are vast even by DOD standards.[14] Moreover, every year some $20 billion is spent upgrading this network, whereas President Reagan's largest initiative to improve strategic nuclear command and control, announced in October 1981, totaled only $18 billion (in 1982 dollars) over a period of six years.[15]

But if the commercial system is to be strengthened to serve *military* as opposed to *civil* needs in wartime, policymakers must be absolutely specific about what the network is supposed to accomplish under each level of hostilities. The price of confusion on this point can be seen by considering NETS—the Nationwide Emergency Telecommunications System, a program of the NCS.

After the AT&T breakup the NCS won the charter not only to plan, coordinate, and cajole, but also to establish spending programs to support national security telecommunications. Through these NCS programs, the government can in principle fund carriers directly to do things that the Bell System might have done under tariffed rates in the past. At the moment these so-called national-level programs are minuscule and few in number. Each government agency that is a member of the NCS is supposed to contribute from its budget in proportion to its interest, with the result that the great bulk of the funding for survivability is borne by Defense.[16] Still, the total amount is in the low tens of millions of dollars. By far the largest of the survivability efforts, accounting for almost all of the budget of these programs, is NETS.[17]

NETS arose from an examination of how the phone system would perform under massive nuclear attack. The public switched telephone system has a hierarchy of switching centers—five layers in all. Using the predivestiture AT&T nomenclature, there are roughly 19,000 class-5 centers, some 950 class-4 centers, 175 class-3 centers, 50 class-2 centers, and only 10 class-1 centers.[18] Today the class-5 centers and some of the class-4 centers would be said to constitute the local exchange component, and the other centers would be interexchange carriers' points of presence

14. United States Telephone Association, *Telephone Statistics 1986* (Washington, D.C., 1986), pp. 2–3.

15. Ibid., p. 7; and "Why C^{3}I Is the Pentagon's Top Priority," *Government Executive*, vol. 14 (January 1982), p. 14.

16. NSDD-201, "National Security Emergency Preparedness (NSEP) Telecommunications Funding," December 17, 1985.

17. The other two national-level programs of the NCS, called the Commercial Network Survivability and Commercial Satellite Communications Interconnectivity programs, fund other technical measures to make portions of the commercial networks usable after massive nuclear attack. They are much smaller than NETS. Many of the issues associated with NETS apply to these other programs as well.

18. R. F. Rey, ed., *Engineering and Operations in the Bell System*, 2d ed. (Murray Hill, N.J.: AT&T Bell Telephone Laboratories, 1983), p. 109.

(POPs) and long-haul network nodes.[19] Though the network is changing rapidly, the old AT&T classification still affords an adequate model for the nationwide public network.

The analysis begins with a hypothesized Soviet nuclear attack aimed at U.S. missile silos, submarine and bomber bases, military installations of all sorts, transportation nodes and other economic infrastructure, key industries, and—crucially—the command centers, warning sensors, and communications links serving the national civilian and military leadership. Most of the class-1, class-2, and class-3 centers are in or near areas that would be targeted by the Soviet Union in a massive attack. Many class-4 and class-5 centers, on the other hand, would survive.[20] The rich telecommunications system could be partially reconstituted if these surviving class-4 and class-5 centers could be reconnected.

The NETS program would add software and hardware "call controllers" to selected class-4 (and possibly a small number of class-5) centers expected to survive attack. Centers chosen to receive call controllers that are not now connected by trunk lines to other chosen centers would be so connected, with the cost borne by the NETS program.

NETS is supposed to serve the national civil and military leadership in managing retaliatory operations.[21] Surviving leaders, if in an area served by one of the NETS class-4 offices, would dial up a call controller through one of the surviving class-5 offices it served. By entering an encrypted password, they would gain access to the call controller, which would attempt to contact other call controllers, establishing a network among the surviving centers. NETS would explore more paths to route a call than the phone system ordinarily explores. Normally, a phone call is not routed through many links because the quality of the connection suffers. Instead, the caller is asked to redial.[22] But NETS would incorporate special features

19. For a comprehensive description of the modern telephone system, see Peter W. Huber, *The Geodesic Network: 1987 Report on Competition in the Telephone Industry* (Washington, D.C.: U.S. Department of Justice, Antitrust Division, January 1987); and Bell Communications Research, *Notes on the BOC Intra-LATA Networks–1986*, Technical Reference TR-NPL- 000275 (Livingston, N.J., April 1986).

20. Lt. Gen. William J. Hilsman, director of the DCA, testified in 1981 that under the massive Soviet attack modeled by his office, almost three-quarters of the Bell system class-4 offices would survive. *Telecommunications Competition and Deregulation Act of 1981*, Hearings before the Senate Committee on Commerce, Science, and Transportation, 97 Cong. 1 sess. (Washington, D.C., 1981), pp. 151–52.

21. On procedures for continuity of government during nuclear war, see Paul Bracken, "Delegation of Nuclear Command Authority," and Ashton B. Carter, "Assessing Command System Vulnerability," in Carter and others, eds., *Managing Nuclear Operations*, pp. 352–72, 555–610.

22. These routing limits are being relaxed as digital transmission pervades the networks.

to improve the quality of voice transmission and would route calls over any path that was physically intact. NETS could carry ordinary clear voice, digital data, or digital encrypted voice at 2.4 kilobits a second.[23] In this way national leaders, including successors to the presidency, could learn of one another's existence and location, receive intelligence on what had happened, order retaliation, and try to put an end to the war.

Under this scenario, the NETS class-4 offices are *assumed* to survive because they are not located near valuable military targets. But there is a tautology involved in declaring a military asset "survivable" simply because it is outside the target zones of current interest to the enemy. Presumably if NETS became a truly valuable asset, the Soviets *would* target it. In the scenario hypothesized for NETS, there would be little reason for them not to. This scenario postulates a determined attack aimed at eliminating the United States as a military opponent and avoiding, to the extent possible, retaliation by U.S. strategic nuclear forces against the Soviet Union. Urban areas come under heavy attack. That is why the class-1, class-2, and class-3 centers and the non-NETS class-4 centers are destroyed in the first place. Once the threshold into nuclear violence on this scale was crossed, the Soviet Union could hardly expect the United States to show much restraint in its own retaliation. Thus Soviet targeters devising this attack plan would go all-out to achieve the express objective of disrupting communications from U.S. national leaders to the military forces. The only limits on the Soviets' ability to nullify NETS would be physical, not strategic.

The Soviet Union possesses roughly 10,000 strategic nuclear weapons capable of striking the United States. The number of class-4 offices ultimately to be furnished with NETS call controllers is unknown, since the project is scarcely under way and will suffer from the usual budget pressures.[24] The number cannot be more than the number of class-4 offices located outside the hypothesized "target areas," that is, about 500. Costs, including the need to connect centers that are not already connected with trunks, could bring the number considerably lower. What is more, only some of the NETS nodes, perhaps half, would need to be destroyed to render the network useless or to chop it into small, disconnected islands. The effect of NETS is thus to create a new target set of about 100 locations

---

Regenerators of digital signals eliminate noise in retransmission, whereas analog amplifiers merely pass noise on to the next node.

23. NETS itself would not provide encryption. The call controller would encrypt the telephone number being called, however, to keep the whereabouts of the national command secret.

24. Planning numbers, for what they are worth, are classified.

on top of the thousands of targets already struck in this massive attack. Though the identities of the class-4 offices provided with call controllers would be kept secret as long as possible, Soviet targeters could probably devise an adequate strike plan by studying old AT&T network maps. Barring very deep cuts in the superpower strategic arsenals as a result of arms control, NETS will be neutralized the moment it achieves, in Soviet eyes, significant capability to aid U.S. postattack military operations. Locating telephone facilities outside "target areas," like hardening them to collateral blast damage, belongs to an era when the Soviet Union had many fewer strategic nuclear weapons than it has now.

Despite the richness of the telephone system, it would not be an easy matter to make it work with any confidence after a massive nuclear attack. The prospects seem dimmer still when one considers that the phone system receives its electrical power from the national grid, which itself would be targeted.[25] And the NETS class-4 centers, class-5 centers, and their interconnecting transmission plant are all susceptible—to a degree unknown but worrisome—to electromagnetic pulse damage.

Fortunately for Defense planners, the key *military* capability called for in this circumstance of massive attack is rather rudimentary—simple and massive retaliation by surviving U.S. forces in an entirely prearranged operation triggered by a low-volume traffic of "go-codes." The Defense Department operates a wide range of custom communications systems for this purpose: UHF, SHF, and soon EHF satellite communications; HF, LF, and VLF broadcasts; and UHF and VHF line-of-sight radio. These systems are designed, within the limits of technical feasibility, for high-confidence low-data-rate communications under the most hostile circumstances. It is difficult to see how NETS can compete with them in performance. Surviving remnants of the phone network, aided by NETS, might provide an uncertain backup to the dedicated radio and satellite systems upon which nuclear deterrence rests. In this role they might add further to the Soviet Union's doubts about its ability to blunt a retaliatory strike. NETS would in small measure further raise, by about 100 warheads, the "attack price" the Soviet Union must pay to destroy U.S. military communications in a major attack and would thus reinforce the gulf between a massive nuclear war like the NETS scenario and a truly "limited" nuclear war of incomparably smaller scale. The *necessity* for the Soviet Union to strike key nodes in the commercial phone network as part of any attempt to "decapitate" U.S. nuclear weapons, with all the destruction

25. Some telephone facilities could run for a short period on generator power.

of civil society such a strike entails, is one reason that such an attempt is tantamount to all-out war. And the remnants of the telecommunications system, including NETS, would also be crucial to survivors attempting to rebuild civil society. But giving the commercial telecommunications network a *military* role in a *massive* nuclear war is impractical. Similar arguments of technical practicality apply to private networks, satellite networks, and other commercial systems. Yet the scenario of massive attack lies single-mindedly behind virtually all the national security telecommunications programs of the National Communications System.

Instead of attempting to contribute to retaliatory operations after large-scale nuclear attack, national policy would do better to focus on situations where the damage to the network is much smaller—including crises, conventional war, terrorism, or even limited nuclear wars—and possibly also on aiding the recovery of civil society in the aftermath of a major nuclear attack. These two goals lead to quite different, and much more practical, demands on the network than those envisioned in current NCS programs. These goals will also be taken more seriously within the national security community, within industry, and in Congress.

Telecommunications Needs for Limited Nuclear War and Other Emergencies

The notion that strategic nuclear war might be waged at levels much lower than the NETS scenario over a period of weeks or months is controversial. Because massive nuclear war cannot serve any rational definition of military or political purpose, it is hard to see how one could get started through the deliberate actions of rational leaders. It seems much more plausible, at least insofar as rational decision processes are still at work on both sides, to envision leaders deliberately resorting only to limited and selective use of nuclear weapons in pursuit of military advantage, probably in the course of a war that is already under way with conventional weapons. Over the past two decades, U.S. strategic nuclear doctrine has become increasingly attracted to the abstract logic of contingency planning for limited, protracted nuclear operations.

Unlike the NETS scenario, which envisions a relatively crude and massive retaliation, truly limited contingencies would call for much more from the military and civil telecommunications systems. These systems would need to coordinate limited retaliatory strikes without accidents or loss of control that could ignite all-out war; deter a massive attack by continuing to maintain the unused bulk of the strategic nuclear arsenal at high readiness; continue to prosecute the conventional war from which nuclear war erupted; and communicate with allies and opponents to end the war before it escalated further. Though these telecommunications tasks are much more demanding than the NETS scenario's doomsday retaliatory strike, the national networks carrying them out would be—by hypothesis—largely intact.

It is not hard to grasp why preparing for limited nuclear war is controversial. Skeptics doubt whether the onset of nuclear war *would* be dominated by rational calculations of military purpose and national interest. They foresee instead miscalculations, error, the pace of events, and the fear of escalation itself creating pressures on both sides to hurl the limited nuclear war out of control. They are also not reassured by the fact that the mission of limited strategic nuclear operations is taken seriously mostly by a small group of civilians and is far from being backed up by realistic plans, the needed flexibility in command and control systems and arsenals, or an accepting attitude by the military commands that operate the strategic arsenal. A sign of ingrained military skepticism about the need to plan realistically for truly limited war is the tendency for military planners (as distinct from civilian theorists) to apply the term "limited war" to attacks that are not all-out but still massive, and "protracted war" to the months following such attacks. Skeptics worry that, at a minimum, such a caricature of limited war places impractical and wasteful demands on nuclear forces, diverting attention from more realistic ends. At worst, planning for limited strategic nuclear war only opens another door to massive nuclear war rather than showing a practical possibility of truly purposeful use of nuclear weapons.

Suppose, whatever its plausibility, that a truly limited strategic nuclear war *does* begin. It is then clearly desirable for the national telecommunications networks to function as normally as possible to support the war effort, including the central government's efforts to direct and control U.S. forces and to seek a negotiated end to the fighting. Unlike the NETS scenario, any attack that deserves the adjective "limited" allows one to assume almost all of the network would survive. Thus, for example, Washington, D.C., and key switching, network control, and transmission facilities in major cities would be spared. Damage would be isolated and local, not pervasive. Just as it defined target zones in the NETS scenario, the NCS would need to make detailed and specific targeting assumptions corresponding to each of these hypothetical limited attacks.

For example, a scenario popular among civilian theorists, and influential among proponents of a Strategic Defense Initiative system to intercept ballistic missiles even if it cannot defend population, posits a Soviet attack on the ten or so ports and airfields on the East Coast through which the United States would resupply the NATO theater.[26] With ten detonations, or one one-thousandth of its arsenal, the Soviet Union could temporarily cripple the allied war effort. Attacks such as these, which are limited but have high military leverage for the Soviet Union, are at least as likely, according to these theorists, as massive attacks like the NETS scenario, and national policy must plan for them. The degree of network "survivability" sought here is much less than in the NETS scenario: the network should remain functioning in areas not directly struck, and telecommunications service in the target areas should be reconstituted as soon as possible. The network should be able to route calls around target areas; it should not rely on a small number of key nodes whose destruction would disrupt service over a wide area; it should give preferential treatment to defense-related traffic; and it should be resistant to electromagnetic pulse.

Survivability of this kind is essential for other types of national emergencies: natural disasters (floods, hurricanes, tornadoes, earthquakes, and incidents like nuclear power plant accidents that involve evacuation or quarantine), civil disturbances, vandalism, terrorism, sabotage, or nonnuclear attack. Steps to improve network survivability for these other emergencies would probably receive wider support than they would receive for the scenario for limited nuclear attack, which must surely appear rather exotic to most people. Notwithstanding its preoccupation with nuclear war, the NCS defines the category of national security and emergency preparedness telecommunications to encompass all of these emergencies, and the NCS has responsibility for planning for all of them. The telecommunications carriers take many pertinent survivability measures as a matter of operational routine. Some effects of a limited nuclear attack on East Coast targets, for example, would bear a bizarre resemblance to network operations on Mother's Day morning, when a burst of mid-morning calling on the East Coast overflows north-south trunks, forcing network controllers to route north-south traffic through midwestern centers that have not yet grown busy.[27]

Trends after divestiture in the resilience of the network to limited damage are mixed, some improving survivability and some detracting from

26. See, for example, U.S. Department of Defense, *Ballistic Missile Defenses and U.S. National Security: Summary Report*, prepared for the Future Security Strategy Studies (Washington, D.C., October 1983).

27. Rey, ed., *Engineering and Operations*, p. 180. The wave of morning calls moves across the time zones, followed by a second wave in midafternoon.

it. In general the networks are becoming more intelligent and hence more adaptable to sudden change. They are also growing physically more extensive and more diverse. Yet in some cases the intelligence is concentrated in a few nodes. An important example is the Common Channel Interoffice Signaling (CCIS) system, a separate communications system overlaid on the phone system. "Signaling" refers to the nonvoice communications within the network that set up, route, and terminate calls, deliver dial and busy tones, and so on.[28] In the past signaling was accomplished on the same communications circuit that carried the telephone conversation itself. Increasingly, signaling is accomplished over separate circuits like CCIS. A signaling network might have far fewer nodes than the communications network it controls. The result could be an overall system more vulnerable than its predecessor to destruction of a few offices. Competition might also result in a fragmented network consisting of pieces owned by different carriers that could not be interconnected in emergencies. The points of presence, or POPs, at which all the exchange carriers must provide their services within each Local Access and Transport Area (LATA) concentrate network vulnerability. Overall, it is difficult to generalize about whether trends in the network make it more or less survivable, and assessments depend on a host of engineering details in each LATA.

The NCS has undertaken a number of measures to increase the survivability of the public switched network. Though these measures tend, like NETS, to be couched in terms of survival under massive nuclear attack, they are actually practical only for limited attack and for other limited emergencies.

RESTORATION PRIORITY. Since 1968 the NCS and the Federal Communications Commission have jointly administered a program requiring common carriers to give priority to restoring key communications like the SAC PAS if they are interrupted. The U.S. government does not compensate the carriers for these services, drawing its authority from the Communications Act of 1934. The program has been restricted, however, to intercity dedicated private line circuits. The existing system for priority in restoration of service is widely recognized as obsolete. As FCC rule-making for the postdivestiture restoration priority system approaches, the carriers are attentive to its technical, operational, and cost burden.

AVOIDING DEPENDENCE ON A FEW NETWORK NODES. The federal government is not willing, or able, to require network engineering for survivability through FCC regulation. Thus it is unclear how measures

28. Digital networks also require time synchronization.

to make the emerging networks more survivable will be funded. A strong possibility is that government procurement specifications will require carriers providing essential services to national security agencies to take some or all of the following measures:

—Provide multiple, prearranged routings for priority national security circuits like SAC PAS. It is possible that this can be accomplished with the kind of network controls against congestion and blocking that already exist in the carrier networks.

—Distribute signaling and network control over a large enough number of offices that there are no "Achilles' heels" in the network.

—Be capable of interconnection with other carriers in an emergency. Where interexchange carriers share the same point of presence, for example, they could be connected without adding expensive trunking.[29]

—Adhere to federal standards. The NCS administers the U.S. government's program for telecommunications standards, establishing uniformity among all government agencies. Generally, the NCS follows industry standards. The National Bureau of Standards administers federal computer standards.

—Provide transportable switches, microwave relays, and other equipment to serve as backup for destroyed nodes.

—Provide backup power and physical security at network nodes.

HARDENING TO ELECTROMAGNETIC PULSE. EMP figures among the "cheap shots" that might cause extensive disruption of the networks without extensive physical destruction to the United States. The most important type of EMP to the terrestrial telecommunications system would be caused by a nuclear detonation at high altitude (above 40 kilometers).[30] Such a burst would cause no direct blast or heat damage to the ground below. But gamma rays from the burst would knock electrons out of air molecules, and as the electrons turned under the effect of the earth's magnetic field, they would generate powerful radio signals in the frequency range from extremely low frequency to very high frequency. The pulse would extend over the part of the earth's surface visible from the burst point, allowing a single detonation at 300 kilometers to blanket virtually the entire continental United States. Conductors exposed to the pulse would act like antennas, collecting the radio energy, transforming it into potentially destructive currents, and conveying it into electronic equipment. At the same time it produced the EMP, the detonation would distort the

29. The U.S. government has also taken steps to interconnect the Defense Department's AUTOVON and the government-wide Federal Telecommunications System (FTS).

30. Carter, "Communications Technologies and Vulnerabilities," pp. 273–75.

earth's magnetic field lines. Long cable runs would develop voltages in this changing magnetic environment in much the same way as armature windings in an electric generator. These voltages would induce additional harmful currents in telecommunications equipment.

The amount of disruption EMP would cause in the telephone network is uncertain. The strength and frequency spectrum of the pulse are not fully known, since experimental detonations in the atmosphere are forbidden by treaty. Switching and transmission components have in many cases not been exposed even to simulated pulses in test facilities. Even if the pulse characteristics and component vulnerabilities were known, it would be difficult to calculate the effects on the network as a whole. The electrical power system, and not the telecommunications network itself, might be the weakest link. Lightning protectors, supplied by carriers as a matter of course at repeaters and at ports where cables enter central offices, might offer some protection against some EMP effects, although the pulse is much more rapid than lightning. Fiber-optic cables, which do not conduct current themselves, nonetheless might have conducting members along their length for physical strength that would collect EMP energy. Some EMP protection is part of good engineering practice and is incorporated into normal equipment design.[31]

EMP would probably cause more intermittent disruption than permanent damage. Data processors "hung up" by the pulse would need to be restarted, relays tripped by the pulse reset, and so on. Since the purpose of NSEP policy with regard to limited nuclear war is to keep the telecommunications system working in support of the ongoing war effort, transient disruption is probably acceptable, provided it is brief. Because the nation will not, by hypothesis, have suffered extensive physical destruction, work crews can be expected to make rapid repairs.

Even transient EMP disruption is troublesome in one specialized but quite important context: the early minutes of a large-scale nuclear attack. Soviet intercontinental ballistic missiles take a half hour to arrive at the United States, but submarine-launched missiles could cause EMP within only five minutes of the start of the attack. Unfortunately, the strategic forces need to take some key steps in the intervening twenty-five minutes, and these steps rely in part on the terrestrial communications system, which would be physically intact during this time. First, messages warning of the attack need to travel to command centers from the sites that process

31. For example, AT&T's 5ESS digital switch has been exposed to electromagnetic pulses in simulators, and circuits found particularly vulnerable have been "hardened" in normal production models.

radar and satellite sensor data. Next, orders to take off must go out to bombers, aircraft-carrying cruise missiles, tankers, airborne command centers, and airborne radio relays before they can be destroyed on their runways. Finally, the United States has stated that it wishes to preserve its ability to launch silo-based missiles under attack, though this step is obviously dangerous in view of the risk of erroneous launch. For all these reasons, the circuits that serve the Strategic Air Command, the North American Aerospace Defense Command, and other key parts of the nuclear command system need EMP protection that prevents even *transient* disruption.[32] Presumably the Defense Department will allow the costs of hardening such circuits to be included in its contracts with the carriers.

Recovery from Massive Nuclear Attack

As already noted, military operations in the aftermath of massive nuclear attack would be best achieved with the Defense Department's dedicated command and control systems, not the commercial networks. The only remaining NSEP policy goal for the circumstance of massive attack is aiding the "recovery" of civil society. Many NCS initiatives are better suited to this mission than to aiding military operations, though they are frequently couched in war-fighting terms. Civil defense is the province of the Federal Emergency Management Agency, leading to some tension between that agency and the NCS.

From the outset civil defense planning confronts large and ineradicable uncertainties, ranging from nuclear effects to human behavior. In the face of these uncertainties, many Americans have questioned the value of making preparations for recovery, seeing such measures as trivial and hopeless against the background of large-scale destruction and so many unknowns. The NCS could therefore find it difficult to garner support in the telecommunications industry for civil defense plans, let alone for technical modifications that are not fully funded by government. It is perhaps for this reason that the NCS sometimes tries to associate its initiatives with strategic military objectives rather than with civil defense.

The telecommunications network is a basic infrastructure like the transportation, fuel, food, water, and power systems. The network therefore has logically a high priority in any plan to reassemble the rudiments of material life after a massive nuclear attack. Priority among users clearly

32. The Defense Department is constructing a special purpose low-frequency radio system called the Ground Wave Emergency System (GWEN) to perform these vital "transattack" functions, lessening reliance on the common carriers.

ought to go to public officials coordinating recovery. The needs of these officials are rather different from the needs of national command authorities coordinating retaliatory operations. National commanders need nationwide (and worldwide) narrowband communications in the hours after attack.[33] Recovery focuses on the needs of *local* officials coordinating recovery in their own states, counties, and municipalities in the days, weeks, and months after attack. A National Research Council study of national security telecommunications policy emphasizes that recovery is primarily a local task, not a federal task.[34] Much of the federal government's role is providing technical guidance to local government in the way that FEMA provides guidance for general civil defense planning. The main efforts under way in FEMA and NCS are:

THE EMERGENCY BROADCAST SYSTEM. The Emergency Broadcast System was established in 1964 to give the president a means of communicating with the populace if war seems imminent, presumably to urge cooperation in civil defense measures like evacuation. The system should also function under limited attack and should be capable of reconstitution after massive attack. It consists of 9,500 participating radio and television stations linked by commercial carrier. The carrier network, together with some other FEMA networks, is exempt from many of the provisions of the modified final judgment and of Computer II.

ORDERWIRE NETWORKS. In NSC planning the first step after massive attack is the establishment of low-capacity "orderwire" networks to allow surviving officials to learn of one another's existence and to exchange traffic necessary to plan reconstitution. For example, the NCS has noted that U.S. government agencies own about 1,200 high-frequency radio stations in the continental United States alone. The NCS is making a directory of these stations and their operating frequencies and devising procedures for passing critical traffic among them after attack. It might also enlist the 12,000 amateur radio operators ("hams") in the effort.

RECONSTITUTION OF THE TELEPHONE SYSTEM. The NCS urges common carriers to designate evacuation sites for key technical personnel, to establish caches of supplies away from target areas, and to store backup copies of network maps, operating software, and equipment inventories.

33. Retaliation would be most efficient if ordered and executed within hours of a massive attack, since the command and control systems and some of the weapon systems would not function for much longer. See Carter, "Assessing Command System Vulnerability," pp. 573–95, 605–10.

34. National Research Council, *The Policy Planning Environment for National Security Telecommunications: Final Report to the National Communications System* (Washington, D.C.: National Academy Press, 1986).

These measures for recovery from massive attack, unlike measures for limited nuclear war and other limited emergencies, have little overlap with the carriers' normal safeguards against network disruption. The FCC will apparently not mandate such measures; they will be voluntary for the carriers or included in contracts for government circuits. The National Coordinating Center has a plan to relocate in emergencies from its offices near the Pentagon to a site outside Washington.

CREATIVE USE OF OTHER TELECOMMUNICATIONS SYSTEMS. The NCS is studying the use of a host of other systems besides the switched terrestrial networks for postattack recovery: satellite television for emergency broadcast, modified cable television networks for two-way communications, cellular radio, the private landline and microwave systems owned by the railroads and electric power utilities, and so on.

A particularly interesting example is provided by domestic communications satellites. The United States would probably be unable to build and launch communications satellites for years after a massive nuclear attack, so all reconstitution would have to be done with satellites on orbit at the time of attack. Since the current Soviet antisatellite vehicle cannot reach geosynchronous orbit in which U.S. communications satellites reside, the satellites would not be directly threatened. Nonetheless, nuclear detonations at high altitudes or in space might damage them. Direct radiation from the bursts reaching the satellites could cause ionization in electronic circuits and resulting current surges.[35] Space bursts would also create artificial radiation belts that might be harmful to satellites. Military satellites are designed to withstand these effects, but commercial satellites are not. The NCS has considered funding commercial satellite owners to incorporate such hardening, since the owners are unlikely to do it for commercial reasons. Hardening would serve the goal of limited nuclear war as well as recovery from massive nuclear war.

Assuming the satellites survive physically, the next concern is the telemetry, tracking, and control (TT&C) uplinks that send stationkeeping (and other routine "housekeeping") commands to the satellites. The satellites do not need frequent commands, so they would remain useful for some time after attack even if their TT&C stations were destroyed. Backup or mobile stations could be built for survivable TT&C. All commercial satellites belong to one of two families: cylindrical satellites built by Hughes, and winged satellites built by RCA.[36] Any surviving Hughes

35. This effect is called system-generated electromagnetic pulse (SGEMP) to distinguish it from ordinary EMP.

36. The Hughes satellites spin about their line of symmetry, the other two axes remaining

TT&C station could control all the Hughes satellites, and an RCA station could control RCA satellites, though the two families are incompatible. TT&C links on some satellites are encrypted (with government help) to prevent a hostile party from sending the satellites disruptive, false commands.

If the satellites survive and are on station, all that remains is for earth stations to communicate through them. There are hundreds of satellite earth stations in the continental United States that use these satellites, though the stations are not hardened against EMP. Some would not survive the attack physically. Many of the surviving earth stations could point their antennas at any one of the satellites in that part of the geosynchronous orbital arc allocated by the FCC to domestic satellites. The next question is whether surviving fixed earth stations could communicate with one other (or with truck-borne mobile stations) through the satellites. The commercial satellites operate either at C band (6 gigahertz uplink, 4 GHz downlink) or at K band (14 GHz up, 12 GHz down), and the military satellites operate at X band (8 GHz up, 7 GHz down). Ground stations designed for different frequency bands cannot easily be modified to communicate with one another. The transponders on all C-band satellites use exactly the same bandwidths and center frequencies. But Hughes and RCA satellites use different multiplexing schemes, so Hughes C-band ground stations can communicate with each other via Hughes satellites but not via RCA satellites, and vice versa. At K band, interoperation is limited even within the same company because of differing transponder bandwidths and center frequencies. In practice, then, Hughes and RCA C-band earth stations can communicate with other C-band stations designed for the same company's satellites, but interoperability is otherwise impractical. Within these constraints, some communication among surviving satellite ground stations after a massive nuclear attack is technically feasible. Government telecommunications policy for the recovery from nuclear attack would aim to establish the plans and make the minor technical modifications necessary to allow satellite communications to participate in national recovery for as long after the attack as the satellites could be kept on station.[37] The government would have to provide trunks from

fixed with respect to the earth by TT&C command; they are called "two-axis stabilized" spacecraft. The RCA satellites do not spin, and all three of their axes remain fixed with respect to earth; they are "three-axis stabilized."

37. Jamming of communications transponders and TT&C links is a remaining problem. Presumably, Soviet jammers in the Western Hemisphere having line of sight to the domestic satellite arc would themselves come under attack by American forces in the course of the U.S. retaliation.

the satellite ground stations to toll switches of the terrestrial networks in order to use satellites to bypass interrupted cable and microwave trunks.

Mobilization

Like other industries, telecommunications participates in planning for war mobilization. The president draws clear authority from the Communications Act of 1934 to direct the industry in time of war. Initially, mobilization would change the priority assigned to various services by the carriers, and eventually network construction and equipment manufacturing would also be shaped by wartime needs. Planning, rather than tangible and costly steps like maintaining excess manufacturing capacity (as is done in some other military-related industries), is all that is contemplated at this time. The National Security Telecommunications Advisory Council has established a Panel on Telecommunications Industry Mobilization to analyze "surging," or quickly increasing, services to government; surging the manufacture of military equipment; stockpiling vital equipment; mobilizing key repair and managerial personnel; coordinating with other infrastructures such as electrical power and transportation systems; and untangling the host of involved federal, state, and local jurisdictions.

The Global U.S. Military Mission

With all the attention in the NCS and FCC to divestiture and deregulation in the domestic telecommunications industry, it is easy to forget that the worldwide military mission of the United States poses important telecommunications problems far from American shores. The United States has long-standing commitments in Europe and the Pacific and growing commitments in the Middle East and Latin America. Strategic nuclear war involving the U.S. homeland would probably be the culmination of a long and intense conventional or even nuclear war in these theaters. Military, diplomatic, and intelligence operations all depend to some degree on local telecommunications systems. Telecommunications in Europe and Japan are undergoing structural changes analogous to the AT&T breakup and deregulation in the United States. The same problems that plague the NCS in the U.S. homeland recur in these theaters. In fact, they are probably worse. Foreign industry and regulatory bodies are likely to be less cooperative in national security planning than are U.S. carriers. Overseas communications rely on a relatively few undersea cable shore points and satellite ground stations. Armed conflict in these theaters is more likely than strategic nuclear war, and the number and variety of possible contingencies are great compared with the relatively stylized "set-piece" strategic nuclear war. These needs are increasingly serviced by an internationalized information industry without the close ties of American companies like IT&T to the U.S. national security establishment, making more difficult such sensitive activities as assured diplomatic communications or access to international traffic by the National Security Agency.

Technology Policy and National Security

While the National Communications System concerns itself almost exclusively with the national security obligations of the commercial telecommunications network, elsewhere the Department of Defense is taking a much broader look at the relationship between military and commercial information technologies. The watchwords here are "spinoff" and "dependency." Spinoff has traditionally referred to a process by which technologies developed for defense find their way into the commercial sector. The word itself suggests that this process is both automatic and one-way, from defense to commerce. But both these features can be questioned today. Dependency signals anxiety that the United States' forty-year monopoly on advanced technology is ending and that the Defense Department is having to choose between fielding the very best technology or being independent of foreign supply.

Underlying both the spinoff and dependency issues—and causing considerable analytic confusion—is the fact that the United States has no federal technology or industrial strategy. American political traditions do not easily accommodate such central planning in economic life. It challenges both free-market laissez-faire *and* fair-market antitrust regulation. Curiously, technology policy thrives in state governments. But the federal government is as yet unable to give political legitimacy to overt technology policymaking.

In a pattern established after World War II, almost all federal funding for research and development is made in pursuit of a specific social goal, passing through "mission agencies" like the Defense and Energy departments and the National Institutes of Health. The National Science Foundation, the only agency whose "mission" is science per se, funds basic research neglected by the mission agencies and supports science and engineering education. There is no mission agency charged with advancing the nation's commercial technological welfare, and the National Science

Foundation has no traditional role in applied or commercial technology. Efforts to create an American cabinet-level department analogous to the Japanese Ministry of International Trade and Industry (MITI), creatively dubbed DITI, fell prey immediately to the argument that it was not the American way.

Industrial and technology policy thus tends in the United States to be accomplished in "bootleg" fashion through the Department of Defense. Recently, this tendency has exhibited itself under the banner of "competitiveness," but the tradition is much older. For example, the Sputnik shock of 1957 allowed not only science and technology, but education and even interstate highways, to be justified on national security grounds. Appeal to defense to accomplish broader technological aims has some compelling attractions. Though overt industrial policy is immediately challenged, it is widely accepted that the United States should be militarily unsurpassed. Moreover, about 70 percent of federal R&D spending, accounting for about a third of the nation's total innovative effort (public plus private), passes through the Department of Defense.[38] And the department's efforts are technically broad, technically sophisticated, and well managed by federal government standards. It is tempting to turn the nation's broader technology goals also over to Defense.

Since the Defense Department plays a shadow role in the nation's technology strategy, it follows that not everything it does can be fully justified on the basis of national security. Inevitably, some of the security "imperatives" adduced as the basis for DOD spending on technology are exaggerated, if not entirely fabricated. Though these fictions are created with the best of intentions, and no doubt often with the best of outcomes, they could prove dangerous if the nation's true economic or military interests get lost in the fog. Conflating economic and military security makes it hard to sort out where the two compete and where they complement one another. Yet sorting out the two is especially important in the information industries today, where both security concerns are pressing.

Spinoff

The spinoff tradition derives from the successful stimulus DOD spending gave to aircraft, computers, and microelectronics in the 1950s and

38. National Science Board, *Science Indicators: The 1985 Report* (Washington, D.C.: National Science Foundation, 1985), pp. 218, 226. Despite the sharp upturn in Defense Department R&D spending in the past eight years, it only recently surpassed, in real terms, the level it reached in the mid-1960s. Furthermore, this spending takes place against the background of a much larger GNP and technical community than in the 1960s. Ibid., p. 39.

1960s.[39] But a growing body of opinion questions whether these models apply to defense technology overall today, and especially to information technologies.[40] One supposed factor is a growing divergence between defense and commercial technologies. Military components need thermal, mechanical, and radiation hardening; ultrareliability; and other features not needed by commercial components. Military systems like ELF, VLF, and LF radio, and features like jam-resistance, also have little commonality with commercial systems. Sometimes commonality seems to arise because the Defense Department is following trends in commercial technology rather than asserting its own, rather different, needs. For example, the new MILSTAR (Military Strategic and Tactical Relay) satellite system incorporating advanced communications technology is perhaps less well suited to its mission of surviving massive nuclear war than a constellation of more numerous, hardened, simple transponders in lower orbit would be. But a rugged and crude system lacking the latest technology would simply fail to attract the support of either the satellite industry or military technologists. More important than divergences in technical substance might be the engineering atmosphere in which military technology develops. Military technologists, this argument goes, have become accustomed to designing to performance and not to cost and schedule, making them poorly adapted to commercial work. Development cycles as long as a decade are common for DOD systems but would be lethal for commercial firms. Many companies eventually choose to do either military or commercial work but not both, and companies that do both sometimes segregate their different divisions.

Secrecy is also a barrier between the defense and commercial worlds. Bell Laboratories delayed informing the Department of Defense of the discovery of the transistor until shortly before they made a public announcement for fear the department would classify it.[41] Sterner controls

39. See Richard C. Levin, "The Semiconductor Industry," and Barbara Goody Katz and Almarin Phillips, "The Computer Industry," in Richard R. Nelson, ed., *Government and Technical Progress: A Cross-Industry Analysis* (Pergamon, 1982), pp. 9–100, 162–232. David C. Mowery, "Federal Funding of Research and Development in Transportation: The Case of Aviation," paper presented at a Workshop on the Federal Role in R&D sponsored by the National Academy of Sciences, November 21–22, 1985.

40. Nathan Rosenberg, "Civilian 'Spillovers' from Military R&D Spending: The American Experience since World War II," paper presented at the Conference on Technical Cooperation and International Competitiveness, Lucca, Italy, April 2–4, 1986; and Jay Stowsky, "Beating Our Plowshares into Double-Edged Swords: The Impact of Pentagon Policies on the Commercialization of Advanced Technologies," Berkeley Roundtable on the International Economy Working Paper 17 (University of California, Berkeley, April 1986).

41. Levin, "Semiconductor Industry," p. 58.

on information and exports, should they continue to appear, could well further segregate military from commercial technology.[42]

A crucial factor in the early development of the semiconductor industry was strong and predictable demand for components by the Defense Department (and NASA). But since the 1960s the department's purchases have gone from being nine-tenths of U.S. semiconductor sales to being less than a tenth.[43] In fact, the other federal agencies together spend slightly more than Defense on information technology.

Above all, when it comes to information technology, Defense is in general a technology follower, not a leader. The department's very high speed integrated circuit (VHSIC) program acknowledges this fact, aiming to make some crucial military circuits the equal of their commercial counterparts in processing capability, size and weight, and power consumption. The long DOD development cycles virtually guarantee a follower status in all areas of technology that have commercial commonality.

Economic analyses of spinoff usually turn on what other economic activity defense R&D displaces. If it displaces no other technical activity elsewhere in the economy, that is, if defense does not take its "slice of the pie" at the expense of commercial R&D, but only makes for a bigger overall pie of national R&D, then the spinoff appears positive. But if the defense spending is compared with an equivalent amount spent directly on commercial technology in the commercial sector, implying dollar-for-dollar displacements, then it appears detrimental. Actual DOD technology efforts are a mixture of positive and negative spinoffs. Indirect effects of military spending are also important: to the extent that defense technical activity bids up the price of engineering talent in the commercial sector, it is also harmful. Economic analyses that average many sectors and many technologies obscure the diversity of effects in pursuit of a single bottom line. They can also lead to embarrassing results for the spinoff model, such as the conclusion that R&D-intensive federal programs do not create more income multiplication or jobs than income redistribution programs

42. Department of Defense, *The Technology Security Program*, report to Congress (Washington, D.C., 1986); Committee on Science, Engineering, and Public Policy, Panel on Scientific Communication and National Security, *Scientific Communication and National Security* (Corson Report) (Washington, D.C.: National Academy Press, 1982); and Committee on Science, Engineering and Public Policy, Panel on the Impact of National Security Controls on International Technology Transfer, *Balancing the National Interest: U.S. National Security Export Controls and Global Economic Competition* (Allen Report) (Washington, D.C.: National Academy Press, 1987).

43. Stowsky, "Beating Our Plowshares into Double-Edged Swords," pp. 48–49.

do![44] About 90 percent of defense R&D is development, not research, making dollar comparisons with the spending of industry and other government agencies misleading.[45] Development would generally be expected to yield less spinoff, dollar for dollar, than basic or applied research.

For all these reasons, spinoff should be assessed case by case. Industries differ greatly, the differences arising from history as well as technology. Aviation has been nurtured on defense spending, whereas the chemical industry, which is also research-intensive, receives almost no government support. The information industries are somewhere in between, coming from the aviation direction and heading in the direction of the chemical industry. Within the information industries, the diversity is also great, from microchips to software development and from satellite transponders to satellite launch vehicles. Analytically well-founded programs to increase spinoff are likely to be piecemeal: they will arise from selecting technologies of genuine common interest to both sectors and managing them within the Defense Department in a style congenial to the commercial sector. As for spinoff from the commercial to the defense sector, the department will need to shake off its ingrained conception that it is the technical leader and that commercial technology will follow in the paths it sets.[46] Instead, it will need to shape its habits and even its systems to the commercial trends if national defense is to enjoy the products of firms at the technical cutting edge.

Dependency

Not only is unquestioned technical preeminence slipping from the defense sector into the commercial sector, but it is diffusing worldwide as well. The fact that the United States must share the technical frontier, while unaccustomed and certainly unwelcome, is inevitable. It feels strange only because American habits of mind were shaped after a war that left the rest of the industrialized world prostrate. Europe and Japan were destined to make progress in relation to the United States, though they

44. Such outcomes in studies of NASA spinoff have been noted by NASA's former chief economist, Henry R. Hertzfeld, "Measuring the Economic Impact of Federal Research and Development Investment in Civilian Space Activities," paper presented at a Workshop on the Federal Role in R&D sponsored by the National Academy of Sciences, November 21–22, 1985.

45. *Special Analyses, Budget of the United States Government, Fiscal Year 1989*, p. J-9.

46. An interesting case where Defense remains a technology leader is information security, ranging from cryptography to preventing tampering with computer systems. Industry's interest in information security is growing to safeguard sensitive records and communications from exploitation by employees, by competitors, and by foreign governments promoting their nation's commercial interests.

are not necessarily bound to pull even with it or to surpass it. Indeed, one cannot regard the position these allies now enjoy as anything but the successful result of a deliberate American policy in the postwar years to rebuild their economies. Yet the fact that the best technologies are not all made in the United States has consequences for national defense, and these consequences are only now becoming clear.

Concerns for these consequences are articulated, for example, in a very explicit 1987 report of the Defense Science Board.[47] The impetus behind the so-called Augustine report is the Japanese capture of almost the entire market for the 256-kilobit dynamic random access memory (256 K DRAM), a basic commodity semiconductor chip. The Augustine report views the Japanese capture as a herald for the passage of the entire semiconductor manufacturing industry to Japan, despite a temporary U.S. retention of its lead in design-intensive custom logic chips and microprocessors. The argument envisions a slippery slope: when high-volume production of commodity chips like DRAMs moves offshore, then specialty components, upstream materials (like silicon) and tools industries, downstream product industries, and even the education of semiconductor engineers will all follow. The Augustine report recommends several remedies, mainly a Semiconductor Manufacturing Technology Institute funded in large part by a $1 billion DOD commitment. The institute would involve industry, academia, and government in developing process technology for volume semiconductor manufacturing, aiming specifically at establishing dominance in 64-megabit DRAMs.

What, in concrete terms, would "semiconductor dependency" mean for military security? The Augustine report gives no guidance on this point, but there are several possibilities. One is that Japan would not side with the United States in a major war, that the war would last long enough to require manufacturing of weapons after expenditure of existing stocks, and that the United States would be unable to make all the kinds of equipment it needs. A second possibility is that Japan might oppose some future U.S. policy, for example, a war of the Vietnam type or deployment of some high-profile, controversial system like an antisatellite or missile defense weapon or nuclear weapons in the Pacific theater. Japanese political forces opposed to the policy might restrict high-tech exports to the United States until the policy was changed. A third possibility is that Japan

47. Office of the Under Secretary of Defense for Acquisition, *Report of the Defense Science Board Task Force on Defense Semiconductor Dependency* (Washington, D.C.: Department of Defense, February 1987). The task force was chaired by Norman R. Augustine, president of Martin Marietta Corporation.

would withhold the latest technology from American defense contractors (who depend on merchant suppliers for their components) until it had been incorporated into Japanese export products. Military systems would then not be at the cutting edge of technology. Japanese suppliers might decline to produce some military components, like radiation-hardened chips, at all. A last possibility is that Japan might sell leading technology to the Soviet Union, or might fail to police exports to the Eastern bloc to the satisfaction of the United States.

Though these possibilities are serious, they need to be placed in context. Most fundamentally, U.S. national security commitments in the Pacific and Persian Gulf are in part for Japan's benefit. In a sense, Japan is the reason for some of America's most challenging military missions, and without its ally the United States would be without many of its commitments as well. Modern wars might well be too short and intense for new manufacturing to make a difference. Japan already has many ways to show dissatisfaction with U.S. policies, and the United States has many ways to show its own annoyance at Japanese disloyalty. U.S. weapon systems already undergo such long development cycles that they do not incorporate cutting-edge technology; until this problem is solved, little additional military penalty would be incurred if the Japanese withheld new components for one or two years. On the other hand, if the Japanese withheld new components from U.S. *commercial* users, the consequences could be serious for American competitiveness in the downstream product lines that use the chips.

The Augustine report's proposed fix to dependency is narrow. Even if the silicon processing takes place in the United States, dependency persists if the packaging and assembly take place offshore and in nations at least as likely as Japan to cut off American supplies. And so extensive are the webs of international commerce that no one quite knows the existing American dependence on foreign suppliers for all sorts of other military equipment. Foreign companies may do manufacturing in the United States, and many U.S. companies have factories abroad under the physical control of foreign governments.

Last, one can question whether the Defense Department is the right platform from which to try to rescue an entire commercial industry. The department accounts for only 10 percent of the U.S. semiconductor end-use market. But 35 percent of the department's combined R&D and procurement budget goes to sophisticated electronics.[48] Defense thus has a

48. Ibid., pp. 31, 49.

strong interest in shaping the industry to its needs. But the industry cannot survive if the department's needs diverge too far from commercial needs in, for example, reliability testing.

The Augustine report's lack of military analysis about dependency is telling. It reveals an effort that is as much, or more, national industrial policy than national security policy. It is a brave attempt to revive a vital American industry using the successful spinoff tactics of the 1950s and 1960s. It wisely incorporates insights of more recent vintage to increase the efficiency of spinoff: focusing on commercialization, not just research; stressing process as well as product technology; guaranteeing a DOD market; and allowing key technical decisions to be made in industry, not the Pentagon. It may well succeed.

But insofar as dependency is a genuine security concern, and not a cover story for industrial policy, the United States will need to make a much more comprehensive study of its many forms of dependency. It must identify the situations in which dependency could seriously affect military security. And it must look at a wider menu of solutions than remaining preeminent in all technologies. Above all, the United States needs to adapt its security policies to the inevitable end of its postwar monopoly of technology. It is fortunate that its main opponent has far more serious problems with technology. But until these analyses are done, it is impossible to justify such a highly selective industrial policy on *military* grounds.

Pressures for Change in Global Markets

International Telecommunications in Transition

Eli M. Noam

THE TERM *international telecommunications* encompasses a complex matrix of countries and issues. It is possible to view international telecommunications as merely a hodgepodge of national systems, each reflecting its society's history and economics, and each happily self-contained except for collaborations on technical issues. It is equally possible to see national developments as variations on a single global theme, inexorably driven by an underlying technology that is destiny, such as the convergence of telecommunications and computers. Or one can argue, as this article does, that a common development is changing all the institutions of international telecommunications today: the rent-seeking coalition that provided links of shared economic interests across frontiers is steadily breaking down. In this light the turmoil of telecommunications should be understood as nothing more than a normalization—one of the most tightly controlled sectors is becoming more like the rest of the economy, not necessarily deregulated but more "normal."

For a long time the traditional arrangement was remarkably stable, successful, and undisputed. For a number of years, however, it has been subject to forces of disintegration. While it was at first possible to dismiss changes as policy initiatives of a conservative American regime, later events in Britain, Japan, the Netherlands, Denmark, and other countries suggest that broader forces are at work. And now as third world countries such as Pakistan and Malaysia, and even a mainstay of the traditional system such as Germany, are seriously contemplating change, a trend can be discerned. I have discussed the causes of this change and its regulatory implications elsewhere.[1] The purpose of this paper is to provide a factual analysis of the international scene, survey the battles, link them to the

1. Eli M. Noam, "The Public Telecommunications Network: A Concept in Transition," *Journal of Communication*, vol. 37 (Winter 1987), pp. 30–48.

broad forces of change, and observe the defense strategies of the traditional institutions.

Origins of the Traditional Network

For almost a century the key institutional feature of traditional telephony around the world has been a ubiquitous network operated by a monopolist. The operating entity was usually a government administration known generically as a PTT (post, telegraph, and telephone administration). The United States split these three functions among three near-monopolists: American Telephone and Telegraph, Western Union, and the U.S. Postal Service.

Public telecommunications were not merely a technical system, but social, political, and economic institutions. One must go back to their origins, to the emergence of European postal monopolies in the sixteenth century. Much later the monopoly system was rationalized as based on technical economies of scale, strategic necessity, cross subsidies, or public infrastructure needs, but the early creators of the postal monopoly were quite forthright in their primary mission—to make profits for the state and its sovereign.[2] The postal system was a major source of revenue, just at a time when absolutist European rulers had insatiable needs for money. This goose with its golden eggs was ardently protected through the centuries against encroachment by private competitors and by other states.[3] When the telegraph and later the telephone emerged in the nineteenth century, they were rapidly integrated into the postal monopoly system and guarded by the same protective policies. Together they became the PTTs.

The PTTs were supported by a broad political coalition, a "postal-industrial complex." It included the PTT itself and the equipment industry as its supplier, together with residential and rural users, trade unions, the political left, the newspaper industry (whose postal and telegraph rates were heavily subsidized), and affiliated experts. The system worked in no small measure to the benefit of the equipment industry. The PTTs through their huge procurements, especially after World War II, provided large markets for the industry. Even better, buy-domestic policies substantially protected these markets from foreign competition and production. Within most advanced countries, domestic equipment manufacturers

2. Heinrich von Stephan, *Geschichte der Preussischen Post von ihrem Ursprunge bis auf die Gegenwart* (Glashütten im Taunus: Verlag Detlev Auvermann, KG, 1976; originally printed in Berlin: Unveränderter Nachdruck der Ausgabe, 1859).

3. Martin Dallmeier, *Quellen zur Geschichte des Europaischen Postwesens, 1501–1806*, pt. 1 (Verlag Michael Lassleben Kallmunz, 1977).

often collaborated with each other in formal or informal cartels that set prices and allocated shares of the large PTT contracts.

Political Télématique: The New Ideology of the Postal-Industrial Complex

For a long time the mission of the postal-industrial complex in developed countries centered on achieving universal penetration of basic telephone service. With this goal largely reached in effective collaboration of PTTs and industry, a new organizational ideology needed to be articulated, both to instill a sense of purpose internally and to legitimize the continuation of the institutional regime externally. An expression of the new direction was the influential 1978 Nora-Minc report commissioned by the French government.[4] This report broadened the range of telecommunications issues to encompass vital questions of national technological capabilities and sovereignty. It concerned itself at length with France's lack of control over the industry of the future, electronics. IBM, the electronics paragon, was viewed as a threat to French sovereignty by its control over technology. "As a controller of networks, the company would take on a dimension extending beyond the strictly industrial sphere; it would participate, whether it wanted to or not, in the government of the planet. In effect, it has everything it needs to become one of the great world regulatory systems."[5]

How does a government deal with "one of the great actors on the world stage"? The growing interaction between computer technology and telecommunications, what the authors termed *télématique,* provided the answer. Governmental influence over the computer industry was limited, but the industry's overlap with telecommunications—over which the state traditionally had tight control—provided the government with a lever of power. Governments need to "strengthen their bargaining position with a solid mastery of their communications media." Importantly, this needs to be coordinated with other governments because "the difficulty lies even more in the fact that no country can play that role alone."[6] This political analysis, which may be described as *political télématique,* became extraordinarily influential. PTTs embraced its notions, which assigned to them a central role in high-technology policy and in the preservation of

4. Simon Nora and Alain Minc, *The Computerization of Society: Report to the President of France* (MIT Press, 1980).
5. Ibid., p. 72.
6. Ibid.

the national interest against America (and later Japan). The equipment industry was similarly supportive, since political télématique notions created a presumption in favor of government subsidies and technological protectionism as a matter of national sovereignty.

Political télématique's defense of the telecommunications monopoly was carried forward in a series of lengthy articles assessing U.S. deregulation. Published in the influential French daily newspaper *Le Monde,* these articles described the United States as engaged in two wars: a military war against the Soviet Union and an industrial war against Japan. The advanced technologies of computers and communications were viewed as vital factors in both battles.[7]

To win this international war the United States deregulated and divested AT&T. At first glance this move may be surprising. "Why smash this power [AT&T] in the middle of a war against Japan?" The answer, said *Le Monde,* is that "deregulation of communication in the U.S. has as its main goal to give American industry a good 'kick in the pants' in order to get it to start a conquest of the rest of the world."

Given such energizing effects of deregulation, one would expect the United States to prefer to be the sole custodian and beneficiary of such a deregulated system. Nonetheless, *Le Monde* viewed the United States as proselytizing the rest of the world; opening the American equipment market to foreign imports was really part of a U.S. export offensive. International liberalization would give the United States several advantages. It would reduce the communications costs of its internationally active firms and pry open the European equipment market. Once "liberated," European telecommunications would be captured by American firms. "Would not abandonment of state control over communications cause them to fall under the control of IBM?" Having posed the issue in such a way, the analysis advocated domestic restrictions and international agreements. Liberalizing change was viewed as a profound threat to French and European economic and sovereignty interests; energetic containment was recommended.

Forces of Disintegration

Political télématique is colored by a Spenglerian pessimism about the ability of major European countries—despite their proud scientific and technological traditions, well-functioning research and development infra-

7. Eric le Boucher and Jean-Michel Quatrepoint, "La Guerre Mondiale de la Communication" (four parts), *Le Monde,* January 11–14, 1984.

structures, sophisticated users, and large financial markets—to succeed in the electronic field against rivals in the United States and Japan. It marshals new arguments in the service of an old cause, the preservation of a state-controlled monopoly in telecommunications.

And yet, for all its political strength, the traditional system has been subject to forces of disintegration. Technology is one of the reasons, though one should not exaggerate its contributions. A chief driving force has been the phenomenal growth of user demand for telecommunications, which in turn is based on the shift toward a service economy. Information-based services, including headquarters activities, emerged as a major comparative advantage of developed countries. These activities were reinforced by productivity increases in information transactions through computers and advanced office equipment. Consequently, electronic information transmission—that is, telecommunications—became of ever-increasing importance. It also became a major cost item. Specialized managers emerged whose function was to reduce telecommunications costs for their firms and enhance their internal infrastructure, and who for the first time established sophisticated telecommunications expertise outside the postal-industrial coalition. Once users on a large scale emerge, it becomes easier and more desirable for them to organize in a political pressure group of their own.

Traditional PTTs provided standardized and nationwide solutions, carefully planned and methodically executed. In the old days sharing a standardized solution was generally acceptable to users because the loss of choice was limited and outweighed by the benefits of the economics of scale gained. But as the significance and diversity of telecommunications services grew, this balance shifted, providing the incentive for private and group network solutions.

The globalization of commerce also created forces of centrifugalism. If one country's PTT exercises restrictive policies, its firms will be disadvantaged internationally, and foreign firms may choose not to locate there. Similarly, acquaintance with options available elsewhere creates pressures for change across borders. For example, in country after country the international electronic funds transfer network Swift (Society for Worldwide Interbank Financial Telecommunications) was able to force PTTs to change their rules on group networks as applied to Swift operations, or else the country's banks would have been left out.

For satellite transmission in particular, the marginal cost with respect to distance is low. Communication flows can be routed in indirect ways to circumvent regulatory barriers and restrictive prices. Arbitrage becomes possible and with it the incentive for a country to become a "communications haven" by liberalizing its regulatory regime. This undermines the stability of administratively set rules for prices and service conditions.

The Technology Gap of the Traditional System

Meanwhile, traditional telecommunications firms were also losing their hold. Insulated from competition and secure in their profits, they were not particularly successful in technological terms relative to their resources. Almost all missed out on the development of computers or failed to stay on the industry's leading edge. Siemens A.G. in Europe and several Japanese firms are the main exceptions. This lack of success came despite major national efforts and subsidies.

The development of microelectronic components illustrates how the traditional equipment manufacturers fell behind technologically, and how they permitted the emergence of a "second" electronic industry that is now undermining them. Contrary to popular belief, the American advantage in electronic component development did not result from European devastation in World War II. The war had provided an impetus for innovation in Britain, Germany, Italy, and the Netherlands, and though many production facilities were destroyed, the technical know-how remained. European firms were as advanced in tube technology as their American counterparts, and they were doing sophisticated research in solid-state technology, such as the work that led to the development of semiconductor diodes.

In late 1947 the transistor was invented at Bell Laboratories, and its superiority soon became apparent. The large, established telecommunications suppliers moved into transistor manufacturing, and although the Americans had a head start, European companies managed to keep up with the new developments. N.V. Philips (with its various European subsidiaries), Siemens, AEG-Telefunken, Plessey, Ferranti, GEC, and Lucas were all doing quite well.[8]

But in the next stage of microcomponents—integrated circuits—different market structures evolved on the two sides of the Atlantic. The new technology was based on silicon instead of germanium, and on planar fabrication, which made mass production easier. And it made possible substantial component integration within one chip. In the United States these innovations were met not so much by traditional manufacturers as

8. Franco Malerba, *The Semiconductor Business: The Economics of Rapid Growth and Decline* (University of Wisconsin Press, 1985), p. 62.

by new firms. Furthermore, American computer manufacturers themselves went into component production. In Europe, much of the development of the new technology was left to the traditional manufacturers who were larger and slower to innovate than their counterparts in America—and later, Japan.

The integrated circuit period lasted from 1959 until the 1971 beginning of a new stage—large-scale integration (LSI) and microprocessors. Very large-scale integration (VLSI) began in the early 1980s. During the LSI period, American firms, mostly nonexistent or hardly known before 1945, were dominant. By then European public policy had focused on microelectronics and encouraged finished goods producers such as telecommunications and consumer electronics firms to integrate vertically into component manufacture. Government development projects provided investment funds. On the whole, however, none of these efforts significantly challenged the Americans and the Japanese. With the advent of the VLSI stage, Japanese firms took the lead in mass manufacturing of components.

Many attribute European firms' poor showing during the 1970s and 1980s to their lack of research and development funds. However, a survey found that research and development expenditures by the European computer and component industry were about $3.7 billion in 1982, compared with about $1.7 billion by Japanese firms and $4 billion to $5 billion by U.S. firms.[9] A high-technology and telecommunications specialist for the Organization for Economic Cooperation and Development (OECD) observed: "Per unit of output, and especially exports, the R&D spending of [European] high technology firms—notably that part of it financed by public money—vastly exceeds that of its trading partners. Whatever the cause of Europe's difficulties may be, it is not that too few resources are devoted to R&D."[10]

The Emergence of the "Second" Electronics Industry

Given the importance of electronics, elements of an independent computer and component industry evolved in most developed countries, forming a "second" electronics sector in contrast to the established telecommunications supply firms. In Europe, Nixdorf Computers and Ing. C. Olivetti are probably the best known among them. These firms are

9. Jim Kraus, "EEC Computer Manufacturers Tie R&D to Compete with U.S., Japan," *Electronic News,* September 10, 1984.

10. Henry Ergas, "Can Europe Catch Up? Exploding the Myths about What's Wrong," *Financial Times,* June 26, 1985, p. 15.

used to direct relations with the users without the mediation of the PTTs. They are an element of the new coalition that is challenging the postal-industrial complex—the alliance of large service users with the second electronics sector in a "services-information coalition." Examples in the United States are American Express, IBM, Time, TWA, Silicon Valley firms, and Citicorp. The traditional system has been defended primarily by AT&T—not enough to stem the tide. Hence the victory of the services-information coalition over the traditional forces was inevitable in the United States; Judge Harold Greene and antitrust chief William Baxter in the AT&T case merely fixed the details of a historic trend.

In Britain the new coalition was slower to gather, and the defense of the traditional industrial sector was more tenacious and ideological. But the balance of power swung in the 1970s. The British electronics industry was not successful internationally, particularly once one subtracts the United Kingdom's former colonies as a market. GEC, Plessey, and Standard Telephone and Cable (STC) were solid performers, but they were not successful in mass production of novel technology. British service industries such as banking, insurance, trading, publishing, and media were doing well, however. London is the preferred European headquarters of non-European firms and, along with New York, is a major center for international services. The Thatcher government advocated the deregulation of telecommunications largely to make British high technology more competitive, but the most important effect was to help make London the convenient center for European business transactions. Given its traditions, Britain was comfortable and familiar with this role.

In Japan the telecommunications equipment industry transformed itself into the new information industry better than anywhere else, and the changes were smoothest. Reform was accomplished as a continuation of industrial policy. Nippon Telegraph and Telephone (NTT) was privatized; and competition was introduced, under the prodding of the Japanese Ministry of International Trade and Industry (MITI), and without the public conflict that occurred in the United States and Western Europe.

In several other European countries the service sector is politically weak compared with manufacturing. French banks have long been nationalized and do not play the same role in international business as do London banks. At the same time the industrial sector has been a particular darling of the political left. This can be partly explained by a traditional socialist emphasis on the production of goods with its proletarian connotations, in contrast with the more middle-class-based, white-collar service activities. It also reflects the electoral base of socialist parties in the working class and the trade union movement. And the emphasis on high technology fits neatly into France's traditional concern with national autonomy, which appeals to the political right, too.

Besides the external challenges, the traditional coalition weakened internally where its constituent parts began to redefine their advantages. A good example is the Netherlands. The 1985 report of the Steenbergen Commission led to a functional separation of the PTT's telecommunications activities into the basic network "social services" of the PTT, the PTT, and a "competitive services" complex of PTT and private suppliers, which included user group networks, value-added networks, and complex terminal equipment.

Perhaps the most significant aspect of the Dutch reform was the attitude of the PTT and its labor union. The PTT concluded that some change was in fact in its own self-interest. In particular, younger managers preferred the greater independence possible outside of the government civil service. The PTT labor union, for its part, concluded that wages, salaries, and especially pensions would be improved by a switch to an independent corporation status. Under the old system, employees were paid as civil servants and tied to the pay scale of the entire bureaucracy rather than of the electronics industry.

The Dutch example shows that a transition can be smooth when it is not enacted as a PTT-busting measure, but rather when the PTT embraces it as an opportunity. In Japan, too, the trade union Zendentsu went along with privatization in order to uncouple from the lower pay scale of the civil service. The union gained above-normal increases after privatization, while achieving job security in the reform legislation. Indeed, as the example of British Telecom will illustrate, invigorated and entrepreneurial PTT successor organizations may gain more power than before—and create a whole new set of problems.

Liberalization and Industrial Policy in the United Kingdom

Margaret Thatcher won the 1979 general election with the slogan "It's Time for a Change" and took a personal interest in applying it to telecommunications.

In its advocacy of reform the government emphasized industrial policy. It hoped to influence the structure of the telecommunications industry on the assumption that structure determines conduct, which in turn affects performance, the classical paradigm of industrial organization economics. The five distinct elements of government policy were (1) a formal separation of telecommunications from the Post Office and establishment of British Telecom (BT) as an independent but regulated entity, (2) establishment of competition in services by permitting rival carriers and value-added network services, (3) privatization of the public network by selling a majority of British Telecom, (4) liberalization of the market for peripheral equipment, and (5) establishment of the regulatory body Oftel.

Thatcher and her advisers' plan of restructure was based on two partly conflicting policy goals: to encourage the service sector and to reverse the decline of British technological leadership. The British share in the world market of telecommunications equipment had fallen from 25 percent in 1960 to 5 percent by 1980. The government wanted to encourage industries with a future—electronics, information, biotechnology—industries that were, conveniently, closer to the interest of Tory followers than to traditional smokestack firms. A minister of state for industry and information technology, serving under the secretary of state for industry, was named, the first such appointment anywhere.

On the whole, Britain's encouragement of the service sector has succeeded, particularly in conjunction with other steps such as liberalization of financial services. It has proved more difficult, however, to change things for the electronics industry. Despite efforts to provide a competitive environment and public money, the electronics industry has continued to slide. Ironically, one factor that has pressured the industry is the increased cost consciousness of British Telecom and of the Ministry of Defense, by far the largest customers of electronic equipment.

The traditional British electronics firms were not doing well. At Standard Telephone and Cable, the chairman and chief executive, Sir Kenneth Corfield, resigned in August 1985, partly because Britain's major computer firm, ICL, which STC had acquired in 1984, performed poorly. STC also experienced problems in its traditional telecommunications equipment market. Meanwhile, STC's parent, International Telephone and Telegraph Corporation (ITT), was itself in need of money and reduced its ownership share successively. Eventually, ITT brought its telecommunications interests into the French-dominated Alcatel N.V., with the notable exception of STC, which therefore was left potentially stranded without a strong technology supplier.

The other main telecommunications equipment makers, Plessey and GEC, had their own problems. Development costs for their flagship "System X" digital switch were much higher than expected, production was delayed, and export orders were not forthcoming. In 1985 Plessey's profits fell dramatically, and the company also had to carry the losses of its American acquisition, Stromberg-Carlson.

The response of the ailing firms was to seek a merger that would further reduce the already limited competition. In December 1985 GEC made a takeover offer of Plessey of about $1.7 billion. Plessey rejected the offer and counterproposed to take over all of GEC's System X digital switch operations. Each company argued that the other was attempting to end duplication.

Meanwhile, the smaller and newer semiconductor and computer companies were also performing poorly. Acorn Computers was largely taken over by the Italian firm Olivetti. Sinclair Research had to sell out. Inmos, Britain's most important developer of semiconductor technology, was an acquisition target for AT&T. For industrial policy reasons, a British "white knight" was promoted in the form of Thorn-EMI, which purchased the government's stake. Soon Thorn-EMI itself went through turmoil, with Inmos a major money loser.

Effects of Liberalization on End-Users and on British Telecom

If the traditional equipment industry has not greatly benefited yet from the reorganization of British telecommunications, who has? The answer is, not surprisingly, large service industry users and, much more than expected, British Telecom itself. The latter observation suggests a future general weakening of the alliance between network operators and equipment industry.

As a supplier, British Telecom became much more sensitive to its customers, particularly its business customers. Examples of improved service include a speedup in installation of private lines in the business district of London. Within the company, independent profit centers were established to control performance. Management employment contracts began to include performance clauses, and at high levels were limited to three years.

The United Kingdom's liberal telecommunications policy and low international telephone rates attracted traffic. One large user, the Ford Motor Company, set up its European communications center in the United Kingdom partly because of its operating flexibility. In 1986 about 40 percent of all North American private-line traffic to the European continent was routed through the United Kingdom.

In the meantime rates increased for small users; the rules established in the British Telecom license were to limit aggregate price increases on domestic calls for a five-year period to the rate of inflation minus 3 percent. In the first year after the license, long-distance call charges were increased for five types and reduced for four. Overall, the weighted average price increase was 3.7 percent. For low-volume users, rates went up 7.1 percent and for high-use residents, 5.7 percent. For a "moderately high" business user, telecommunications rates went up 2 percent.[11]

The following year, the inflation rate was 2.5 percent, which therefore required a price roll-back by British Telecom in absolute terms, somewhat mitigated by a small accumulated credit from the previous period. Overall rate reductions totaled 0.3 percent. But residential rental rates increased 3.7 percent, ahead of inflation.

Movements toward Vertical Integration

The strength of the postal-industrial complex derived from the strength of its members, and this strength existed even after liberalization. It would be naive to expect a newly reorganized telecommunications organization to strive only for an improvement in its efficiency. It finds itself in a double bind: if it takes seriously the exhortations to entrepreneurialism and expands, it is criticized for power grabbing and unfair competition; if it fails to embark on new activities, it is dismissed as hopelessly stagnant.

In Britain, British Telecom began to pursue several avenues of expansion by vertical integration. In the equipment field it purchased the Canadian private branch exchange (PBX) manufacturer Mitel. This led the regulatory body Oftel, and nearly the entire British equipment industry, to argue that the acquisition would not be in the public interest because it would strengthen British Telecom's power in terminal equipment and would threaten British PBX manufacturers. The Monopolies and Mergers Commission, though acknowledging the problem, accepted the merger with some tough conditions attached. But even those conditions were waived by the government when it approved the acquisition.

Another attempt at vertical integration by British Telecom was in advanced services. In June 1984 British Telecom and IBM/U.K. Ltd. announced their intention to establish a joint value-added network services (VANS) venture for data network management service, and they applied for a license. The plan set off strong domestic protests. About a hundred computer and communications companies registered their opposition. Some

11. "British Telecom's Price Changes," Oftel press notice, December 16, 1985.

were concerned with the reliance on IBM's Systems Network Architecture (SNA), while others feared the linking of two dominant firms in closely related markets. In the face of such pressure, the British government rejected the application but left the door open for either company to offer such services on its own.

British Telecom is not unique in seeking to expand its market power vertically in an (old) AT&T-like fashion. In Spain the telecommunications monopoly Compañia Telefónica Nacional de España (CTNE), which is partly private, has increasingly been involved in manufacturing. It holds a large interest in Alcatel's Standard Electrica, which is by far the largest electronics manufacturer in Spain, and owns a majority of the stock of twelve equipment firms and minority interests in seven others. Their aggregate output accounts for about a third of total Spanish telecommunications production.[12] CTNE also linked up with AT&T in semiconductor manufacturing, with Corning Glass in fiber optics, and with Fujitsu Ltd. in microcomputers.

Vertical integration also occurred in Sweden and Italy. In Italy the predominant telephone carrier Società Italiana per l'Esercizio Telefonico (SIP) is not run as a government administration but is largely owned by the Società Finanziaria Telefonica (STET), a company controlled by the government holding organization Instituto per la Recostruzione Industriale (IRI). STET also owns several major manufacturing firms, including Italtel, the country's largest telecommunications equipment firm, and several leaders in semiconductor components and robotics. In Japan the newly privatized NTT formed within a year almost seventy subsidiaries or new ventures, which have only begun to be active in new products, services, and marketing.

These instances of vertical integration by the network operators indicate how liberalization can transform the relationship between state PTTs and private equipment firms from one of partnership into one of conflict or, alternatively, can lead to an AT&T-style vertical integration of the two.

Liberalization of Services: The New Generation of Telecommunications Carriers

Liberalization has led to carriers outside the traditional PTTs. These can be either operators of new facilities, as in Britain or Japan, or enhancers

12. U.S. Department of Commerce, National Telecommunications and Information Administration, *Telecommunications Policies in Ten Countries: Prospects for Future Competitive Access,* NTIA-CR85-33 (Washington, D.C., 1985), pp. 131–45.

of regular PTT transmission capacity, as in the case of value-added service networks. To meet the challenge, the traditional networks have been upgrading their technology, but they have had increasing difficulty keeping together the traditional international cartel in the face of arbitrage, resale, breaches in cartel solidarity, and new international links.

Mercury and Japan's New Common Carriers

The British government encouraged three major companies—Cable and Wireless (C&W), British Petroleum Company (BP), and Barclay's Merchant Bank—to form an alternate long-distance carrier. Although this consortium, Mercury Communications Ltd., was modeled on the American MCI Telecommunications Corporation as a competitor to AT&T, there were great differences. MCI was an entrepreneurial maverick that entered the market by opposing federal authorities and prevailing in court. Mercury was born with three silver spoons in its mouth and the government as its godparent. (Within a short time, however, Mercury became wholly owned by C&W.) Mercury is less a response by entrepreneurs to the market than a government blueprint. The Conservative government staked the credibility of its telecommunications program on the effectiveness and survival of this particular enterprise. In 1982 it made Mercury the only licensee for the foreseeable future, thus giving it a monopoly on competition.

Mercury's permanent license, granted in 1984, permits it to run, install, and operate an independent national telecommunications system for at least twenty-five years. The license is similar to British Telecom's, but with several important differences. Mercury does not have to fulfill British Telecom's universal service obligations, and thus it need not operate a national system.

Mercury quickly established a microwave network within London in 1983. It later constructed a figure-eight fiber-optic trunk system centered in Birmingham. Full-scale operations started on May 15, 1986, with long-distance rates about 15 percent to 20 percent lower than British Telecom's, despite that firm's anticipatory tariff reductions. Mercury's goal is a 5 percent market share by 1990, but obviously a much larger share of large-user business. Price advantage is not the only reason why large users are likely to allocate part of their use to Mercury. Another is simple diversification. In a country as prone to strikes as Britain, dependence on one supplier seems unwise.

Mercury's main contribution in its initial phases was to lead British Telecom to reduce its long-distance rates. For example, anticipating Mercury to be a potential competitor, British Telecom cut its prices to North America up to 20 percent in mid-1985.

As in the case of MCI in America, the issue of interconnection with the British Telecom network was central. In 1985 Oftel decided on a framework for the interconnection, largely in favor of Mercury. British Telecom must provide Mercury with local interconnections at both ends of a telephone conversation, and the compensation that Mercury must pay British Telecom will be set, in the absence of agreement, by regulation.

The issue of fair interconnection is complicated. In the United States it led to two decades of dispute and was a main cause of the AT&T divestiture, when the Justice Department, and with it Judge Greene, concluded that one could not expect genuinely nondiscriminating access by a local monopoly to its long-distance competitors. The divestiture established the principle of complete and equal access and of user choice of a "primary long-distance carrier." The issue of the cost for such access to the local network, whether by AT&T or its competitors, precipitated one of the fiercest battles between long-distance carriers and local-exchange companies, and among long-distance carriers themselves.

Interconnection is British Telecom's lever to control the competition. Thus the regulations governing that interconnection were critical. They affected how much British Telecom could charge, a murky area of conceptual and accounting issues, as well as the numbering system, the points of interconnections, the quality of service, the number of digits to be dialed, and other technical matters. In all these issues British Telecom has an understandable reason to be uncooperative. Even when British Telecom's proposals are fair by some objective standard, however, Mercury has an incentive to cry wolf and seek an advantageous interconnection arrangement. It can argue that, as an "infant" competitor, it needs a period of protection. Without protection Mercury could not compete with British Telecom, and this would undermine the entire basis of government policy on competition.

Hence Mercury has leverage over British policymakers out of proportion to its economic power. With the Labour party in power, the reverse is possible: Mercury could be choked to death by "technical" regulatory decisions rather than by policy decisions debated and passed by Parliament. Probably some form of tacit collaboration will occur. Mercury does not have much of a support base. At the same time, Mercury's existence provides BT with a useful argument against government interference in its operations. British Telecom is not necessarily interested in totally winning this contest.

In Japan seven domestic carriers, including NTT, received operating licenses in the first year of the new law that became effective in 1985. They are affiliated in several instances with major institutions—Japan National Railways, the Public Highway Authority, the Tokyo Electric Company, Mitsubishi/Ford, and Itoh/Mitsui/Hughes. Of these carriers two plan to use satellites, two fiber-optic cable, and the other microwave. The first alternative service began in August of 1986 by the National Railways' Japan Telecom Company in the Tokyo-Osaka corridor. Under government prodding NTT reached interconnection agreements with the "new" common carriers (NCCs) that were favorable to the latter.

In light of the American experience, it is questionable whether all these licensees will be profitable. In another liberalized service, mobile telephony, the PTT ministry already has been restrictive. The law suggests that telecommunications facilities should not exceed demand.

Value-Added Network Services: Hybrid Communications

Specialized network services can be offered by specialized providers, as well as by international and domestic carriers. Value-added network services play a much greater role in policy discussions in Europe and Japan than they do in the United States, and it is important to understand why. The fundamental interests of the traditional network operators and not technical or business reasons have led to the new regulatory category of value-added network (VAN) services. It is an intermediate step in the liberalization of services, and just as inevitably it is not the last one. The key problem that prompts the licensing of value-added networks is the potential resale of leased transmission to third parties. This form of arbitrage by a service reseller leads to loss of control by the basic network provider, to competition, and to a reduction in revenues, at least in the short term.

In the United States such resale is possible and widely practiced. Lessees can do almost anything they want. The regulatory constraints that do exist are largely to prevent the basic carriers from extending their market power over the network downstream into the applications stage by internal subsidies. But in other countries resale is prohibited, although it seems to exist unofficially in several instances. Some of these countries have realized, however, that the use of leased lines can provide communications applications of a sophisticated nature for use by third parties, and they do not wish to prevent these services from emerging. Thus they lean toward permitting the provision of "value-added" services, where something has been added to basic transmission. This technical addition legally transforms into a sale what otherwise would have been a resale. (Another alternative is to establish usage-sensitive pricing to eliminate the incentive for retailing of services. But this creates other efficiency problems in pricing.) As in any attempt at price discrimination that is not cost-based, one should not underestimate the ingenuity of arbitrageurs. Those who wish to resell basic transmission (or switching services) but can sell only value-added service may try to add a trivial amount of value or an entirely unnecessary amount, solely to become legal. To prevent this, it is necessary to license value-added networks, after scrutinizing the nature of their "value added." A formal approval process is therefore needed, together with some form of ongoing monitoring, to protect the system of price discrimination. This restricts the range of services and limits the licensed network's operating flexibility. It is deemed to provide some protection to the monopolistic "basic" service, although the stability of this protection is illusory over time.

In the United States, as mentioned, such procedures do not exist. Value-added network services are merely a functional description and not a regulatory category. Since they are undefined officially, they mean different things to different people and often simply refer to packet switching networks. There is a regulatory distinction in the United States affecting VANs between "basic" and "enhanced" services, but it serves an entirely different purpose. Whereas PTT countries seek regulation of VANs to prevent the resale of leased capacity (that is, to protect the PTT service monopoly), the U.S. categories are to prevent the cross subsidization by a dominant carrier of its value-added services through revenue gained in those dominant activities. The United States distinguishes between basic and enhanced service in order to *prevent* the dominant carriers' exercise of market power—not to protect those carriers from competition.

The British Telecommunications Act of 1981 authorizes the secretary of state for industry to license value-added services. A general license is required that prevents interconnections except to a public telecommunications system. It also requires the use of approved equipment. Resale or shared use is not permitted.

New legislation governing value-added network and data services was passed in 1985. By that time there were 688 VANs operating under a general license, operated by 164 different companies.[13] Of these VANs the most popular were store and retrieve systems (89); mailbox service

13. "Future Licensing of VANs," *Oftel News*, no. 1 (December 1985), p. 7.

(71); protocol conversion between incompatible computers and terminals (71); customers' data bases (54); deferred transmission (50); user management packages (46); view data (49); word processor and facsimile interfacing (40); multiaddressing routing (49); and speed and code conversion between incompatible terminals (43). Other VANs include automatic ticket reservation, conference calls, long-term archiving, secure delivery services, telesoftware, retrieval, and text editing.

The "liberalized" licensing system quickly showed itself to be overly rigid. This led to still newer rules in 1986 and 1987 that substantially simplified and liberalized procedure and made, in effect, the resale of capacity for computer data transmission fairly unrestricted. Large VANs are subject to rules that prevent the establishment of a dominant market position. These limitations were aimed at British Telecom and IBM, which had unsuccessfully applied to establish a joint venture in value-added network services. Unlike its competitors, BT must provide these services nationwide. It is also subject to rules that prevent a cross subsidy out of other services.

In Japan two types of VANs were established by the 1984 reform: "special" type II carriers and "general" type II carriers. Special type II carriers resemble large packet data networks such as Tymnet and Telenet in the United States. Several networks of this kind have been established by the computer firms Hitachi Ltd. and Fujitsu Ltd. and by NTT itself jointly with IBM Japan (Japan Information Service), to name a few.

More than two hundred general type II carriers emerged after the law went into effect in 1985. By far the largest category is order networks of retailers and wholesalers (for example, for food and used cars), followed by credit card verification, financial networks, electronic mail services, voice mail services, and transportation. Also in this regulatory category are a dozen resale carriers, of which the largest are Recrute and K-VAN. They offer rates 20 to 30 percent lower than those of NTT. This resale, however, cannot include connection with the NTT public switched network, and therefore it makes economic sense only for larger users with private-line needs.

Japanese providers or VANs found themselves unable to offer international service under the recommendations by the International Telegraph and Telephone Consultative Committee (CCITT). To do so they had to be awarded by the government the status of a "recognized private operating agency" (RPOA). As both the British and Japanese experiences indicate, it is difficult to permit competition in new and advanced services while protecting the traditional services.

International Carriers

Cable and Wireless is the prototype for the new generation of international carriers. The company once operated telecommunications services for Britain's overseas colonial possessions. C&W was nationalized by the Labour government in 1947, and it remained state controlled for almost thirty-five years. In the mid-1980s it was still operating public telecommunications services in twenty-eight countries and territories on behalf of their national governments, with a major operation in Hong Kong.

In 1981 the Conservative government reprivatized C&W. Privatization made it possible for the firm to expand rapidly and to transform its somewhat sleepy image into what is arguably today the most interesting telephone company in the world.

C&W's announced goal is to become the first global telecommunications carrier. It aims to link the four major financial centers in the world: London, New York, Tokyo, and Hong Kong. Already it is a dominant presence in Hong Kong, where it owns the local telephone company. It is a major participant in a joint venture with Nynex for a private submarine fiber-optic cable to the United States, to be operational in 1989. C&W also hopes to participate in a transpacific fiber-cable venture. In Britain C&W has become the sole owner of Mercury, which provides it with a long-distance capability within Britain.

To compete with the former monopolist Kokusai Denshin Denwa (KDD), two Japanese consortia applied for a license to provide international service. One of them is International Telecom Japan (ITJ), owned by fifty-three large users including Mitsubishi, Sumitomo, Mitsui, the Bank of Tokyo, and Matsushita. ITJ planned to commence service at first on circuits leased from KDD. The second consortium is International Digital Communications, in which C. Itoh and Company and C&W, the largest partners (each with 20 percent), are joined by thirty-three others, including Toyota. The Ministry of Posts and Telecommunications tried to nudge the two ventures into a merger. Part of the agreement would have been to reduce C&W's share to 3 percent for reasons of "national security" and to exclude it from a role in management. This had the British and American governments up in arms. The example shows how difficult it is to reconcile the conflicting philosophies and interests in this field.

Integrated Services Digital Networks

The preceding discussion of new carriers should not lead one to believe that the traditional telecommunications coalition is technologically passive in defending its position against challenges. One form of this defense is expansion into new and adjoining fields of activity such as cable television transmission and videotext. Another strategic move is the upgrading of the network in a way that raises barriers to entry. Monopoly, it is argued in the spirit of political télématique, is a condition for technological innovation. The primary initiative is known as the integrated services digital network (ISDN). At its most elementary, ISDN is an integration of voice, data, and telex networks into a unified "superpipe." Though hundreds of papers on ISDN have been published—almost all of them from a technical perspective—virtually no public discussion of the ISDN concept has taken place. Part of the problem is that the term *integrated services digital network* encompasses several subconcepts. As a move toward more *digitization* of the network, it is squarely and positively within the trend of technology. As an upgrading of the networks to *higher transmission* rates, it similarly responds to the need for greater data communications, particularly those of larger users; for residential users the need is less clear except to create the proverbial egg (the network) for future chickens (the applications).

The third element of ISDN is *integration,* and its rationale is much weaker. To put separate communications networks into one superpipe is elegant from a technologist's view, but from a user's perspective what count are the cost, performance, and choice of services. Integration is a standardization process, which is always a trade-off between the cost reduction of streamlining and benefits of diversity. Integration usually reduces options. Users are interested in choice for selection, at a price, while network operators may be more interested in providing standardized options.

The implicit assumption in the justification for the nonduplicating superpipe is that cost functions (for example, for telephone and telex networks) are static. Yet different services under rival control usually create a dynamic downward shift of the cost curves, because of the extra efforts of competitors, in contrast with the monopolistic situation of unified services. The effects of these downward shifts in costs can offset, partly or totally, the economies of scope of integration.

ISDN as a technical concept does not prevent multiple ISDN networks and networklets from coexisting, competing, and interconnecting. There is no notion of exclusivity in the technical integration, but anything less than exclusivity is almost impossible for ISDN's PTT promoters to accept. After all, eliminating duplication is the primary rationale for ISDN.

To permit multiple integrated networks would defeat the purpose.

For the equipment industry, ISDN is welcome. After several decades of enormous public investments to expand the public network, growth has declined. Export markets are limited because many of the larger ones are protected against imports. Therefore, one way to activate the sagging domestic market was to launch an ambitious program of upgrading.

Rate Differentials and Arbitrage

The traditional collaborative system has faced three main challenges: how to maintain control over the international segment of communications, how to deal with the new American carriers, and how to resolve disputes involving international satellite service.

The area of international communications has been a major contributor to PTT profits, but it is more vulnerable to rival service provision than is domestic communications, where both ends of the link tend to be under the control of the same PTT. International telecommunications is the soft underbelly of the traditionalist system. The need for international coordination and agreement has illuminated the different perspectives of the traditionalist coalition and its opponents. In consequence, disputes have been frequent and harsh. Profits on international service are high because costs have dropped faster than rates. The investment cost per transatlantic cable circuit has dramatically decreased, from $133,000 in 1940 and 1941 to a projected $670 for the fiber-optic cable TAT-8. For satellite circuits, costs have dropped from $86,000 on Early Bird to $450 in the Intelsat VI satellite generation. In 1981, according to one study, the yearly cost of a direct broadcast-grade connection between London, New York, and Frankfurt was $53,000 a year, but British Telecom charged $750,000.[14]

A U.S. Federal Communications Commission (FCC) study showed that the average rate from Europe to the United States exceeded that from the United States to Europe by 34 percent in 1981. After AT&T's 1981 price cut, the weighted-average foreign tariff was almost 95 percent higher than the American.[15] Another report shows that in 1982 a daytime telephone

14. Barry Stapley, "Managing Communications: The Value of Choice," quoted in Organization for Economic Cooperation and Development, *Telecommunications: Pressures and Policies for Change* (Paris: OECD, 1983), p. 106.

15. Evan Kwerel, "Promoting Competition Piecemeal in International Telecommunications," Working Paper Series 13 (Washington, D.C.: Federal Communications Commission, Office of Plans and Policy, December 1984), pp. 18–19; and U.S. General Accounting Office, "FCC Needs to Monitor a Changing International Telecommunications Market," GAO/RCED-83-92 (Washington, D.C., March 14, 1983), p. 17.

call from New York to Munich, Germany, cost $1.38 per minute, while the same call made from Munich to New York cost $3.03 per minute.

High profits encouraged the emergence of arbitrage. A Telex message from Germany directly to the United States cost $2.58 a minute in 1981, but only $1.76 if it was routed via the United Kingdom. This difference led to a substantial transatlantic traffic from the European continent via London telex bureaus. European PTTs fiercely tried to stamp out this arbitrage (citing CCITT rules on ''golden handcuffs'' of a high-priced cartel), but they were harshly rebuffed by the European Commission and the European High Court of Justice.

The legal foundations of the cartel have thus been shaken. One single country that breaks ranks—for reasons of economic ideology or a desire to profit as a ''communications haven''—can undermine the profitable arrangements that have endured for a century.

The second area of challenge to cartel solidarity concerns the new American carriers. In 1984 the European PTTs reaffirmed their policy on the control and limitation of other common carrier (OCC) entry. The PTT organization (the European Conference of Postal and Telecommunications Administrations, or CEPT) advised its members to open their markets only to the traditional seven U.S. carriers (AT&T and the six international record carriers). The new guidelines stated that new carriers would have to provide better technical service at a lower cost than at present in order to be permitted entry into the European markets. New carriers would be permitted for new types of communications service (such as videotext, teletext, facsimile, and packet switching), but they would be restricted to one carrier for each new service.

The CEPT guidelines effectively limited the number of carriers PTTs were able to choose from. Normally, it is to the advantage of any party in a transaction to be able to pick and choose among competitors, and particularly if the party is a monopsonist. The PTTs, however, acted to restrict entry to prevent competitive bargaining between countries. A PTT presumably would not enter into an agreement with additional American carriers if it were not in its own self-interest, economically or technically. CEPT recommendations aim at collective forbearance from any future bidding for new carriers' entry by establishing cartel solidarity. They aim at preventing the establishment of telecommunications hubs such as Britain.

Given the hostile reception, it stands to reason that the new American carriers, in order to be admitted, must offer significantly more attractive deals to the PTTs than AT&T does. But the FCC, concerned about PTT monopolists squeezing U.S. competitors, set rules against "whip sawing" that hinder the new carriers' ability to compete with AT&T and with each other for PTT business. Since uniformity tends to benefit the incumbent, AT&T's competitors were unhappy.

MCI tried to reduce the barriers to entry by buying from Xerox an existing carrier, Western Union International (renamed MCI International). This created a convenient international outlet for MCI's U.S. involvement in electronic mail and provided it with an already established relationship with the PTTs.

MCI actively sought to provide end-to-end international voice traffic, as had AT&T. It concluded agreements with several countries, in particular the United Kingdom and Hong Kong, which became its major international hubs. Like U.S. Sprint, MCI reaches other countries by transfer through these hubs. Overall, the other common carrier procedures can be complicated enough to prevent profitability in operations to many countries served by the OCCs. However, the OCCs require a full international service to compete with AT&T on equal footing in the United States.

The third type of dispute involves international satellite service. In an extension of its domestic "open skies" policy, the FCC accepted applications from a group of private entrepreneurs for a license to operate the private international satellite system Orion. This was followed by similar applications by International Satellite Inc. (ISI), PanAmSat, and Cygnus. All were fiercely resisted by Intelsat and the PTTs.

Ironically, the opponents of liberalization of international satellite communication are partially responsible for its emergence. Several regional and intercontinental satellite systems have been established outside the Intelsat organization. They include a Scandinavian satellite consortium, Arabsat, Eutelsat, and a French system that is "domestic" but stretches that term to encompass communications with French possessions in the Caribbean and South America. Several countries believed that they could follow their telecommunications goals better if they had more control over satellite communications. This is one reason for the emergence of these satellite projects. Moreover, various industrial policies promoted electronic development projects. These industrial policies undercut the argument that an international satellite system must be controlled by one organization for reasons of economical and technical efficiency.

Fearful of satellite competition undermining their highly profitable international service, the PTTs pursued several defenses. An "up-link" strategy was intended to prevent the FCC from granting private licenses as a violation of the Intelsat agreement. A "preemptive" strategy sought to cut rates and offer new service options as a way to deter potential entry. A "down-link" strategy tried to prevent new satellite carriers from connecting into national networks. And a "third world" strategy rallied the less developed countries' PTTs, fearful of losing the cross subsidies to low-traffic routes.

In the end PanAmSat got a limited Intelsat approval, while the other applicants were stymied. But this was a rearguard action. Whatever one may think of the desirability of a single global system with its economies of scale, the simple fact is that a distance- and border-insensitive technology such as satellite transmission cannot be successfully restricted for long. Even without competing satellites, rivalry from private submarine cables threatens the Intelsat arrangements.

Equipment and Trade

In the past the traditional coalition was fairly successful in holding the line on equipment imports from the United States. Once divested, however, AT&T emerged as a competitor in international markets, a sharp break with the past. The divestiture received much attention. It was portrayed as part of an American telecommunications equipment offensive into the rest of the world. But for all the publicity, actually the opposite has happened: American equipment makers recently have been repulsed and almost expelled from the international markets, with ITT, GTE, and Honeywell mostly departing, and AT&T largely unsuccessful. Meanwhile, foreign manufacturers have rapidly gained a fairly large aggregate share of the market in the United States.

Containment of American Equipment Manufacturers in Europe

For more than fifty years AT&T had stayed out of international equipment activities, despite its position as the largest manufacturer of international equipment in the world. In the early years of telephony, the Bell System licensed several European equipment manufacturers, acquired others, built its own foreign facilities, and had a substantial manufacturing and distribution presence abroad. Then in the 1920s American critics of AT&T charged that American ratepayers were subsidizing its international operations. For that and other reasons, the company in the 1920s decided to sell its European operations to ITT, then a relatively insignificant firm run by the Virgin Islands entrepreneur Sosthenes Behn. This purchase marked ITT's entry into the big league of telecommunications.

Following divestiture in the 1980s, the international field became interesting to AT&T once again. Because the domestic equipment market had been opened to all comers, its U.S. market share had nowhere to go but down. The rest of the world became its field of growth. AT&T's strategy was to align itself with domestic interests, thus lowering the barriers that an American company would face. A series of joint ventures emerged, some ad hoc, but in the aggregate part of a new orientation. The first such alliance was a joint venture with the Dutch electronics giant Philips, which had run out of steam in telecommunications development. The Netherlands, however, is a very limited market.

AT&T's second major international involvement was with Olivetti, the Italian manufacturer of office equipment and small computers. AT&T purchased 25 percent ownership of Olivetti for $260 million, with the option to acquire another 15 percent after four years. Olivetti's ambition is to become the major European player in world computer markets. Its main rival is IBM, with more than 50 percent of the European market. Olivetti's alliance with AT&T is a great advantage, given AT&T's own technological capabilities and capital.

AT&T, which had looked for a European beachhead and distribution system, found itself making a nice windfall profit on its Olivetti investment. Fueled by Carlo de Benedetti's success as Olivetti's chief executive officer and the rise of the Italian stock market in general, AT&T's investment quintupled in value. Olivetti's private branch exchanges, terminal equipment, and personal computers make it an increasingly strong rival to the STET group. STET, in turn, is collaborating with IBM. Thus the Olivetti-STET rivalry is joined by the American antagonists, AT&T and IBM, each the ally of a major Italian company.

Another major move by AT&T was to try for an agreement with the dominant French firm Compagnie Générale d'Electricité (CGE) group, and this unleashed another round of conflict-laden politics. Behind the story was the unresolved question of what to do with the remaining public switch manufacturer in France, Compagnie Générale de Constructions Téléphonique (CGCT), acquired by the government in 1982 from ITT. CGCT was losing money quite heavily, but it had one major paper asset: it was traditionally allocated 16 percent of the French public switching market, about 300,000 lines a year.

After the government-generated merger of the telecommunications activities of Thomson into CGE (and its Alcatel subsidiary), CGCT was the only remaining second source for the Direction Générale des Télécommunications (DGT), which did not want to confront CGE as the only supplier. CGCT, in consequence, was given the task of manufacturing the Thomson digital switch under a license, but it was in no position to develop new equipment. Who, then, would be the second source supplier to the DGT? With 84 percent of the market, CGE negotiated with foreign firms to determine which of them should be admitted to the market; in effect, it was selling the small share of the French market that it did not hold for political reasons and selecting its own competition. Eventually, CGE agreed with ATT-Philips Telecommunications (APT) that APT would receive CGCT's market share of 16 percent for its 5-ESS PRX switch, which would be manufactured in France by CGCT and adapted to French standards. In return, AT&T would help CGE modify its E10-5 switch for North American use, include it in its product line, and pay certain indemnities if sales for that switch did not reach a specified amount. Furthermore, Philips would transfer the microwave equipment manufacturing of its French TRT subsidiary to a joint venture controlled by CGE, while AT&T would undertake to buy at least $200 million worth of microwave transmission equipment over four years. Lastly, CGE would receive $100 million.

CGCT was unenthusiastic about the deal with AT&T, preferring to deal with Siemens or L.M. Ericsson. But for CGE, AT&T was a more compliant partner. A company such as Siemens, once it had a toehold in the French market, could not be as easily contained, given the close European collaboration between France and Germany, which includes French companies' involvement in German television set manufacturing. In contrast, any AT&T involvement in France would be subject to much greater government scrutiny and future pressure, since public opposition against it could always be more easily organized.

The telephone administration (DGT) was in favor of the ATT-Philips deal because it wanted to use the AT&T Centrex capability. It wanted to get a better bargain, however. It pressured AT&T's equipment price down and called upon other firms to enter into negotiations. Left out were Plessey and GEC after British Telecom decided to use as a second source ("System Y") the Swedish firm Ericsson rather than CGE; this the French PTT minister considered to be un-European, despite the geographical facts to the contrary.

The story now gets a new twist: what started out as bargaining for more favorable terms from AT&T changed when CGE struck a historic deal with ITT and gained control over its telecommunications activities.

The French Conquest of ITT: The High Point of Political Télématique

To appreciate the significance of the acquisition by CGE of ITT's worldwide telecommunications operations, one must go back to the close of World War II, when the French telecommunications industry was almost nonexistent. There was CIT, part of CGE. And there were Le Matériel Téléphonique (LMT) and CGCT, both French subsidiaries of ITT, as well as the foreign-owned Ericsson-France and the Philips group. These companies were licensed by the PTT to manufacture items of foreign design.[16] Other companies that later joined the market were the French Thomson-CSF and AIOP. And yet by the mid-1980s, only one company, the French Alcatel, remained, with CGCT surviving artificially. When the French government expanded the national telecommunications network and therefore the market for equipment, the foreign presence was almost entirely eliminated. First, the government forced the transfer of ITT's LMT, and Ericsson's French subsidiary transferred to Thomson-CSF, a private firm that was later nationalized by the Socialist government. ITT's other subsidiary, CGCT, was nationalized in 1982. During that period CGE and Thomson-CSF also took over AIOP, a workers' cooperative. Later it transferred Thomson's telecommunications activities to CGE.

In July of 1986, CGE entered into an extraordinary transaction with ITT, leading to its taking control over the telecommunications operations of the American-based firm. Despite ITT's American headquarters, the company had only a limited equipment presence in the United States, and in the telecommunications equipment field it was a multinational firm without a home base. A far-flung conglomerate in the 1960s and 1970s under the leadership of Harold Geneen, ITT later came into hard times. It was losing money heavily, and its innovative telecommunications switch, System 12, had technical trouble. The company also had to concede humiliating defeat in adapting the European-developed switch to U.S. specifications.

Meanwhile, CGE's Alcatel telecommunications subsidiary had difficulties of its own, especially in export sales abroad; it had almost no public switch sales in Europe to match its French dominance, and many of its international sales were political deals, particularly with former French colonies, or part of foreign aid packages. Domestically, the golden years

16. Jacques Darmon, *Le Grand Dérangement: La Guerre du Téléphone* (Poitiers/Liguge: Editions Jean Claude Lattes, 1985), p. 79.

of the expansion of the French network were coming to a natural end, and orders were dropping. Its mainstay digital switch family, the E10, developed for the expansion, was showing signs of age. Having failed to penetrate foreign markets through its products, or by French diplomacy, CGE set out to purchase foreign toeholds by acquiring ITT.

Through a complex agreement, ITT merged its telecommunications equipment, office automation, and consumer electronics into the new entity. CGE contributed its own Alcatel equipment subsidiary. ITT kept a 37 percent share of the holding company, while CGE, together with other European firms, controlled the rest. ITT received $1.5 billion for giving up its share, and the holding company assumed $800 million of ITT debt. The new firm was named Alcatel N.V.

Because of the problems inherent in having a nationalized French company own the centerpieces of other countries' electronics industries, it was envisioned for several other entities to have a share in the new venture. But this proved difficult, and in the end only the Belgian holding company SGB participated in a limited way. Alcatel was headquartered, nonetheless, in the Netherlands, to provide for a less French image.

The merged firm became the second largest international telecommunications firm after AT&T, with $7 billion in assets, almost $10 billion in sales, and 150,000 employees. It accounted for 42.5 percent of European public telephone switches, by far the largest share. CGE heralded the agreement as establishing for the first time a large-scale European telecommunications firm; most European telecommunications experts, however, did not get enthusiastic over such a French government–dominated arrangement. CGE claimed that the merger was necessary for reasons of economies of scale. The notion that it takes almost one-half of the European market to be successful is part of the obsession with economies of scale that pervades the industry's thinking.

The deal put into question CGE's separate arrangement with AT&T for CGCT's market share. With CGE inheriting many of ITT's footholds in so many other European countries, a greater reciprocity probably would be expected by those countries, and the 16 percent market share, whose allocation had been anticipated by AT&T, might be needed instead to assuage one or several European countries. In particular, the German government began to be active on behalf of Siemens. It pressured the French government at the highest levels to substitute Siemens for AT&T-Philips in the spirit of European solidarity, as well as in reciprocity for the newly acquired German ITT subsidiary Standard Elektrik Lorenz (SEL). The tug-of-war grew acrimonious. FCC Chairman Fowler pointedly sent inquiries to major American telephone companies about their use of equipment from countries that discriminate against U.S. firms. Within the French government, the DGT preferred APT, but other ministries did not wish to antagonize Germany. The rival companies successively sweetened their bids. In the end as a compromise, the government chose Ericsson, together with the French defense firm Matra, which thus gained a foothold in telecommunications.

Roughly at the same time, another U.S. telecommunications firm, GTE, stopped manufacturing international equipment. GTE had substantial manufacturing involvements in Italy and Belgium until a deal with Siemens transferred 80 percent of these interests to the German company.

The Failure of Political Télématique in Mainframe Computers

It is useful to contrast France's successful empire building in telecommunications with its lack of success in computers or components. Here monopoly leverage did not exist to the same extent as in telecommunications, and the performance of French firms was less impressive. The French government had long been worried about "computer sovereignty." De Gaulle unsuccessfully attempted to veto GE's ownership involvement with the French computer firm Compagnie des Machines Bull. The French had been rightly shocked when the U.S. State Department refused an export license for large scientific computers to the French atomic energy commission for use in H-bomb research. When the GE-Bull deal could not be prevented for financial reasons, the government formulated in 1966 its "Plan Calcul." Among the plan's projects was a merger between the two remaining French-owned computer manufacturers into the firm Compagnie Internationale de l'Informatique (CII). Bull was left out of the Plan Calcul because of its American links.

Plan Calcul established targets for development of scientific computers, leaving much of commercial office computing to IBM. This projection completely misjudged the explosive growth of business applications of computers and, in any event, never touched the predominance of the American Control Data Corporation in this field.

The product strategy behind the Plan Calcul missed other developments. Observing time-sharing use of computers in which terminals were linked to a powerful central computer, it predicted a future with a few giant mainframe computers only and began developing them. But the trend was almost the opposite: minicomputers and microcomputers proliferated. Also unanticipated was the main benefit from the Plan Calcul. It provided technology skills to Thomson and to CGE that they later applied in their development of digital telecommunications switches.

General Electric, in the meantime, experienced major headaches with Bull. Later it sold its interest to Honeywell when it left the computer business altogether.

By 1976 neither CII nor Honeywell-Bull was doing well. With government pressure and financing, they were merged into CII-Honeywell-Bull, which was 53 percent French (private and governmental) and 47 percent Honeywell. During the next two years, the French government subsidized the merged firm with about $300 million and arranged loan guarantees for about $1 billion. In 1982 the Socialists nationalized the computer firm, and Honeywell was forced to reduce its involvement substantially. Because of its access to advanced technology and the American market, Honeywell was left with 19.9 percent of Bull and a ten-year marketing and technology agreement with its former subsidiary. In 1986 Honeywell exited from computer manufacturing altogether and sold its French interest. Bull's affiliations with American technology also included a 7 percent share in the U.S. company Trilogy, founded by Gene Amdahl in 1981 to advance the state of high-speed mainframe computers and VLSI components. Trilogy, however, was unsuccessful in its development efforts. After RCA, GE, and Honeywell, this was Bull's fourth luckless marriage with an American firm.

The costly subsidies to computer development were part of a large effort in electronics. Following its victory in 1981, the Socialist government nationalized twelve big industrial groups at a cost of about $6 billion. The electronics firms included CGE, Thomson-CSF, CGCT, Matra, and CII-Honeywell-Bull. The total losses that the government subsidized grew from $226 million in 1980, to $4.6 billion in 1982, to $4.2 billion in 1983. The subsidy of losses was more than two-thirds as high as the initial costs to the government of taking over the companies! Direct government aid for the electronics industry was $1 billion in 1983, $1.2 billion in 1984, and another $1.2 billion in 1985.[17] This does not include indirect support through the telephone administration, DGT. In 1985 Bull received $100 million in new equity, the same amount as the year before. Thomson got $100 million in 1984 and another $130 million in 1985, primarily for its electronic components division.

Eventually, ending the state companies' deficits became a government priority, and the nationalized companies began to cut jobs rapidly. By

17. *Electronic News*, February 25, 1985, p. 13.

1984 French unemployment was at a postwar high of 10 percent. The electronics companies, too, laid off workers during 1984 and 1985. For example, Alcatel cut 1,700 jobs, Thomson 4,000.

Does the French example mean that the industrial strategy of government subsidy is doomed to failure? Not necessarily, for there have been successes, too. Furthermore, the British example does not indicate that the free-market approach is superior. At the root of the problem is the inability of institutions to transform themselves and the considerable hold they have over public policy. If a free-market approach is to be superior at all, it cannot be simply an opening up from above as part of a policy blueprint. It requires a vigorous entrepreneurial element from below and support by the educational and financial institutions. Conversely, the efforts of the French government to finance and guide the electronics industry will create self-sustaining growth only if a dynamic rather than bureaucratized environment results. Large, established firms have benefited, but there has been no notable emergence of small and innovative firms.

Equipment Imports

Despite its criticism of U.S. liberalization, the traditional coalition benefited in trade at the same time that it resisted change in domestic procurement practices. This created problems of reciprocity in trade that spilled into the political arena. The models for American exporters, once one goes beyond official assertions of openness, have already been described. This section deals with trade to the United States.

The U.S. market is not only the largest domestic market in the world by a large margin, but it is also relatively free and has many independent telephone companies (roughly 1,500). There are more potential customers in the United States than in the rest of the world. (Many of these firms are, of course, quite small.)

The U.S. liberalization provided non-U.S. manufacturers with exciting opportunities. Before the divestiture of AT&T, the Bell operating companies relied largely on Western Electric for their equipment, thus giving AT&T a captive market of 80 percent of the total U.S. equipment market. After divestiture, they were free to buy equipment from other suppliers. They have actively done so, primarily from Canada's Northern Telecom, but also from Siemens.

Even before the AT&T divestiture, the Swedish firm L.M. Ericsson had been an active supplier to American independent telephone companies and to MCI, and the British company Plessey had purchased the public switching business of the well-established American manufacturer Stromberg-Carlson.

The opening of the American market was some of the best news that many non-U.S. manufacturers had had for a long time. Other industrial countries' markets are largely closed to imports, even within the European community. Demand from the OPEC countries declined because of the fall in the price of oil and the fact that the initial large equipment orders had already been placed. Likewise, in the third world markets fewer funds than before were available for telecommunications investments, and many countries encouraged the development of a domestic telecommunications industry to spur their own industrial development. Thus open markets for telecommunications equipment were limited to less than 15 percent of the world market, according to a 1982 estimate by the Organization for Economic Cooperation and Development (OECD). Since then the United States has opened its market, more than doubling the total.

As Georges Pebereau, then president of CGE, declared,

> It is obvious that no European company, French or not, can remain a world company if it does not have a significant position in the American market, which represents 40 percent of the world market and, in addition, is from the point of view of technology the best testing grounds one can imagine. Fortunately, we have a historic opportunity to develop ourselves in the U.S., with the deregulation of ATT. . . . If, unfortunately, CGE's presence in the U.S. failed, we would need more than a decade to regain the confidence of our American customers. Thus the interest in finding a partner in place which would permit us to penetrate the American market faster, more surely and at a lesser cost. Of course, if such an occasion presented itself, we would seize it.[18]

Some of the strongest advocates of protectionist policy in telecommunications procurement subsequently began to seek their fortune in the newly liberalized U.S. market. But this asymmetric situation unavoidably created tensions. The U.S. government would not stand by as others sold freely in America but shut out U.S. manufacturers. It is therefore not surprising that the FCC took a first step in December 1986 and invited comments on whether there should be restrictions on the approval of equipment exported from countries that discriminated against American equipment.

18. "Pebereau Joue Quitte ou Double," *L'Expansion* (June 7–20, 1985), p. 73.

Thus the opportunity to enter the U.S. market is ultimately a double-edged sword, threatening to bring about a reduction of European and Japanese firms' own protected positions. It has the tendency to split the telecommunications industries of other countries. Strong and advanced manufacturers who can compete successfully in the American market and at home on the merits of their products could accept American entry in their home base, but weak firms in need of protectionism could have little to gain and much to lose from the lowering of the barriers.

Services Imports

Just as in the equipment market, deregulation of U.S. domestic telecommunications services gave foreign organizations new opportunities to enter the American market. Cable and Wireless established with Nynex a joint venture for fiber-optic transatlantic and transpacific cables. The liberalized environment makes it possible for European carriers to acquire American domestic long-distance carriers. Cable and Wireless owns TDX System Inc., an American discount long-distance carrier servicing business users. Likewise, France Cable et Radio, a subsidiary of the French DGT, in 1983 took a share in Argo Communications Corporation, a newly formed American interexchange, or long-distance, carrier that early on offered an ISDN-type service. The relative ease with which services can be offered in the United States contrasts strikingly with the barriers that prevent American carriers from even reaching international markets.

Cable and Wireless also attempted to acquire Pacnet Communications Corporation. Pacnet requested a data network identification code that would enable it to provide overseas customers with a U.S. resale packet switched network. Under the *Computer II* decision, Pacnet, as an enhanced service provider, would not have had to file with the FCC and could even have acquired satellite circuits from the Communications Satellite Corporation (Comsat) without authorization. Thus a PTT could set up its own unregulated distribution network in the United States. As a staff memorandum to the FCC concluded,

> It is fair to say that the ability of foreign telecommunication entities to enter the U.S. international telecommunications market is in large measure unprecedented and raises serious issues not presented by foreign entry into the U.S. domestic market. . . .
>
> In the U.S. international telecommunications market, an unregulated foreign enhanced service provider would have the ability to both prevent

> the entry of additional U.S. entities into the market for service between the U.S. and the home country of the foreign entity and to remove existing U.S. carriers competing in that sub-market, at least where the service involved is classified by the foreign country as a common carrier service to be provided by the telecommunications entity of the country. Such action by foreign entities would run directly counter to the U.S. policy of fostering increased competition in the provision of international telecommunications services.[19]

Although the Pacnet application was withdrawn, similar actions are a possibility. Argo provided long-distance service in the United States and also served as the sole connection for all American competitors to AT&T who wished to be routed to France. Just as in the case of the opening of the American market to European equipment sales, this potential of European service provision within the American domestic market, linked to a domestic monopoly position, raises issues of reciprocity of entry. It highlights again the problems inherent in coordinating a system of communication links when its two ends are controlled by fundamentally different concepts of the nature of telecommunications.

International Collaboration in Telecommunications

From the beginning, telecommunications have been highly internationalized. For a long time international organizations were used to shore up domestic arrangements and protect PTTs by creating welcome international restrictions. What started as technical collaboration across borders almost immediately became deeply involved in economic arrangements and the protective regulations of a cartel. This tradition goes back to the early period of postal systems when the checkered map of central Europe often permitted alternative routes and thus made intergovernmental agreements desirable for states in need of the postal revenue. The maintenance of stable international arrangements is a central policy concern for the postal-industrial complex. But times have been changing. Other international organizations have begun to disturb the established harmony, and private collaborative ventures also have affected the compartmentalization of national markets.

19. GAO, "FCC Needs to Monitor a Changing International Telecommunications Market," pp. 27–28.

The Traditionalist International Institutions

In 1865 several telegraph administrations founded the venerable International Telegraph Union (now the International Telecommunication Union, ITU). From the beginning the ITU was controlled by the major European powers. Not only were these countries at the forefront of telegraph technology and usage, but they also provided for themselves voting membership through the colonial telegraph administrations of their overseas colonies. In 1925 France, Great Britain, Italy, and Portugal all had seven votes in the ITU.

Technical coordination was only one aspect of the ITU's activities. The issue of international rate making, that is, of economic collaboration, was important from the beginning, and much time at ITU meetings was spent establishing uniform rates and agreeing upon the charges for coded messages.[20]

For years the United States regarded the ITU with benign neglect. It did not send delegates or observers to the International Telegraph Union and did not participate in the international consultative committees when they were formed in the 1920s. In the late 1920s the United States opposed the creation of the unified telecommunications ITU because it extended the potential for an international cartel. Following both world wars, the United States became more interested in international collaboration. This led to the 1947 Atlantic City conferences that reshaped international communications into arrangements that have lasted until today.

A majority of ITU members are against any form of liberalization. But the victories of the conservative majority would be hollow if the minority consists of the United States, the United Kingdom, and Japan, major telecommunications countries that may not abide by the recommendations.

Of particular importance in the telecommunications field is the International Telegraph and Telephone Consultative Committee, which has a subsidiary relationship with the ITU. The role of the CCITT is to harmonize operational, technical, and tariff issues of international telecommunications. It functions primarily through expert groups that deal with specific questions. It issues recommendations, but it has no enforcement power. It is not a treaty organization with binding resolutions. Instead, it functions as a de facto standards-setting organization, for which there is often a need. The emphasis on tidy standards was most appropriate in an

20. George A. Codding, Jr., and Anthony M. Rutkowski, *The International Telecommunication Union in a Changing World* (Dedham, Mass.: Artech House, 1982), p. 7.

era in telecommunications when technological change was relatively slow. In the present rapidly changing environment, however, standards can be used to establish artificial stability and to protect favored firms. CCITT recommendations can and do clash—not only with the liberalization of individual countries (as in the case of the Japanese VANs' ability to operate internationally), but also with other international agreements.

The European Community's Treaty of Rome provides for the elimination of restrictions in trade of goods and services among European countries. During the 1970s telex bureaus emerged in Britain to route telex messages from Europe to the United States through London (at a considerably cheaper rate than that charged by the European countries' PTTs for direct service). At that time CCITT recommendations prohibited such third-country traffic, and the PTTs sought to enforce them. The telex bureaus sued under the Rome Treaty and eventually won before the European Commission and the European High Court.

The European Conference of Postal and Telecommunications Administrations (CEPT) is another PTT organization. Established to represent the interests of twenty-six European countries, it harmonizes European positions for CCITT and ITU discussions.

Modernist International Institutions

Today telecommunications policy issues are addressed not only by the ITU, CCITT, and CEPT, but also by the OECD, the General Agreement on Tariffs and Trade (GATT), and the European Community. This is a reaction to the often narrow perspective of the PTTs in their own international bodies. Domestic conflicts in many advanced countries between the PTTs on the one hand and the ministries of economics or industry and of antimonopoly agencies on the other have been extended to the international level.

Yet one should not overestimate the divergence of interest. The ministries of industry, the OECD, and the EC primarily focus, not on services or on user interests, but on equipment issues. They are allied primarily with the "second" electronics industry rather than with service-industry users.

The OECD has taken a leading role in identifying and discussing the issue of transborder flows of data, privacy, and national sovereignty. With regard to the GATT, the United States has proposed to extend the GATT code to cover trade in services, including telecommunications services.

This was to match the GATT's liberal trade regime for goods and commodities, which did not apply to services.

For a long time telecommunications matters were outside the reach of the European Community and its commission. In June of 1983 the European Commission concluded that the fragmentation of European telecommunications into a nationally protected environment was an important element in Europe's falling behind the United States and Japan. This led to the formation of the European Strategic Program for Research in Information Technology (Esprit) development program, which excluded, however, telecommunications. In 1985 the Race program was added specifically for telecommunications, and Eureka was established in 1986.

The commission has pursued several cases against member states for discrimination against each other's equipment. It also has created links between firms in different countries, to dilute the notion of nationality, to strengthen Europeanism, and to encourage European high-technology development.

Private Collaborations

Even without EC subsidies numerous international joint ventures have been undertaken.[21] Olivetti owns a major part of the British firm Acorn Computers; Philips, besides its links with Siemens and AT&T, entered into joint ventures with Ericsson in Sweden and with Bull, Alcatel, and Thomson in France. In 1984 it took control of the German consumer electronics firm Grundig when that company had financial problems. Similarly, Thomson acquired in Germany the consumer electronics division of AEG-Telefunken and the consumer electronics firms Nordmende, Saba, and Dual. Siemens and ICL distribute Fujitsu computers under their labels. Siemens has a good number of U.S. ventures, including a joint one with Corning Glass for fiber optics. CGE and the Belgian holding company SGB acquired a majority share in the Belgian electronics company ACEC. SGB is also a part owner of the Alcatel venture. Alcatel, Siemens, Plessey, and Italtel have a joint research effort for telecommunications switches and transmission.

A major cooperative effort for the development of semiconductor components is the $600 million "Megaproject" of Siemens and Philips, to which the German and the Dutch governments have contributed about

21. Jonathan David Aronson and Peter F. Cowhey, *When Countries Talk: International Trade in Telecommunications Services* (Ballinger, 1988).

$100 million each. Another project has joined Siemens with ICL of Great Britain and Bull of France.

International joint ventures are often difficult. Besides the obstacle of incompatible products, they must overcome problems of selecting the physical location of a project, the language to be used, the composition of management, and labor sensitivities. Siemens, ICL, and Bull, in their collaborative effort to develop fifth-generation supercomputer technology, had an R&D lab in Germany, a French director, and English as the operational language. For the same reasons that firms like to see duplication of efforts reduced by joint ventures, trade unions are suspicious of such efforts. They are fearful of employment reduction and of the ability to deflect the effects of strikes in one country.

Because of difficulties in cooperation, several joint projects have collapsed. In the mid-1970s Unidata, a data processing venture of Philips, Bull, and Siemens, fell apart after bitter disputes about the French government's alleged overaggressive involvement in its affairs.

Multinational cooperation and mergers can also be hampered by some countries' promotion of high-technology companies as ''national champions,'' making it difficult to have these firms as junior partners in a collaborative effort, as is often necessary. This was one of the factors that prevented the British firm GEC from acquiring its German counterpart AEG-Telefunken when the latter was having financial problems.

Joint ventures increase the importance of standardization. In March of 1984 the twelve leading European computer and communications firms agreed to draft common standards for the interconnection of their products. In 1985 six European computer manufacturers—STC, Nixdorf, Siemens, Olivetti, Philips, and Bull—decided to base their future computer systems on AT&T's Unix operating system and to develop software for such uses. The following year they agreed to collaborate on Open Systems Interconnection (OSI) standards. Membership was open to other European firms, but American and Japanese companies were pointedly excluded.

Outlook

Communications are becoming too varied, complex, and significant for one organization, together with a handful of favored suppliers, to cover the entire field well. The old arrangements may have been effective for earlier and simpler times, but circumstances change and so must institutions. Some contend that all communications flows should pass through one superpipe controlled by a single organization. This notion, however, is hard to entertain on technical, economic, or political grounds in the information age and in societies operating largely on the market principle, except by reference to the present balance of power. But this condition is not likely to prevail. The traditional arrangement is being challenged from a multitude of centrifugal forces. Demand conditions are changing because the information-based service sector is growing. Technology is changing and merging and propelling the telecommunications industry into the broader electronics sector, with less cozy relations with the PTTs, more competition, and weaker protection. Moreover, the greatly increased volume of international transactions creates pressures of interjurisdictional competition. If one country's PTT exercises restrictive policies, its firms may be disadvantaged internationally, and foreign firms may choose not to locate in that country.

These forces, while having different manifestations in each developed country, are not peculiar to any of them. Consequently, the breakdown of the system of domestic monopoly and international cartel will continue and spread to other industrialized countries.

This does not mean that PTTs will cease to exist or to predominate. They will still function as the core of telecommunications service provision. And indeed, as has been argued above, their role may actually increase through vertical integration into equipment supply. But the exclusivity of their monopoly will become a thing of the past, and they will have to contend with domestic and international rivals. Such rivalry is likely to be the strongest for advanced services and to reach basic telephony later. As this process takes place, the telecommunications network changes from a hierarchical model built on the concept of a star (with PTT control at its center) to one of a matrix (with numerous connecting points between networks that partly collaborate and partly compete, and with software-defined, value-added networks superimposed). Such a configuration cannot be contained within the nation-state. It will require new forms of international cooperation. Because the traditional institutions are not ready to lead but rather will retard this development, new arrangements will emerge. There will be greater involvement of non-PTT international institutions and greater bilateralism. As in air transport, a loose international regime with numerous and specific bilateral agreements is likely to emerge as a transitory system.

Such a new network system will considerably lower terminal and central office equipment prices and reduce the profitability of the postal-industrial coalition. The equipment market will become much more open to foreign manufacturers as well as to members of the "second" electronics industry.

Traditional telecommunications firms will accelerate their rate of innovation or lose out; equipment itself will become unbundled, modular, and specialized so that multiple suppliers can seek their niches. The conventional wisdom that there is room for only six to eight switch manufacturers worldwide will prove as nearsighted as the "mainframe thinking" in the computer industry of the 1960s. Over time, developments will push telecommunications rates toward cost, particularly in the highly profitable international services. These prices will be unstable, since the excess capacity will lead to periodic price wars, and they will be de-averaged among routes of different traffic density and competitiveness.

Once the notions of the traditional network are breached in some respects, the dynamic process of change will be hard to contain; each step of liberalization will lead to a challenge of the next. In international communications the absence of an effective centralized regulatory mechanism leads one to expect the breakdown to be fastest. The growing complexity of the system will make it increasingly difficult to formulate consistent rules. And these rules are not likely to be enforceable. The subject of the regulation—streams of electrons and photons and patterns of signals that constitute information—are so elusive in physical and conceptual terms, and so fast and distance-insensitive, that a regulatory mechanism, to be effective, must be draconian, and for that the traditional system has neither the will nor the political support. Regulatory oversight under which networks and users will interrelate is still needed, but with less control than in the past.

These developments are inevitable, not because they lead necessarily to a superior result, but because the traditional centralized and protectionist network and its international extension into a cartel is an anomaly, though one almost too familiar to be noticed as such. As long as the economic system of Western industrialized democracies is based on markets supplied by private firms, the exclusion of major economic parties from a major field is an unstable affair at best. It is hard to keep a moat between telecommunications and the rest of the economy. To differentiate it as an infrastructure service is conceptually too vague to be useful. Telecommunications, unlike a lighthouse, is not a public good in the classic sense: users can be excluded, and charges can be assessed, breaching some of the criteria for a public good.

The traditional system was international in the sense of a collaboration on the level of government organizations. It held together well because of a similarity in perspective—the values of engineering and bureaucracy—and because of a common interest to protect the domestic arrangements. For a long time national PTT administrations participated almost joyfully in the international sphere because they could return home with an international agreement that would buttress their domestic position. But in the age of satellites, internationalism has become a threat, since it is more and more difficult to reconcile the traditional arrangements with it. And there is much more change to come. For example, we still think of international telecommunications as a federation of networks that are legally, operationally, and territorially based on the nation-state. But the breakdown of the system will not stop at national borders. In the long run telecommunications networks will transcend the territorial concept, and the notion of each country having control over electronic communications may become archaic in the same sense that national control over the spoken and later the written word became largely outmoded in open societies.

As this process of normalization takes place, those identified with the traditional system, who are rightly proud of its technical and social accomplishments, will defend it as best as they can. The transition will therefore be a difficult one. The United States is at the leading edge of the long-term change in international telecommunications—no place to make many friends. Hence one should expect the future to be full of discord as the telecommunications of the developed world move reluctantly toward normalcy.

Information Industries in the Newly Industrializing Countries

Ashoka Mody

In the past two decades some newly industrializing countries (NICs) have shown clear signs of catching up to the industrialized nations in technological capability and productivity levels. They have been formidable competitors in several industries—garments, textiles, shipbuilding, and steel—that are labor-intensive or use mature technologies. Most analysts expect that their capabilities will continue to grow and that they will become serious competitors in knowledge- and capital-intensive industries also.

This transition can be seen in the inroads of the NICs into the international markets for electronics products. Table 1 shows the performance of the major electronics exporters. Exports from newly industrializing countries are small relative to those from Japan, the United States, and the European leaders, which are Germany and the United Kingdom. But Taiwan, Singapore, and Korea are approaching or have exceeded the levels of the other European countries. There is a real possibility that some NICs may leapfrog to the technology frontier. The most striking pointer has been the Korean effort to produce dynamic random access memories (DRAMs). According to a forecast by the Integrated Circuit Engineering Corporation, Samsung of Korea will be the world's ninth-largest merchant producer of integrated circuits by 1996.[1] In both the United States and Japan, the NICs are being viewed as major competitors.

In this analysis I shall be concerned with the information industries in the NICs. I use the term *information industries* to include data processing and communications services as well as electronics hardware. Though diverse, most firms in these industries face common challenges: uncer-

1. See Samuel Weber, "The Look of the Industry in 2000," *Electronics*, vol. 60 (April 2, 1987), p. 60. A *merchant producer* sells its output to others; a *captive producer* manufactures only for its internal needs. Another high-technology area, the light commercial aircraft industry, has seen inroads made by Brazil. See Ravi Sarathy, "High Technology Exports from Newly Industrializing Countries: The Brazilian Commuter Aircraft Industry," *California Management Review*, vol. 27 (Winter 1985), pp. 60–84.

Table 1. *Major Exporters of Electronics Products, 1979, 1983–85*
Billions of dollars

Country	*1979*	*1983*	*1984*	*1985*
Developed countries				
Japan	13.77	26.78	35.50	36.26
United States	14.32	23.09	27.29	26.49
Germany	7.60	8.56	9.25	10.72
United Kingdom	4.75	5.90	7.34	8.77
France	3.96	4.51	5.02	5.86
Netherlands	2.93	3.00	3.55	3.67
Canada	1.31	2.44	3.41	3.47
Italy	2.01	2.51	2.82	3.62
Ireland	0.51	1.45	1.99	2.28
Sweden	1.58	1.76	1.85	2.24
Newly industrializing countries				
Taiwan	2.15	3.23	4.55	4.50
Singapore	2.07	3.36	4.34	4.19
Korea	1.59	2.67	3.65	3.75
Hong Kong	1.35	2.04	2.80	2.25
Malaysia	0.93	1.75	2.30	2.10
Mexico	0.91	1.49	1.79	1.94
Philippines	0.30	0.89	1.15	0.81
Brazil	0.21	0.30	0.44	0.45
Thailand	0.05	0.18	0.31	0.26

Source: General Agreement on Tariffs and Trade, *International Trade, 1985–86* (Geneva, 1986), p. 178.

tainty in technological and market development, high initial investment requirements, and knowledge-intensive production technologies. Although there is no generally accepted definition of a newly industrializing country, one key characteristic is industrial literacy. I shall draw on examples from Brazil, South Korea, Taiwan, India, and Singapore. What these countries share is a relatively developed industrial base, a large stock of educated manpower, and a commitment to research and development and to the promotion of information industries. Despite these common features, specific policies have varied in important respects, as has performance. A comparative assessment, therefore, allows one to identify successful strategies.

As "latecomers," the NICs have had the benefit of access to technology developed in the industrialized countries. There is some evidence that the international diffusion of technology has speeded up in the past two decades, spurred by several factors.[2] First, the technology of transmitting

2. Edwin Mansfield, "R&D and Innovation: Some Empirical Findings," in Zvi Griliches, ed., *R&D, Patents, and Productivity* (University of Chicago Press for National Bureau of Economic Research, 1984), pp. 127–48.

and distributing information has improved. Second, the internal capabilities of the NICs have increased. And finally, the NICs have benefited from intensified international competition among developed countries over the past several decades. More intense rivalry has meant that in seeking markets, developed-country firms have been willing to transfer technology to and form alliances with firms in the NICs. It has also led to cost rationalization and movement to low-cost production sites. Though the influx of new businesses has not always benefited the developing or newly industrializing country, in some instances it has provided exposure to new technologies and has helped train the domestic work force.

As a group the information industries have high barriers to entry: the physical and human investment requirements are often large and probably will become even greater. The fierceness of international competition that has made technology more available also means firms must learn quickly and price aggressively. Even firms in developed countries have often been unable to cope with these pressures and have sought government support in various ways. The NICs have similarly tried to overcome entry barriers by active government participation.

In particular, governments have promoted firms that they believe have the potential to overcome capital and information market imperfections; these include transnational firms, domestic conglomerates, and publicly owned firms. Singapore has almost entirely depended on transnational corporations (TNCs). Taiwan also has encouraged foreign investment, though because it is larger than Singapore, it has made a conscious effort to promote its domestic firms. Brazilian computer and telecommunications industries have historically been dominated by transnational firms; as the Brazilian market has grown and indigenous capabilities have developed, this domination has been unacceptable to some groups, leading to a series of policy measures designed to increase local participation. The two countries that have made the strongest effort to retain domestic control over production are Korea and India. Korea has promoted the growth of privately owned conglomerate firms, whereas India has relied on public sector firms, though private initiative by large business houses has been encouraged in the last decade. To support the operation of the firms, each country has established specialized credit and research institutions.

Each of the big four Korean conglomerates (Samsung, Hyundai, GoldStar, and Daewoo) spans a large range of production activities: construction, shipbuilding, chemicals, heavy machinery, and electronics. Consequently, Korean firms rank by far the largest among the nonpetroleum firms in the NICs. In fact, Samsung and Hyundai of Korea are among the fifty largest industrial corporations in the world.[3] All four Korean conglomerates have major and growing interests in electronics and information industries.[4] The main economic virtue of a conglomerate is that it provides an internal substitute for an external capital market; where external capital markets are not working efficiently, the conglomerate form increases the ability of the firm to undertake large and risky investments.[5] The conglomerate is not a panacea, as can been seen from the often dismal performance of conglomerates in the industrialized countries. What the Koreans may have going for them is that their conglomerates are for the most part still managed by first-generation entrepreneurs; in the next couple of decades, as a new generation takes over, the ability to maintain the current momentum will determine the longer-term prospects of Korean industry.[6]

In the next section, I describe the experience of the NICs with regard to their choice of products and timing of product introduction and evaluate the manufacturing and design capabilities they have achieved. International tension over strategic government intervention and strategic behavior of firms is affecting at least two areas of international policy: trade relations and intellectual property legislation. In the third section, I consider the rationales for stricter intervention and the instruments that have been used in the NICs. The concluding section highlights some emerging issues in the international trade and intellectual property regimes.

The Evolution of Information Industries in the NICs

Information industries are highly diverse. Many information products have a low technology content, while others embody extremely high research input. In table 2, I have classified the industries into four sectors by technology and investment levels. The scale of investment pertains to the minimum required for efficient production. By choosing the correct technology, producers in low-wage countries may reduce the required investment somewhat; the difference, however, is unlikely to be large because the scope for substitution between inputs is not high.

3. Alan Farnham and Carrie Gottlieb, "The World's Largest Industrial Corporations," *Fortune* (August 4, 1986), pp. 170–71.

4. In 1986 Samsung had an electronics-related export target of $1.5 billion; the export targets of GoldStar and Daewoo were $1 billion and $400 million. Hyundai, a relative newcomer to electronics, has already set up considerable production facilities in semiconductors and personal computers. See David Lammers, "Business Is Booming for South Korean Electronics Industry," *Electronic Engineering Times*, October 20, 1986.

5. See Oliver E. Williamson, *Markets and Hierarchies: Analysis and Antitrust Implications—A Study in the Economics of Internal Organization* (Free Press, 1975), pp. 158–59.

6. See Paul Ensor, "Seoul Succession," *Far Eastern Economic Review* (June 18, 1987), pp. 90–91, for a discussion of the transition at Hyundai.

Table 2. *Technological Structure of the Information Industries*

Sector	Products	Level of technology		Minimum investment for efficient production (millions of dollars)
		Design	*Manufacturing*	
Advanced	Advanced semiconductors, computers, telecommunication equipment	High	High	100 and above
Design-intensive	Mini or supermicro computers, software, simpler telephone switching equipment	High	Medium to low	5–25
Medium-technology	Color televisions, video cassette recorders, disk drives, microcomputers	Low to medium	Low to medium	5–50
Low-technology	Black-and-white television sets, passive components, simpler semiconductor devices	Low	Low	1–20

I have distinguished between manufacturing and design in considering technology levels. This distinction is important because barriers to entry are at first glance lower in the design-intensive sector. Countries such as India, Singapore, Taiwan, and the Philippines are hoping to become major players in international design services, particularly software. There is at least some evidence of their success. On the other hand, reputation, credibility, and the ability to keep pace with changing technologies do create significant entry barriers even in design services. I shall examine these conflicting forces in the context of emerging medium-sized firms.

Product Choice and Sequencing

The main effort in NICs has been toward manufacturing capability, but the approaches followed have been quite different.

Electronics production in Taiwan and Korea was started in the late 1950s by Japanese and U.S. firms seeking sites for low-cost production of components. Most of these components embodied elementary technology. Domestic entrepreneurs were soon attracted to this activity, establishing many small firms. Even today component manufacturers are a major sector in both the Korean and Taiwanese electronics industries. Though unable to grow or diversify into other products, these firms have served at least two useful purposes. First, they brought about the development of a capital goods industry that produced the machinery needed for the assembly of components. Second, to feed the development of consumer electronics in these countries, they ensured that a ready and cost-competitive supply of components was available. The ready availability of good-quality and low-cost components continues to give Korean and Taiwanese producers a competitive edge as they move into industrial electronics and office automation products.

The consumer electronics sector in Korea and Taiwan has been the main driver of growth. The Korean conglomerates GoldStar and Samsung, and the large Taiwanese firms Tatung and Sampo, became involved in this sector in the late 1950s and 1960s. Until the 1980s the product choice in the consumer sector was determined essentially by demand trends in the United States conditioned by the perceived degree of Japanese competition. As Japan continually differentiated its products and hence generated markets that were relatively price-inelastic, Korea and Taiwan responded by moving into the price-elastic (mass) market left behind by the Japanese. In the 1980s, however, Korean and Taiwanese firms have attempted to close the gap with the Japanese. They have sought to differentiate their products and to enter markets for more advanced products.

Thus the Koreans and Taiwanese followed a *sequential approach* to product selection for about two decades. By the early 1980s, they had a strong component base, a trained work force, and large firms with experience in manufacturing and marketing consumer goods. The Koreans, in particular, and the Taiwanese, to a lesser extent, are trying to use these advantages to increase their competitive strength in the high-technology sector. The current Korean and Taiwanese approach may be characterized as "big-push," inasmuch as the firms and the governments are engaged in promoting products and technologies across the technology spectrum described in table 2.[7]

Korea's entry into production of dynamic random access memories reflects the commitment to high technology. Semiconductor production, in a broad sense, consists of three stages. First, a substrate (mainly silicon, but increasingly gallium arsenide) is processed. The substrate is in the form of a wafer, and the processing involves imprinting several "dies" on the wafer. Once the wafer has been processed, the dies are sawed off and packaged. Finally, the packaged product is tested. At all times, wafer processing has been significantly more capital-intensive than packaging or testing. In 1984 Korean producers began to develop a capability in 64K DRAMs; in late 1985, they began to produce 256K DRAMs and now are making sample quantities of 1M DRAMs. Because of the fierceness of the competition in DRAMs, they have also shifted to other semiconductor products using similar technologies. It seems clear that the Koreans are moving into semiconductor wafer processing neither for reasons of static comparative advantage nor for reasons suggested by a simple product cycle theory. Instead the efforts represent an attempt to anticipate their comparative advantage. Since organizing semiconductor facilities, recruiting a critical mass of engineers, and coming down the learning curve can be agonizingly slow, one must start early. Reportedly Korean producers at first had significant losses: the slump in international demand and the large international capacity drove prices down faster than had been generally expected. With the appreciation of the yen and the signing of the U.S.-Japan semiconductor pact, however, Japanese competitive strength has declined somewhat and has given the Koreans an opportunity to gain international market share. On the other hand the 1987 finding by the U.S. International Trade Commission that Samsung has infringed on

7. Though I have stressed the similarities between Korea and Taiwan, there are important differences, which I have discussed in "Recent Evolution of Microelectronics in Korea and Taiwan: An Institutional Approach to Comparative Advantage," Discussion Paper 36 (Boston University, Center for Asian Development Studies, 1986).

patents held by Texas Instruments is going to reduce Samsung's revenue streak, at least temporarily.[8]

At the other extreme, India has had no sequencing strategy. For the last two decades, production has been divided more or less equally among components, consumer electronics, industrial electronics (including defense), and communications. As a consequence, India has a very poor component base. Furthermore, institutional linkages between the different electronics sectors are weak, making coordination extremely difficult. Brazil has had similar problems. The institutional linkages are probably even weaker there. Consumer electronics has grown in isolation from the computer or "informatics" sector (the computer hardware and software industry), and despite similar treatment under government policy, the communications and informatics sectors do not seem to have formed linkages. Thus India and Brazil have sought to develop all parts of the electronics complex, but without capturing technological linkages between its sectors and without fostering institutional linkages.

Because Korea and Taiwan paid greater attention to appropriate sequencing than did Brazil or India, they developed better internal linkages. Attention to sequencing allowed Korea and Taiwan to establish manufacturing facilities that exploited technological economies of scale; as a result the facilities created enough domestic demand for upstream components and subsystems to allow the upstream industries to produce at economic production scales. In contrast, because Brazil and India lacked well-thought-out sequencing policies, downstream sectors failed to develop adequate production scales, and that failure in turn restricted the development of the upstream sectors. The Indian electronics industry, for example, has been caught in a trap of small scales of component output. Component prices have been high, resulting in high prices and low demand for final products. That in turn has meant low demand for components: hence the underdeveloped state of the component industry.[9]

Economies of Scale and Competition

Cost-competitiveness in the NICs has been influenced by several interrelated factors: product sequencing, choice of markets (export and domestic), institutional development, and degree of competition.

8. Louise Kehoe, "Samsung Faces Import Ban for Infringement of U.S. Patents," *Financial Times*, September 23, 1987.

9. Because of the poor availability of components, many systems producers must fabricate some required components in-house. The small internal requirements led to loss of scale economies and spread the limited technical and managerial resources more thinly over the activities of the firm.

Economies of scale are hard to achieve if firms spread their efforts among too many product lines or if there are too many competitors for the same product line. In the NICs the existence of a large number of product lines has affected the development of the component industry.

The effects of production scale in the NICs may be understood further by comparing Brazil, Taiwan, and Korea. Frischtak has suggested that the number of firms in the Brazilian informatics sector has been excessive: "Unrestricted entry of national firms in a market of limited economic size may have been responsible for the large unit costs of data processing equipment."[10] The competition has largely been among domestic firms. In Taiwan both foreign and domestic firms have competed in most electronics sectors. According to Frischtak, Brazilian data processing firms have been unable to achieve *technological* economies of scale. By contrast, because Taiwan's economy is export-oriented, at least a few large domestic firms have been able to set up plants that capture technological scale economies. What the Taiwanese firms lack, in comparison with Korean conglomerates, are *organizational* economies of scale. In the buying of inputs and more so in international marketing efforts, Korean firms have been able to do much better than Taiwanese or Brazilian firms because they sell large volumes of single product lines and several products. For example, in the early 1980s Taiwanese firms exported more personal computer systems than did Korean firms. Since 1985, however, Daewoo and Hyundai have begun marketing very large volumes of personal computers, outdoing the smaller Taiwanese firms. Reportedly the Koreans "have poured $100 million into new computer factories to improve efficiency and have been quoting prices that the Taiwanese are struggling to match. To keep up with the South Koreans, the Taiwanese have cut prices. Although a few Taiwanese companies can approach Korean production capacities now, the South Koreans can quickly outpace them by switching factories from TV and VCR production to computers if necessary."[11] Similarly, Korean firms have begun marketing many of their

10. Claudio Frischtak, "Brazil," in Francis W. Rushing and Carole Ganz Brown, eds., *National Policies for Developing High Technology Industries: International Comparisons* (Westview Press, 1986), p. 51.

11. Dori Jones Yang and Laxmi Nakarmi, "The Clone Wars: Frenzy on the Asian Front," *Business Week* (September 29, 1986), pp. 90–91. The competitive strength of Korean PCs has been described widely in the commercial press. See, for example, *Wall Street Journal*, November 6, 1986; and *Business Digest of Southern New Jersey*, May 1986, pp. 18, 21. Korean producers have been successful in more sophisticated markets also. An example is a highly price competitive color graphics and publishing workstation designed and manufactured by Samsung Semiconductor and Telecommunications and distributed by MicroDirect. See MicroDirect, Inc., *News Release*, July 29, 1986, p. 1.

products in the United States under their own brand names; Taiwanese firms are just beginning to do so.

The need for economies of scale conflicts with the need to maintain a competitive environment. An attempt can be made to balance these requirements, as, for example, in the Brazilian telecommunications equipment industry, where four firms have been allowed to participate. It was believed that four firms would be enough to provide a high degree of competition without reducing the potential for achieving economies of scale. It has been difficult, however, to achieve either a significant amount of competition or cost reductions. Ericsson do Brazil controls about half of the Brazilian switching equipment market but produces at a cost 25 percent greater than its parent company in Sweden. At least part of the reason, as in India, is the high cost of components.[12]

So far the only NIC that has been able to achieve a successful balance is Korea, where consumer and industrial electronics are dominated by three firms—Samsung, GoldStar, and Daewoo. The oligopolistic market structure has been accompanied by severe competition. The firms have competed, both in the domestic and export markets, through price, quality, and marketing strategies. Each company has chosen to produce entire product lines (for example, each firm typically produces an entire range of television sets rather than specializing in a few models), making competition even more intense. A number of fierce battles have been fought in the domestic market for video cassette recorders and personal computers.

Thus, while the outlet provided by an export market helps, it is clearly not sufficient, as Taiwan demonstrates. Competitive ability in international markets requires large resources for marketing. Only the Korean firms have become large enough to be major international competitors.

Design-Oriented Sector

For many electronics products, design can be effectively separated from actual production. At the same time, design opportunities are continually evolving. For example, a large company may develop a product for a mass market. But because a wide variety of components are now available and new ones are continually introduced (especially semiconductors), it

12. Bo Göransson, "Enhancing National Technological Capability: The Case of Telecommunications in Brazil," Technology and Development Discussion Paper 158 (Lund, Sweden: University of Lund Research Policy Institute, 1984), pp. 19, 30, 38. Göransson has also noted the importance of economies of scale in purchasing and marketing in the Brazilian telecommunications equipment industry. Ibid., pp. 7–9.

has become possible to make variations in design to meet the requirements of a more focused customer group (often referred to as a "niche" market).

The NICs have a possible advantage in this area because of their far-sighted policies on technical education. The entrepreneurs among the engineers have begun to form medium-sized firms. Typically, such firms become viable when their sales reach the range of $25 million to $50 million and they employ a few hundred people. A large fraction, about 25 percent, of the employees are engineers, and about 10 percent of sales goes toward R&D. The comparative advantage of these firms is based on design capability rather than on manufacturing strength. In contrast, large NIC electronics firms for the most part mass produce complex components and consumer and industrial electronic products; small firms mainly produce simple components. The medium-sized design-oriented firms engage in a variety of tasks: they design sophisticated computer and telecommunication hardware, they engineer systems (such as those to computerize airline reservations), and they write systems and applications software.

Some large national firms and a number of multinational firms have also attempted to take advantage of the availability of relatively cheap engineers and scientists in the NICs. In Korea, Samsung Semiconductor and Telecommunications has engaged a few hundred engineers to design integrated circuits, and GoldStar has links with Olivetti and Hitachi for the development of software. Texas Instruments is reported to have set up an integrated circuit design center in Bangalore, India; Citibank has a software development center in Bombay. Tata of India has been involved with Elexi of the United States in a multinational venture in Singapore. A few multinationals have set up R&D centers in Singapore on a trial basis. IBM has been involved in collaborative software and design ventures in Taiwan.

Utilizing their engineering talent in design activities is a good strategy for the NICs. For some of them, in fact, it is a better use of resources than producing hardware. The technology underlying hardware production is becoming more capital-intensive because of the trend toward automation. This trend applies not merely to advanced semiconductor devices but also to computer peripherals, such as floppy disk drives and printers, that require precision engineering.

Still, realizing this potential comparative advantage in design-oriented activities poses problems. Medium-sized firms face problems of growth. As was noted earlier, they rely on niche markets. Over time, however, a niche market changes character: it either disappears or it grows into a mass market. If the latter happens, large firms with superior resources are attracted. The ensuing competition is usually hard on the smaller firms. Some of the firms survive the competition: Multitech of Taiwan has managed to create international production and marketing links. But most medium-sized firms are unable to withstand the competition from firms with larger resources. The Korean computer industry at first consisted mostly of small firms, but with the entry of the giants, the smaller firms have been going through a difficult phase.

Most observers presume the potential manpower supply in NICs is huge, but this may not be correct. The design sector is currently small (the aggregate annual turnover in India is about $100 million). If domestic and foreign firms are to enlarge this sector severalfold, the manpower requirements will be very high, and it is not certain that the supply will keep pace.

Domestic firms, particularly the smaller ones, face the problem of becoming obsolete. For example, Indian exports of software have consisted largely of "porting," or upgrading and modifying, existing software to match changes in hardware. There are signs that this market is unlikely to grow and may even dwindle because hardware designers have been made more conscious of maintaining compatibility with existing software; moreover, automated tools for "porting" software are being developed. Similarly, the increasing use of artificial intelligence techniques in applications software is going to strain the manpower resources of the NICs.

Strategic Policies

Several arguments are traditionally used to justify government intervention for promoting specific industrial sectors. Lawrence and Litan have classified these into three categories: equity-based rationales (maintaining a balance in wages and profits between different industries), efficiency-based rationales (preserving essential production, protecting infant industries, and compensating for private costs that do not reflect real social costs), and rationales based on political efficacy.[13] The efficiency rationale has been used widely to promote high-technology industries, which have often been regarded as "leading," or associated with strong pecuniary and learning externalities.

Not everyone accepts the arguments for strategic intervention. The main case against intervention is that governments have limited ability to identify sectors that need to be promoted. Some writers also question the existence

13. Robert Z. Lawrence and Robert E. Litan, *Saving Free Trade: A Pragmatic Approach* (Brookings, 1986).

Table 3. *Measures to Promote and Protect Software Industries in Selected Developed and Newly Industrializing Nations*

				Government subsidies	
Country	*Mandatory registration*	*Market reserve*	*Procurement preference*	*Domestic firms*	*Foreign suppliers*
Developed countries					
France	No	No	Yes	Yes	No
Japan	Proposed	No	Yes	Yes	No
United Kingdom	No	No	No	Yes	Yes
Newly industrializing countries					
Brazil	Yes[a]	Yes	Yes	Yes	No
India	No	Yes	Yes	Yes	Yes
Singapore	No	No	No	Yes	Yes
South Korea	No	Proposed	Proposed	Yes	No
Taiwan	No	No	No	Yes	Yes

Source: U.S. Department of Commerce, International Trade Administration, Office of Computers and Business Equipment, *A Competitive Assessment of the United States Software Industry* (GPO, 1984), p. 54.
a. Registration (licensing proposed).

of externalities and hence question the very need for government involvement in industrial promotion.[14]

A new argument in favor of government intervention has appeared in the recent literature.[15] In a world of increasing returns to scale and oligopolistic competition, governments can improve national welfare by subsidizing domestic firms. Unlike perfectly competitive industries, oligopolistic industries are characterized by the prevalence of considerable "economic rents." By subsidizing national firms, governments encourage these firms to behave aggressively (through pricing and other instruments) and thereby gain a larger share of the pool of available "rent." This theory can be applied to high-technology industries also, since they often are oligopolistic. But again, such arguments have been challenged, on both theoretical and practical grounds.[16]

Despite theoretical critiques of strategic government intervention, the role of most governments in industrial targeting has increased in recent years in both developed and newly industrializing countries. The similarity of government policies to promote and protect the software industries is displayed in table 3. None of the developed countries has an explicit

14. For a discussion of this point, see ibid., p. 27.

15. See contributions to Paul R. Krugman, ed., *Strategic Trade Policy and the New International Economics* (MIT Press, 1986), and the literature cited there.

16. One of the strongest critiques has come from Grossman. There is a trade-off between promoting firms in international markets and protecting domestic consumers. Subsidization of one production sector can be an implicit tax on other sectors. There is also no guarantee that firms will use subsidies actively to promote exports. See Gene M. Grossman, "Strategic Export Promotion: A Critique," in ibid., pp. 47–68.

market reserve policy; however, a procurement preference in effect provides a market reserve.[17]

In the rest of this section, I shall describe the policies of NIC governments on import protection, foreign investment, telecommunications infrastructure, and education and research policies.

Import Protection

Protection of the domestic market from imports has been a major policy tool for most NICs. Market reserve has been practiced by Brazil, Korea, and India in a significant manner. Singapore and Hong Kong are possibly the only two states that have had open economies.

Korea has used import protection at successive stages to promote its domestic firms. During the years that Korea was becoming a major producer of consumer electronics, it banned imports of its chief export, television sets. When in the 1980s Korea began to produce computers, the government intervened to protect the domestic market against foreign competition. Because Korean producers cannot yet make high-end computers, the government has been selective in its protection. In 1983 it passed legislation prohibiting the import of most microcomputers, some minicomputers, and selected models of disk drives, printers, terminals, and tape drives. Under the regulations, some exemptions are allowed for imported products destined for use in process control, R&D, or other specialized applications. But exceptions are granted only in the most extenuating circumstances.[18] A similar situation exists in the telecommunications sector. The Electronic Industries Association of Korea subjects almost all important telecommunications equipment imports to a case-by-case recommendation. Before a license can be issued, this association must certify that any given import license application involves a product not currently manufactured or not capable of being manufactured in Korea.[19]

The only major sector with a low degree of import protection in Korea is semiconductors. Several things may explain this anomaly. The initiative for the production of semiconductors came from domestic firms rather than from the government. In fact, the official view on these ventures

17. Nelson has argued that U.S. producers had a large protected domestic market because of government procurement policies. See Richard R. Nelson, *High-Technology Policies: A Five-Nation Comparison* (Washington, D.C.: American Enterprise Institute for Public Policy Research, 1984), p. 68.

18. Based on Ashoka Mody, "Korea's Computer Strategy," Harvard Business School case study, 1985, p. 5.

19. P. S. Kim, "CMP Industry Sector Analysis: Telecommunications Equipment" (Seoul, Korea: U.S. Embassy, October 15, 1985), p. 6.

was hostile, at least until recently. Moreover, the firms obviously felt that they could manage without market protection. A large market exists within the conglomerates and so protection was less necessary than in the earlier cases. Further, the firms have developed the ability to sustain losses over significant periods and so depend less on subsidies in the form of domestic market protection. Though this is an achievement, the Koreans may yet be in trouble. The transfer of subsidization from government control to conglomerates with "deep pockets" is not viewed favorably by U.S. trade administrators. If Korean semiconductor firms start gaining significant market shares, they could well come under scrutiny for "dumping."[20]

Despite its potential problems with "dumping" issues, Korea is the only country among the NICs that has made effective use of its import protection policies. Korean firms have used the time to gain experience and move down the learning curve. They have also made effective use of the domestic market when world demand for their products has been weak. In contrast, Brazil and India have had more stringent import protection policies but have been unable to produce internationally competitive products even after long periods of protection. Thus, although market protection can be a useful device for promoting domestic firms, it is clearly not a substitute for making hard choices about products and institutions.

Foreign Investment

Once again, the three countries that have followed the most restrictive policies are Brazil, India, and Korea.

Domestic producers are generally unable to compete with transnational corporations. The corporations have a number of advantages: better reputation, greater access to technology, and greater economies of scale in production, financing, and marketing. Governments seeking to foster domestic entrepreneurship are therefore inclined to put restrictions on transnational corporations. On the other hand, where the gap between the capabilities of domestic producers and those of the corporations is very large, the costs of promoting domestic entrepreneurship can be high. There is thus a conflict involved in promoting domestic entrepreneurs; the manner in which the conflict plays out depends on the balance of forces within the country.

20. For the first seven months of 1986, Samsung made only $25 million profit on electronics-related sales of $1.23 billion. The low profits are partial evidence of cross subsidy. Dumping charges might come not only from the United States but also from the European Community. See Lammers, "Business Is Booming," p. 10.

Two contrasting models are those of Brazil and Korea. Brazil has sought to create large national computer and communications industries of high standards. The computer industry was promoted from within the state apparatus by technically trained bureaucrats.[21] In the early 1970s, Brazil had a substantial stock of engineers and scientists who had been trained in its own universities; some of them had also done graduate work in the United States. They had a strong commitment to the creation of a domestic computer industry. In 1972 an interministerial commission under the Ministry of Planning was set up to regulate the computer industry. Known as CAPRE (Commission for the Coordination of Electronic Data Processing Activities), it became the main vehicle for the realization of domestic technical interests. In the mid-1970s the focus was on preventing foreign competition in the minicomputer market. Production capabilities never reached international standards, but significant local learning took place. Domestic computer firms did more R&D and employed more engineers relative to sales than did the transnational corporations. In 1979 CAPRE was dissolved and SEI (Special Secretariat of Informatics) was set up to guide computer policy. SEI personnel had much less technical competence than did the CAPRE engineers. The CAPRE policies, however, were carried forward and even strengthened. At the same time, a new focus of attention emerged: the microcomputer.

During the early phase, large business interests in Brazil showed limited interest in the development of these technologies. Evans has noted that the traditional names in Brazilian industrial capital—Votorantim, Villares, Bardella—were not directly involved in the industry as producers.[22] Financial interests were also slow to get actively involved. By the late 1970s and particularly in the 1980s, however, the Brazilian banking sector had become heavily involved in the domestic computer industry. Bradesco and Itau, Brazil's largest banks, developed major interests in the largest local computer firms.

21. There is a large literature on the Brazilian computer industry: Peter Evans, "Varieties of Nationalism: The Politics of the Brazilian Computer Industry," in Antonio Botelho and Peter H. Smith, eds., *The Computer Question in Brazil: High Technology in a Developing Society* (Massachusetts Institute of Technology, Center for International Studies, 1985); Peter B. Evans, "State, Capital, and the Transformation of Dependence: The Brazilian Computer Case," Working Paper 6 (Brown University, Center for the Comparative Study of Development, 1985); Emanuel Adler, "Ideological 'Guerrillas' and the Quest for Technological Autonomy: Brazil's Domestic Computer Industry," *International Organization*, vol. 40 (Summer 1986), pp. 673–705; and Ravi Ramamurti, "Brazil's Computer Strategy," Harvard Business School case study, 1985. My description is based primarily on the papers by Evans.

22. Evans, "Varieties of Nationalism," p. 10.

The entry of financial capital has changed the Brazilian strategy. Whereas the technological nationalists have sought to minimize imported technology, financial capital does not have similar interests. Indeed, financial capital has negotiated attractive technology transfer agreements.[23] The big banks have great negotiating strength. They also have an interest in maintaining good relations with transnational corporations. In particular, they have made several attempts to promote joint ventures.[24] Brazilian industrialists are also setting up joint ventures. Recent examples are IBM with Gerdau, a steelmaker, to provide data processing services (IBM will hold a 30 percent equity stake); and Hewlett-Packard with Edisa, a commercial, industrial, and banking automation concern, to produce minicomputers and supermini computers.[25]

In the telecommunications sector also the Brazilian goal was to increase participation of domestic entrepreneurs and increase indigenous technological capability. The strategy followed, however, was different. In 1977 established transnational corporations were asked to transfer majority voting rights to Brazilians to produce in Brazil. (The transfer of voting rights did not imply an equivalent transfer of equity; Ericsson maintains a two-thirds' equity interest in Ericsson do Brazil, the largest producer of switches in Brazil.) At the same time, Telebras, the state-owned telecommunications company, created a Research and Development Center, CPqD, and encouraged producers to do their own research. Telebras sought to influence the use of domestic technology through its purchase decisions. Though there are obvious similarities between the telecommunications and informatics policies, the insistence on local technological effort has been weaker in the telecommunications sector. There has been some conflict between the Ministry of Science and Technology, which has promoted domestic technology, and the Ministry of Communications, which has viewed excessive control over imports of technologies and products as an impediment to the expansion of the telecommunications network.[26]

Several developments suggest that Brazil may loosen its restrictions on foreign capital, technologies, and goods. In late 1986 Brazil agreed to permit up to 40 percent equity participation in joint ventures, up from the earlier 30 percent.[27] Telebras recently awarded Ericsson a $100 million

23. Digital Equipment Corporation has licensed its VAX 11/750, and Data General has licensed its MV4000 and MV8000.

24. See Evans, "State, Capital, and the Transformation of Dependence," p. 36.

25. "Brazil's Informatics MNCs Turn to Joint Ventures to Survive Market Reserve," *Business Latin America* (October 27, 1986), pp. 330, 331.

26. *New York Times*, September 16, 1985.

27. "Management Alert," *Business Latin America* (August 25, 1986), p. 264.

contract to provide local and transit capital AXE digital exchanges for 600,000 subscriber lines.[28] Finally, there are signs that trade barriers will also be relaxed.[29]

Korea has also followed restrictive policies vis-à-vis foreign investment. Nonetheless, there have been important differences between Korea and Brazil. The crucial one has been that major Korean industrial groups have had an active interest in electronics; indeed, electronics has been in the vanguard of their efforts. To a large extent, therefore, technological nationalists have joined with industrial capital in the Korean conglomerates, creating a stronger basis for restriction of foreign investment.

In the 1980s Korean firms have sought greater access to foreign technology. As far as possible, they have tried to buy technology through licensing agreements. Where licensing has been insufficient, they have entered into joint ventures. Most of these licensing agreements and joint ventures have been U.S. firms, since Japanese firms have been reluctant to share technology with the Koreans.[30] U.S. firms have viewed a larger presence in Korea as an important marketing strategy. For example, Robert N. Noyce, Intel's vice chairman, said of his company's link with Samsung: "We believe Samsung's experience in semiconductor manufacturing and their commitment to this sector will help spread the presence of the Intel architecture in the Pacific Basin to our mutual benefit."[31] The Korean conglomerates have been forming joint ventures with U.S. firms from a much stronger position than most developing-country or NIC firms are able to; hence the Koreans have been able to negotiate superior technology transfer deals.[32]

Technology Policy

For the most part, NICs have not engaged in basic research. Their main effort has been to generate the educated work force needed to absorb and modify the technology being developed in industrialized countries. Though

28. *Financial Times*, January 16, 1986.

29. See Jim Van Nostrand, "Brazil to Relax Trade Barriers," *Electronic Engineering Times*, December 29, 1986.

30. One sign of how restricted ordinary technology flows are from Japan to Korea is the large numbers of "moonlighting" Japanese engineers in Seoul over the weekends.

31. See Lee Dong-Geun, "Samsung Makes Electronics Longjump," *Business Korea* (March 1985), p. 62.

32. "Foreign telecommunications companies are welcome, but they've learned that they must work with a local partner and they must meet some very demanding technology transfer conditions." Oles Gadacz, "Reaching for the Information Age," *Business Korea* (May 1986), p. 18.

general educational policies have been critical to forming a base of scientists and engineers, specialized institutions have been found necessary to gain faster access to information technologies. The performance of these institutions has depended on their scale of operation, on their degree of commercial orientation, and on the strengths and weaknesses of domestic firms. In this section, I shall describe some of the principal institutional research efforts in Korea, Taiwan, Brazil, India, and Singapore.

KOREA. The Korea Institute of Advanced Science and Technology (KAIST), the Korea Institute of Electronics Technology (KIET), and the Korea Electrotechnology and Telecommunications Research Institute (KETRI) have been the principal institutions focused on electronics and related technology. KAIST has been charged with producing the several thousand doctorate and master's degree holders that the Korean electronics industry needs. The commercial tasks of product and process development have largely been the responsibilities of KIET and KETRI (operating for the past few years as one unit, the Electronics and Telecommunications Research Institute, or ETRI). KIET was set up in 1979 to demonstrate that semiconductors could be produced in Korea. By the early 1980s, Korean conglomerates had outgrown KIET's capabilities by licensing foreign technology and setting up internal R&D departments and "technology-watch" outposts in Silicon Valley in California. Daewoo negotiated the purchase of KIET's semiconductor production facilities, but the transaction never went through, and Daewoo instead bought Zymos, a U.S. firm, to gain access to semiconductor technology. KIET proved to be a catalyst; however, it could not continue to function as a common semiconductor research center for domestic industry because of strong competition and mutual suspicion among the large Korean firms. ETRI, like KIET, has for the most part played a vanguard role. It is, for example, engaged in developing communications protocol for the integrated services digital network, optical transmission devices, and an earth-station for satellite communications.[33]

TAIWAN. In Taiwan the Electronics Research and Service Organization (ERSO) is the focus of public research, while Hsinchu Science Industrial Park, like Silicon Valley, promotes technological synergism. ERSO has had a strong commercial focus, including one of the two silicon foundries in Taiwan (other foundries are being built by private entrepreneurs). ERSO has had much closer ties with domestic industry than KIET in Korea. The Taiwanese firms are much smaller than the Korean firms, with less ability

33. Ibid., pp. 18–29.

to undertake independent research initiatives. Hence the public role and the degree of cooperation between firms is high in Taiwan.

BRAZIL. Despite the importance accorded to the computer sector, no special institution was developed in Brazil for doing research on computer technology, probably because minicomputer and microcomputer technologies were widely known. Nonetheless some research has been conducted within firms (and to some extent in university departments). Indeed, ratios of R&D to sales have been high in Brazilian firms (8 to 10 percent), and they have also had a large proportion of engineers in their work force.

An institution was created for telecommunications research: the Research and Development Center, known as CPqD. It carries out its own research program and also coordinates research at universities and in the industry. The universities are supposed to do basic research; CPqD and the industry share the tasks of prototype and product development. Brazil has followed a big-push policy in telecommunications, involving an across-the-board attempt at technological competence. There have been both successes and failures. Electromechanical switching has been an example of successful technology development, whereas transmission systems have been more difficult to master. In the area of digital switching, the Brazilians have developed small exchanges, but they have been unable to develop large digital switches.[34]

INDIA. India has several electronics research facilities. In addition, Indian electronics firms, like their Brazilian counterparts, have devoted large shares of their sales to R&D. Still, Indian research efforts have been dispersed and diffuse, even more than those of Brazil. The small scale of the efforts has entailed low research productivity. Moreover the public research institutions have had few links with commercial or market requirements.

The Centre for Development of Telematics (CDOT) is an exception to the general characterization of the Indian R&D sector. CDOT, which is developing small digital exchanges, has had a strong commercial orientation from the start. It has sought to design its products with a common set of components and to work with producers to make component production more efficient. It is, however, too early to judge whether this approach will be successful, since production based on CDOT technology is yet to begin.

SINGAPORE. In Singapore, government efforts have been mainly focused on manpower development. For manufacturing technology, Sin-

34. Personal communication with Bjorn Wellenius, World Bank, Washington, D.C.

Table 4. *Telecommunications Infrastructures in Korea, Brazil, Singapore, China, and India, Selected Years, 1975–85*

Criterion	*Year*	*Korea*	*Brazil*	*Singapore*	*China*	*India*
Telephone sets for 100 people	1975	4.04	2.93	5.24	0.36	0.29
	1983	14.89	7.69	34.24	0.49	0.45
	1985	18.63	8.44	41.73	0.60	0.50
Percentage of main lines connected to automatic exchanges	1975	78	86	100	21	82
	1983	93	99	100	39	86
	1985	97	99	100	45	87
Outgoing international calls (million calls)	1975	1.32	1.28	0.84	0.08	0.39
	1983	5.00	7.48	9.67	4.48	2.56
	1985	9.90	10.40	15.73	12.60	5.35
Number of private leased lines (thousands)	1983	10.0	0.1	0.5	n.a.	0.1
	1985	17.4	9.9	21.2	n.a.	0.4
Telecommunications investments as a share of total domestic investment (percent)	1975	33.4	45.4	13.9	n.a.	12.5
	1983	69.1	22.1	17.9	n.a.	15.7
	1985	68.2	n.a.	23.7	n.a.	16.1

Source: For 1975 and 1983, see International Telecommunication Union, *Yearbook of Common Carrier Telecommunication Statistics: Chronological Series, 1974–83* (Geneva: ITU, 1985), pp. 80–81, 102–03, 114–15, 176–77, 312–13; for 1985, see ibid., *Chronological Series 1977–86* (ITU, 1988), pp. 82–83, 104–05, 116–17, 184–85, 330–31.
n.a. Not available.

gapore has depended on the R&D efforts of multinationals. In keeping with its goal of being a chief center of information technology, Singapore set up three computer training and research institutes in 1981 and 1982: the Institute of Systems Science (ISS), a partnership between the National University of Singapore and IBM; the Japan-Singapore Institute of Software Technology, a joint research project between the governments of Singapore and Japan; and the Center for Computer Studies, a partnership between ICL and Ngee Ann Polytechnic.[35] The emphasis is on developing capability in artificial intelligence techniques, gaining expertise in software production under the Unix operating environment, and increasing software productivity through program generators and other software tools.

Telecommunications Infrastructure

A well-developed telecommunications infrastructure is essential for the diffusion of information technologies. Table 4 provides some indication of public and private efforts at creating such infrastructures in Korea, Brazil, Singapore, China, and India. The most rapid infrastructure development has taken place in Singapore, which, as was mentioned earlier, is positioning itself as a primary information center in Asia.[36]

35. Government of Singapore, *National IT Plan: A Strategic Framework* (Singapore, 1985).

36. The small share of telecommunications investment in total domestic investment in Singapore for 1975 in table 4 is not representative; through the second half of the 1970s, about one-third of domestic investment went into telecommunications.

The country with the most ambitious telecommunications program has been Korea. The share of telecommunications investment in total domestic investment has been high and growing. The number of telephone sets for every 100 persons is second only to Singapore (and about twice as high as Brazil). Of all the countries listed in table 4, Korea has by far the highest percentage of telecommunications investments as a share of total domestic investment.

Korea has also developed a public switched data network, operated by Data Communications Corporation of Korea (Dacom). Dacom is 55 percent privately owned and 45 percent owned by the government. There is no comparable service in the other NICs. The use of the Dacom-Net service has been limited to 600 subscribers (as of May 1986), whereas the leased line service has 17,400 subscribers (up from 10,000 in 1983, as shown in table 4). Still, observers expect that the Dacom-Net usage will grow as value-added services become available.

Emerging International Regulatory Issues

The prospects of the NICs are going to depend a great deal on the way the international trade regime evolves. More protectionism could seriously affect growth of the information industries. This danger looms not only for export-oriented countries, but also for countries, such as India, that have been trying to liberalize imports; to sustain a high level of imports, outlets for exports are required.

One reason for the possible growth of protectionism is the large trade deficits that the United States has been incurring vis-à-vis certain countries. From table 5, one can see that the U.S. deficit in the trade of electronics products has been particularly large with Korea and Taiwan. Such deficits create pressures for protection of domestic industries. In the case of Korea,

Table 5. *U.S. Electronics Balance of Trade, 1985*
Millions of dollars

Country	*U.S. exports to*	*U.S. exports from*	*Balance of trade*
Brazil	546.9	221.6	325.4
India	199.8	28.2	171.7
China	525.2	38.2	487.0
Singapore	1,058.9	2,371.3	−1,312.2
Taiwan	725.8	3,218.7	−2,493.1
Korea	1,003.9	2,521.5	−1,517.7

Source: Electronic Industries Association, *Electronic Market Data Book, 1986* (Washington, D.C.: EIA, 1986), pp. 137–39.

the United States has already levied antidumping duties for color television sets. The general U.S. attitude has been to exert bilateral pressure, asking the particular NIC to open up its market or asking it to restrain exports to the United States. South Korea and Taiwan have been under pressure to revalue their exchange rates. In addition, the United States has been using its privilege of withdrawing favorable terms of export under the "generalized system of preferences" as a mechanism for inducing restraint on the part of countries with large trade surpluses vis-à-vis the United States.[37] Such bilateral pressure, however, only shifts the focus of the international problem. There is an underlying multilateral issue that needs to be addressed. The NICs have much greater difficulty exporting to the two other large markets, Japan and the European Community, than they do to the United States. If the exports of the NICs were spread over the United States, Japan, and Europe, the American trade imbalance would be much smaller and hence protectionist pressures would be much weaker.

Table 6 shows the trade in electronics between the industrialized nations (the United States, Japan, and the European Community), and three newly industrializing countries (Korea, Taiwan, and Malaysia), for 1985. Three features of the trade pattern are worth noting. First, for each of the three categories of products and for each of the countries, Japan has a trade surplus, whereas the United States and the EC have deficits, in eight of the nine slots (the exception being data processing equipment trade with Malaysia). Second, the EC has a much smaller deficit than the United States because its level of imports, as well as its level of total trade, is significantly lower. And third, although Japan's *exports* to the NICs are comparable to those of the United States, Japan has a large surplus because its *imports* from the NICs are negligible. Some might argue that the production pattern of Japan is much like those of Korea and Taiwan, and hence the need to import is limited. But this can only be a part of the story, since Japan imports very little even from Malaysia, a country at a much lower level of development than Korea or Taiwan. The argument is not sustainable, moreover, because Korea and Taiwan import huge amounts from Japan.

South Korea has responded by planning to switch some of its imports from other countries to the United States. Though this has pleased U.S. officials, it is unlikely that the basic multilateral problem will be mitigated. A part of the switch will occur in agricultural products that Korea currently

37. "U.S. Cuts Back on Third World Imports with Duty-Free Status," *Financial Times*, April 7, 1987.

Table 6. *Electronics Trade between Selected Industrialized and Newly Industrializing Countries, 1985*
Millions of dollars

Newly industrializing country	*United States*	*Japan*	*European Community*
		Data processing equipment	
Korea			
Exports to	334	32	173
Imports from	206	215	37
Taiwan			
Exports to	887	42	305
Imports from	229	145	20
Malaysia			
Exports to	36	0	13
Imports from	42	24	21
		Telecommunications equipment	
Korea			
Exports to	1,290	106	220
Imports from	171	311	60
Taiwan			
Exports to	1,820	73	239
Imports from	143	242	44
Malaysia			
Exports to	226	2	124
Imports from	18	167	60
		Other electrical apparatus	
Korea			
Exports to	1,098	232	201
Imports from	592	977	110
Taiwan			
Exports to	1,404	184	396
Imports from	365	932	157
Malaysia			
Exports to	1,217	53	400
Imports from	849	348	134

Source: Organization for Economic Cooperation and Development, *Foreign Trade by Commodities, 1985: Exports*, vol. 1 (Paris: OECD, 1987), pp. 244, 247, 250; and *Foreign Trade by Commodities, 1985: Imports*, vol. 2, pp. 217, 219, 222.

buys from Argentina and China. Korea also plans to switch imports of raw materials and components from Japan to the United States; the value of this switch, however, is expected to be only $250 million. At the same time Korean and Taiwanese firms are increasing their direct investment in the United States. Korean firms have also been voluntarily imposing price floors to avoid charges of dumping.[38]

More specific trade issues relate to transborder data flows and differential degrees of telecommunications regulation. Transborder flows of

38. *Financial Times*, April 27, 1987; Maggie Ford, ''GoldStar Looks Abroad for Growth,'' *Financial Times*, July 28, 1987; Philip Liu, ''Taiwanese Electronics Firms Step Up U.S. Investment,'' *Electronic Business*, vol. 13 (March 1, 1987), pp. 30–32; and *Wall Street Journal*, July 30, 1987.

economic data do not fit into existing regulatory mechanisms. Barriers to transborder data flow arise from legal regulations; from regulations of post, telegraph, and telephone administrations (PTTs); and from practices and technical difficulties, such as standards incompatibilities.[39] And their economic importance is tremendous. Countries that produce the data (mainly the United States) are seeking open international markets; user countries are trying to develop their own capabilities for producing and transmitting data. Observers expect that transborder data flows will be treated under the ongoing negotiations on trade in services.

The deregulation of American telecommunications has led to the removal of a nontariff barrier: foreign companies can now sell telecommunications equipment in the United States, whereas previously Western Electric was virtually the sole supplier to the Bell System. This new openness has created U.S. demands for reciprocity from other countries.[40] The evolution of deregulation, or the lack of it, in other countries will have a great effect on the growth of international trade in telecommunications equipment.

Another primary trend can be seen in the international arena. Intellectual property has emerged as the chief competitive weapon in the 1980s.[41] Manufacturing skills have begun to diffuse quite rapidly, both within the industrialized nations and to the NICs. To keep ahead of the competition, it is becoming more and more important to introduce products embodying superior technology. The higher perceived value of superior technology can be seen in the greatly increased litigation for keeping control over "proprietary technology."[42] Several examples highlight the concern over adequate technology protection and valuation. Matsushita (Japan) paid IBM about $2 million for infringing on software built into IBM's Personal Computer AT. The settlement amount was small, but IBM wished to make plain its determination to protect intellectual property. As one technology analyst has predicted, "Every copyright, every patent right, IBM is going to push to the absolute limit."[43] Texas Instruments has made a number of moves to capitalize on its technology: it has filed an action with the International Trade Commission, seeking relief from injury because of

39. Karl P. Sauvant, *International Transactions in Services: The Politics of Transborder Data Flows* (Westview Press, 1986), pp. 162, 208.

40. Ibid., p. 224.

41. Andrew S. Grove, president of Intel Corp., has stated, "Our very existence is based on the development of intellectual properties." See Andy Grove, "Protecting Intellectual Property," *Electronic Engineering Times*, February 2, 1987, p. 90.

42. Clifford Barney, "Intellectual Property Turns into High-Priced Real Estate," *Electronics*, vol. 60 (April 30, 1987), p. 43.

43. "Matsushita to Pay Firm over Copyrights," *Wall Street Journal*, February 25, 1987.

patent infringement by eight Japanese and one Korean company (Samsung); it has sought to increase royalties on its technologies to 13 percent from the industry norm of 1 percent; and it has been actively patenting in the area of three-dimensional memory cells as a means of creating entry barriers.[44] Intel has charged Hyundai with violating its patents on erasable programmable read-only memories and filed a petition before the ITC seeking prevention of Hyundai sales in the United States.[45] National Semiconductor is suing United Microelectronics Corporation of Taiwan for patent infringement.

Alongside the perceived increase in the value of intellectual property, the U.S. judicial system has been giving it greater protection: "At the patent antitrust interface we deal with two polarities: some people are primarily motivated by their respect for 'property,' others by their abhorrence of 'monopoly.' . . . For several decades, antitrust has been supreme. But recent years should encourage those possessing the 'property' bias, for the pendulum seems to be swinging back."[46]

Existing systems of protection have sometimes been inadequate. Special legislation was passed, first in the United States and then in Japan, to protect the masks that contain the patterns to be printed on silicon for making integrated circuits. There has been some debate on whether software is best protected by patent (which protects novel ideas), copyright (which protects expression), or "trade secret" laws. The courts seem to be favoring copyright; however, there are as yet no firmly established guidelines as to what constitutes software.

The fuzziness regarding the legal status of information technologies extends to the international arena. This is partly because of differing philosophies underlying intellectual property protection and partly because of the lack of an agreement on the best method of protecting the emerging information technologies.

There are two international conventions on copyright laws: the Berne Convention (administered by the World Intellectual Property Organization)

44. Ashoka Mody and David Wheeler, "Technological Evolution of the Semiconductor Industry," *Technological Forecasting and Social Change*, vol. 30 (1986), p. 203. On the past alleged infringements, most Japanese companies have settled with Texas Instruments by paying about $100 million. Samsung, which did not settle out of court, has been found guilty by the ITC of infringing on patents held by Texas Instruments and consequently has been banned from selling DRAMs in the United States. The ban may not take place because Texas Instruments has granted Samsung temporary licenses to its patents.

45. Louise Kehoe, "Intel Suit Could Spark Korea Chip Battle," *Financial Times*, August 7, 1987.

46. David Bender, "Technology Transfer Lessons from Selected Cases," in Tom Arnold and Thomas F. Smegal, Jr., eds., *Technology Licensing* (New York: Practising Law Institute, 1982), pp. 164–65.

and the Universal Copyright Convention (administered by the United Nations Educational, Scientific, and Cultural Organization, or UNESCO). The United States is not a member of the Berne Convention. It is a signatory to the Universal Copyright Convention; however, U.S. withdrawal from UNESCO has prevented it from participating in the UNESCO General Conference, which reviews and approves the various budgets and administrative bodies of UNESCO, including the Copyright Division. The U.S. Office of Technology Assessment notes that, as a result, "the U.S. ability to influence other nations in its favor might be weakened."[47]

Given the limited American involvement in international conventions, the United States has been following a policy of bilateral negotiations to protect intellectual property.[48] It has threatened to withdraw protection of foreign software if U.S. software is not protected in the country concerned; in addition, the NICs and developing countries are in danger of losing preferential import treatment if they do not protect intellectual property in the manner it is protected in the United States. Several countries have in the past year passed new copyright laws, including Singapore, Japan, Korea, and Taiwan. There has been continued friction with Brazil and China, however, and deficiencies have persisted in the Korean patent law. In the meantime, the U.S. policy of bilateralism is creating pressures similar to those described earlier for the international trade regime. South Korea entered into a bilateral agreement with the United States giving retroactive protection as of 1980; for all other countries Korean protection was granted in the main only to trademarks, copyrights, and patents that came into force as of July 1, 1987. Korea's policy stirred up considerable resentment, particularly in Britain, which formally complained to the South Korean government. South Korea has since agreed to give EC firms the same rights as their U.S. counterparts, but many issues are still to be resolved; the status accorded to the rest of the world is not clear.[49]

47. U.S. Congress, Office of Technology Assessment, *Intellectual Property Rights in an Age of Electronics and Information*, OTA-CIT-302 (Washington, D.C., April 1986), p. 223.

48. The U.S. record puts it at some disadvantage in exerting pressure on other countries: "It was not until 1891, when Congress passed the Chace International Copyright Act, that the United States began to recognize international copyright relations. The act, however, provided neither for multilateral agreements, nor for the protection for foreign works manufactured outside the United States. But it did extend copyright relations to nations found and proclaimed by the President to afford adequate protection to American works. This act provided the basis for all of the U.S. bilateral copyright relations for more than the next 60 years." Ibid., p. 215.

49. See Maggie Ford, "Britain Complains Again about Korean Patent Law," *Financial Times*, March 24, 1987; Quentin Peel, "South Korea Gives Patents Pledge," *Financial Times*, May 1, 1987; and Maggie Ford, "EC Set for Row over South Korean Patent Laws," *Financial Times*, September 11, 1987.

Such tensions are likely to accelerate the trend toward greater protection of intellectual property. That would have important implications for NICs. If international protection becomes really effective, NICs will either have a smaller supply of technology or will have to pay more for the technology they receive. In either case, their competitive strength will be weakened.

Conclusion

The linkages between the technologies characteristic of the information industries, and the uncertainties about how these technologies evolve, suggest that a broad (as distinct from a focused) involvement is desirable. On the other hand, economies of scale and large investments suggest the need for focus. The experience of Korea suggests that both breadth and focus are desirable. Focus is needed in the early stages to develop simple production and organization skills. Once such skills have been developed, a broader approach, taking advantage of the technological complementarities and hedging against the uncertainties, is desirable. Hedging against uncertainties does not, however, imply promotion of all sectors or technologies. Diversification should be used to learn the potential of the different technologies; after a period of experimentation or learning, one or a limited number of technologies must be chosen for more intensive development.

Of the countries I have considered in this analysis, India has the least-developed manufacturing technology and telecommunications infrastructure. Korea and Taiwan are at the other end, with the best-developed manufacturing skills and infrastructure. Brazil stands in between. Singapore, because of its small size, is in a category by itself. It has a world-class infrastructure. It also has a highly trained labor force. Domestic entrepreneurship in Singapore is, however, very limited.

For India, and to some extent Brazil, development of manufacturing skills in products of low sophistication (passive components, black-and-white television sets) may be a prerequisite for movement to more complex products. The less-advanced sectors in the information industries tend to have lower capital and skill requirements, making them suitable as transitional sectors. Even if such skill development were not viewed as a prerequisite, India and Brazil would have to pay significant attention to the development of the simpler skills *alongside* the development of more advanced technologies. Similarly, telecommunications development would have to be a high priority.

In Korea and Taiwan, where the prerequisites are in place, the equally hard questions of competing internationally will have to be considered. The high-growth products are also characterized by significant uncertainty, steep learning curves, and short product cycles. To compete internationally, domestic firms must be able to undertake large investments and to sustain losses over time. In this regard, the Korean conglomerates are better positioned than the smaller Taiwanese firms.

Given the externalities associated with information industries and the difficulties in competing internationally, the governments of the NICs have used several mechanisms to promote the industries and the domestic firms. These have included import protection, restriction of foreign investment, promotion of domestic R&D, and the development of telecommunications infrastructure. The role of the government has varied with the needs of the industry. It has typically been stronger when private institutions have been less developed. For example, in the 1970s the Korean government played a strong role in the development of the electronics industry (through import and foreign investment controls and through setting up KIET). In the 1980s the Korean firms have needed less support. In contrast, as the smaller Taiwanese firms have attempted to move to sophisticated technologies, the Taiwanese government has been more closely involved in providing a research infrastructure and in actual production. Note, however, that recently the Korean government has also been taking interest in its semiconductor industry. The government and presumably the Korean firms themselves feel that they are at a stage at which common development of basic technologies could give them a leg up. Thus the strategic role of the government is seen to change over time, responding to the needs of domestic industry. Not all governments are able to effect such "benign" strategic involvement. An understanding of dynamic government involvement, as distinct from static or one-period intervention, should be a clear research priority.[50]

It should be emphasized that government involvement is not a sufficient condition for the growth of high-technology sectors. Clearly a set of finely balanced steps is needed. To gain manufacturing skills, large production scales are desirable, yet to promote competition, scale economies may have to be sacrificed. Such conflicts can partly be resolved by exporting to large markets; in the future, however, trade protection may reduce the access to export markets. Adequate institutional development is needed to overcome market failures; institutional development can also be a means

50. For one such attempt, see Robert Wade, *Governing the Market: Economic Theory and the Role of Government in Taiwan's Industrialization* (Princeton University Press, forthcoming).

of resolving national conflicts. For example, Brazil has had much less political consensus on the development of the computer and telecommunications industry than has Korea; in Korea the interests have been partly coalesced within the conglomerates, and the firms have had close ties with the government.

The serious challenges facing the more successful NICs relate to the evolving international environment. Access to markets and technology will be the two key issues. As the NIC firms have become more competitive, firms in industrialized countries and their governments have become more wary about parting with technology. Japanese firms in particular have followed a policy of minimizing sales of technology. U.S. firms have been more open in this regard; the expectation has been that by collaborating with the NICs, they will be able to develop new markets. At the same time, the U.S. government has made several efforts to exert bilateral pressure on the NICs (and Japan) to follow the U.S. direction in creating an international intellectual property regime to meet the requirements of the emerging technologies. These developments are at an early stage, and their evolution will have an important bearing on international trade and investment.

Growth of the Telecommunications and Computer Industries

Nestor E. Terleckyj

In the early 1960s the communications and computer industries, along with some other industries, began to coalesce into what is sometimes called the information sector. Broadly defined, this sector comprises the development and manufacture of communication equipment, of computers and electronic products, and of software, and the production of telecommunications and computer services as they are performed by distinct firms for sale in the market.

Starting in the early or middle 1960s, the environment of the telephone and telecommunications industry changed as satellite and microwave technologies opened up possibilities of competition in long-distance telephone service markets. The commercial computer industry developed independently, standardized computer languages emerged, and long-distance computer communications were established.

This paper focuses on the growth trends in the computer manufacturing and telecommunications service industries. The growth model developed here sheds light on a number of interesting issues: the determinants of industrial R&D spending, the effects of R&D on productivity, the relationship between productivity change and output prices, the impact of price on demand, interactions between the computer and communications industries and the impact of government on their growth, and the links between foreign and domestic markets. The available data are used to analyze the interdependence of the growth of the two industries over the 1964–85 period. Little evidence of spillovers between the two industries to date is found.[1]

I thank David M. Levy for help in developing and conducting a number of analyses as well as for many suggestions; John Arcate, Martin Neil Baily, Paul Brandon, Robert W. Crandall, Kenneth Flamm, and Harold Furchgott-Roth for many useful comments; and Charles D. Coleman, Houri Ramo, and Jacqueline Rupel for help in research and in the preparation of the paper. The research reported here was supported by the research grant IST-8521100 from the Information Sciences Division of the National Science Foundation.

1. The same conclusion may be found in Flamm's paper in this volume.

A Growth Model of Technology-Intensive Industries

Technology-intensive, or high-technology, industries differ in their growth patterns from other industries largely because of innovations that produce generally high rates of growth in output and productivity and high rates of change in products or in the processes of production. These innovations accumulate into stocks of technological capital. This capital is then employed in the development of new technologies for use in particular industries (though not necessarily the same industries in which the investments are made). Such innovations are particularly effective over intermediate to moderately long periods. Over periods of about one year, other factors may dominate growth. Over very long periods of several decades or more, individual industries may enter and leave the technology-intensive category and move to different growth paths.

As a rule, investments in technology development can be measured by expenditures for research and development. These investments may rival and even surpass the amounts invested in fixed plant and equipment. That is perhaps the most distinctive characteristic of technology-intensive industries.

The social returns from investments in technology can be measured by growth in productivity or by declines in the relative real unit factor cost attributable to these investments. The two approaches—measuring the value of the decline in real unit costs or the value of the increase in total factor productivity—are in principle equivalent. But the social returns resulting from productivity gains attributable to causes other than investments in technology have to be separated from them. The dynamic growth processes that characterize technology-intensive industries can be sustained for long periods, until the industries "mature" or are replaced by later innovations.

The relationships involved in the dynamic growth process of a high-technology industry may be represented by an R&D investment function, a growth function for R&D capital stock, a productivity growth function, a price equation, and an output demand function. In an earlier work I estimated the relationships constituting such a model for the communications industry.[2] Here the model is developed further, and the individual relationships of the model are combined into an interdependent model that is estimated and tested by means of simulation tests, first for the telecom-

2. Nestor E. Terleckyj, "A Growth Model of the U.S. Communication Industry, 1948–80," in Meheroo Jussawalla and Helene Ebenfield, eds., *Communication and Information Economics: New Perspectives* (North Holland, 1984), pp. 119–45.

munications industry and then for the computer industry. The estimation of these dynamic relationships raises some issues that further broaden the scope of the model. Among them are the influence of government-funded R&D on private R&D investments and on private productivity growth, the role of foreign technology development in generating domestic R&D and productivity increases, the significance of international trade and technology transfer, and convergence between the telecommunications and computer industries.

The R&D Investment Function

Investment in the development of technology has been shown to depend explicitly on market opportunities, at least since Schmookler related the rate of patenting to the size of the market for different capital goods.[3] But these market opportunities cannot by themselves produce innovations, however profitable they may be, unless technological opportunities are also present.[4]

In general, R&D investment can be expressed as a function of the expected size of future markets. The technological opportunities are more difficult to identify comprehensively. They may be represented in part by the amount of government-funded R&D in the same industry,[5] by foreign R&D, and by the relevant basic research variables. Like other types of investment, R&D investment may also be influenced by cyclical and other short-term factors, by structural variables influencing the expectations of returns from or the availability of funds for R&D investment, and by policy variables such as regulation and taxation.

The R&D investment function may be written as follows:

$$RD_i = f_i(Q_i, Q_j, GRD_i, X_k), \tag{1}$$

where RD_i is R&D investment in industry, Q_i is the primary market measured by output in industry i, Q_j is the vector of outputs from other relevant markets, GRD_i is government-funded R&D in industry i, and X_k represents all other structural and dynamic variables including the level of aggregate economic activity.

3. Jacob Schmookler, *Invention and Economic Growth* (Harvard University Press, 1966).

4. Nathan Rosenberg, "Science, Invention and Economic Growth," *Economic Journal*, vol. 84 (March 1974), pp. 90–108.

5. David M. Levy and Nestor E. Terleckyj, "Effects of Government R&D on Private R&D Investment and Productivity: A Macroeconomic Analysis," *Bell Journal of Economics*, vol. 14 (Autumn 1983), pp. 551–61.

R&D Capital Stock

Estimation of the capital stock of research and development available for production in a given year requires recognizing an embodiment lag between the time the research and development activity is conducted and the time its results become available for production.

Actually, several time lags are involved. One, an innovation development lag, is the time lag between the average R&D expenditure and the resulting innovation. Another, the innovation diffusion lag, covers the period of the diffusion of the resulting innovations through the production system. A third, the innovation absorption lag, is the time between the adoption of an innovation in production and its operation at full efficiency. Such a lag can amount to several years, as shown by the productivity lag involved in the introduction of jet aircraft.[6] The innovation development and the innovation diffusion lags are also quite long.[7] The following embodiment model intended to reflect all these lags, a model that I developed earlier, can be used for estimating the R&D capital stock.[8]

The embodiment lag is estimated by extending a model of the marginal product of R&D capital, which has been used to estimate the productivity effects of R&D for some time. In this model, output (Q) is a Cobb-Douglas production function of labor (L) and capital (K) with a disembodied productivity growth trend λ, and a factor-neutral contribution of research capital stock, R, with its own elasticity parameter, α:

$$Q = Ae^{\lambda t} L^{\beta} K^{1-\beta} R^{\alpha}. \tag{2}$$

Productivity (F) is defined as follows:

$$F = Q/(L^{\beta} K^{1-\beta}) = Ae^{\lambda t} R^{\alpha}. \tag{3}$$

Differentiating this function with respect to time gives the equation for productivity growth in which the growth rate of productivity is the sum of the disembodied autonomous growth term and growth in the stock of

6. Nestor E. Terleckyj, "The Time Pattern of the Effects of Industrial R&D on Productivity Growth," paper presented at a conference on Interindustry Differences in Productivity Growth at the American Enterprise Institute for Public Policy Research, Washington, D.C., October 11, 1984.

7. See Zvi Griliches, "Hybrid Corn: An Exploration in the Economics of Technological Change," *Econometrica*, vol. 25 (October 1957), pp. 501–22; Edwin Mansfield, "Technical Change and the Rate of Imitation," *Econometrica*, vol. 29 (October 1961), pp. 741–66; and D. Ravenscraft and F. M. Scherer, "The Lag Structure of Returns to Research and Development," *Applied Economics*, vol. 14 (December 1982), pp. 603–20.

8. Terleckyj, " Growth Model," and "Time Pattern of the Effects of Industrial R&D."

R&D capital times the elasticity of output with respect to the R&D capital stock:

$$f = \frac{\dot{F}}{F} = \lambda + \alpha \frac{\dot{R}}{R}. \tag{4}$$

If the depreciation rate for R&D capital is zero (or small enough to be ignored), the elasticity term in equation 4 can be replaced by a term containing the marginal product of R&D capital:

$$\lambda + \alpha \frac{\dot{R}}{R} = \lambda + \frac{\partial Q}{\partial R} \cdot \frac{R}{Q} \cdot \frac{\dot{R}}{R} = \lambda + \frac{\partial Q}{\partial R} \cdot \frac{\dot{R}}{Q}. \tag{5}$$

In this equation the growth rate of productivity is specified as the sum of the disembodied term and the marginal product of R&D capital times the ratio of the increment in the R&D capital to output ($\dot{R}/Q$). The available results of empirical research show that the depreciation rate of R&D as a factor of production is either zero or small enough to be empirically indistinguishable from zero.[9] The data favor zero depreciation of R&D over positive depreciation rates, at least rates large enough (about 10 percent or more) to change appreciably the estimates of the net R&D investment. Thus while R&D capital is subject to obsolescence, it does not wear out by use in production over time and should not be depreciated in productivity analyses.

Equation 5 does not specify the period over which the R&D expenditures become a part of the R&D capital that is available for production. Without an embodiment lag, and with a zero rate of depreciation of R&D, the increment in the R&D capital would equal the R&D expenditure over an interval of time. But the R&D capital resulting from a given R&D expenditure is not immediately available in production. In fact, the available evidence suggests that the embodiment lags for R&D are quite long. For this reason, such lags need to be recognized explicitly in analyzing the effects of R&D on production and productivity. To this end the R&D capital stock in period t is specified as follows for discrete periods (assuming no depreciation of R&D):

$$R_t = W_0\, RD_t + W_1\, RD_{t-1} + \cdots + W_a\, RD_{t-a} = \sum_{s=0}^{a} W_s\, RD_{t-s}. \tag{6}$$

9. Zvi Griliches and Frank Lichtenberg, "R&D and Productivity Growth at the Industry Level: Is There Still a Relationship?" in Z. Griliches, ed., *R&D, Patents, and Productivity* (University of Chicago Press, 1984), pp. 465–501; and Nestor E. Terleckyj, "R&D and U.S. Industrial Productivity in the 1970s," in Devendra Sahal, ed., *The Transfer and Utilization of Technical Knowledge* (Lexington, Mass.: Lexington Books, 1982), pp. 63–99.

Here the coefficients W_s, $0 \leq W_s \leq 1$, are the proportion of the expenditure in period $t - s$, RD_{t-s}, which has already been embodied in the R&D capital stock, and $t - a$ is the first period in which R&D expenditure was made. The values of the embodiment coefficients W_s rise from zero (in the period recent enough for none of the R&D expenditure to be embodied in the R&D capital stock) to one (in all periods far enough in the past for R&D expenditures to be fully embodied in the R&D capital stock). Because the depreciation rate of R&D capital is assumed to be zero, all past expenditures eventually remain fully embodied in R_t. (A positive depreciation rate for R&D would require a relatively simple addition of depreciation terms to the present formulation.)

The change in the capital stock (ΔR_t) in year t is then:

$$(7)\quad \Delta R_t = R_t - R_{t-1} = \sum_{s=0}^{a} W_s RD_{t-s} - \sum_{s=0}^{a} W_s RD_{t-s+1}$$

$$= W_0 RD_t + \sum_{s=1}^{a} (W_s - W_{s-1}) RD_{t-s}.$$

The change in the R&D capital stock over the time interval from $t - 1$ to t is the weighted sum of current and past R&D expenditures with the weights, $v_s = W_s - W_{s-1}$, given by the differences in the embodiment coefficients for the successive past periods. For the current period t, the weight v_0 equals the embodiment coefficient W_0. The positive values of the weights v_s are contained in a finite time interval representing an embodiment period of length a. These weights can be estimated statistically by specifying a dynamic version of equation 5, assuming that the marginal product of R&D capital is constant over time. This version includes the expression for $\dot{R}_t$ ($\approx \Delta R_t$) from equation 7.

$$(8)\qquad \frac{\Delta F_t}{F_t} = \lambda + \frac{\partial Q_t}{\partial R_t}\left[(W_0 RD_t + \sum_{s=1}^{a} v_s RD_{t-s}) / Q_t\right]$$

The assumption of a constant marginal product of R_t is justified on the ground that, relative to the variance in v_s, the trend values of the marginal product of R&D capital are likely to be stable, and the cyclical and other short-term disturbances in productivity can be allowed for in the process of estimation by introducing additional variables. Equation 8 cannot be estimated directly for a single time series because Q_t, which is the denominator of the expenditure-output ratios, would have to change each year with the change in t, and a single series does not provide sufficient observations to estimate the weight structure. (These weights could be estimated if appropriate pooled cross-section time-series observations were available.) For a single-sector model the equation is modified by multiplying equation 8 by the current-year output, Q_t. This nets out the output denominator of the expenditure output ratio of the right-hand side and changes the dependent variable from percentage change in productivity to the dollar value of the productivity change:

$$Q_t \frac{\Delta F_t}{F_t} = \lambda Q_t + \frac{\partial Q_t}{\partial R_t} \sum_{s=0}^{a} v_s \, \mathrm{RD}_{t-s}. \tag{9}$$

This equation is used to estimate the embodiment lag period for R&D and the aggregate stock of private R&D capital.

The Productivity Growth Function

In technology-intensive industries, productivity change is dominated by the amount of R&D capital available for production. The productivity function can be stated as:

$$TFP_i = TFP_i \ (R_i), \tag{10}$$

where TFP_i is the total factor productivity and R_i is the amount of the available R&D capital. In practice, R_i is the stock of R&D investments accumulated for use in production.

The organization of the industry, however, may be important, both for theoretical reasons and for the treatment of data. The R&D stock may be accumulated in the supplying industry, such as communication equipment or aircraft manufacturing, while the productivity growth it generates may be measured in the utilizing industry, such as telephone service or air transportation. The organizational structure of technology development and production varies among industries, over time and internationally. For example, the United States airline industry has been entirely separate from the aircraft industry since very early in its history, and computer technology has been developed both by the manufacturers of computers and in the separate electronics industry. However, the development of technology for U.S. telecommunications was for a long time integrated with telephone service. Industrial organization in telecommunications has changed drastically recently, and the industries providing the technology base for the telephone industry have become much more diversified.

The Price Equation

Over long periods relative prices of individual goods have varied directly with the relative changes in total real unit costs, or inversely with the relative changes in productivity of their industries.[10] This connection is particularly notable in technology-intensive industries. But exceptions have existed because of autonomous shifts in demand, changes in prices of components, noncompetitive conditions, and government intervention, which may have kept the relative prices from following the relative real unit costs. One such exception was the period in the telecommunications industry from the late 1940s to the early 1960s, when the price of the telecommunications service output was only weakly influenced by relative real unit costs and moved largely with the general price level. Since 1964, however, relative prices have corresponded closely with relative real unit costs.

The price of the product relative to the general price level, P_i, is given by

$$P_i = g\,(UC_i, X), \qquad (11)$$

where UC_i is the relative unit cost index, the ratio of the inverse of total factor productivity in industry i to economy-wide total factor productivity, and X represents all other variables.

The Demand Function for Industry Output

The most important variables influencing demand for a product are its price and the income of its buyers. As a starting hypothesis, the demand equation can be specified as a constant elasticity function of the relative price of the product and the size of its market, for example, as measured by GNP:

$$(12) \qquad \begin{aligned} Q_i &= AP_i^{\epsilon}\, GNP^{\eta} \\ \ln Q_i &= \ln A + \epsilon \ln P_i + \eta \ln GNP. \end{aligned}$$

Other variables that may have a significant effect on demand may be shipments by sectors that are heavy users of the output of industry i, and the size of the export market.

10. See John W. Kendrick, assisted by Maude R. Pech, *Productivity Trends in the United States* (Princeton University Press, 1961); and John W. Kendrick and Elliot S. Grossman, *Productivity in the United States: Trends and Cycles* (Johns Hopkins University Press, 1980).

Benefits from Technological Investments

Benefits from technological investments, that is, R&D expenditures, can be measured by the price reduction or (in equilibrium) the productivity gain resulting from these investments. If transitory short-term impacts on price changes or productivity gains average out over the long run, the gain in productivity or the decline in price or unit cost (the direct return to industry R&D) can be separated from other long-term effects.

Application to the U.S. Telecommunications Industry

The history of the telecommunications industry has been characterized by very rapid rates of technological change induced by massive private and government investments in developing the underlying technologies, in particular electronics technology. In the last two decades successive developments in microwave, satellite, and semiconductor technologies facilitated competition in a heretofore monopolistic industry.

The approach taken in this paper follows the analysis I developed in my earlier study of the communications industry.[11] In this approach, industry development is described by the set of relationships discussed in the preceding section, that is, the R&D investment function, the R&D capital stock function, the total factor productivity equation, the price equation, and the demand function.

In my previous study a set of such relationships was estimated with aggregate data for the communications industry and R&D data for the total electrical equipment industry. In this paper, estimates are derived for the telecommunications industry alone, the period of analysis is extended to 1985, and the entire set of the estimated relationships is organized into a predictive growth model of telecommunications. The model is then evaluated through a set of dynamic historical simulations and tested for its sensitivity to the data used.

The telecommunications industry encompasses the telephone and telegraph service industries. The output of telecommunications services is measured by real value added. Output price is the implicit deflator for gross product originating in the telecommunications industry, based on price indexes for the different telecommunications services. The data for real output in 1984 and 1985 are affected by capital write-offs in the

11. See Terleckyj, "Growth Model."

telephone industry occasioned by the breakup of AT&T. They are also influenced by the reduced reporting requirements of the Federal Communications Commission.

Labor and fixed capital inputs are measured by employment and fixed capital stock, respectively. The data are obtained from the national income and product accounts (NIPA), published by the U.S. Department of Commerce, Bureau of Economic Analysis. A measure of total factor productivity for the industry is derived from these data.

The telecommunications industry so defined does not include production of equipment for telecommunications or telecommunications services produced outside the industry in government or in private firms in other industries. Such activities, especially networks involving computers, have become more important in recent years as businesses have begun to operate private communications systems, diminishing the use of services produced by firms engaged primarily in telecommunications services.

The R&D Investment Function for Communication Equipment and Electronic Components

Research and development for the telecommunications industry is conducted almost entirely by the producers of communication equipment and electronic components. The direct cost of R&D is included in the measured output of the telecommunications industry only insofar as the R&D charges of telecommunications companies are reflected in value added. R&D in communication equipment is combined with R&D in electronic components companies not included in the telecommunications industry, in part because fully separate data are not available in sufficient detail, but also because it seems undesirable to exclude R&D in electronic components in estimating productivity gains in telecommunications services attributable to R&D activities. Electronics R&D is as essential to the development of telecommunications technology as it is to the development of computer and other electronics-based technologies. In the future it may be feasible to test for the separate effects of communication equipment R&D and electronic components R&D on productivity in telecommunications, but such a test is not practical here.

R&D expenditures in the communication equipment and electronic components industries are shown in table 12 (at the end of this paper) in 1982 dollars (based on a new price index developed by Mansfield for the electrical equipment industry). The sources of R&D funding in these industries changed substantially from 1957 to 1985. Private R&D spending grew consistently throughout the period, whereas federal R&D expenditures first leveled off in the 1960s and then dropped in the 1970s. Federally funded R&D rose again in the 1980s but did not reach its earlier peak. In 1957 private R&D was less than half of government-funded R&D. By 1985 company-funded R&D in these industries had risen to almost twice the level of government-funded R&D.

Estimates of the investment function for R&D in communication equipment and electronic components for 1964–85 and 1970–85 are shown in table 1. Private R&D investment in these industries depends on the output of both the telecommunications and the computer industries. It also depends on the change in GNP over the previous two years and on the amount of government spending for R&D in the communication equipment and electronic components industries lagged two years.

Two versions of the investment equation shown in table 1 were estimated, reflecting two different procedures used for converting the value of computer shipments to constant dollars. In the first version the computer

Table 1. *Investment Function for Private Research and Development in Communication Equipment and Electronic Components*[a]

	Ordinary least squares			
Variable	*1964–85*		*1970–85*	
Constant	−0.419 (0.86)	−2.030 (2.35)	−1.048 (1.10)	−4.617 (2.04)
Value of telecommunications output	0.043 (8.85)	0.035 (2.33)	0.027 (2.51)	0.084 (1.82)
Value of computer shipments deflated by GNP deflator	. . . . . .	0.077 (3.70)	. . . . . .	0.021 (0.38)
Value of computer shipments deflated by computer price index	0.027 (10.28)	. . . . . .	0.034 (7.20)	. . . . . .
GNP unlagged minus GNP lagged 2 years	0.001 (2.09)	0.001 (2.07)	0.000 (0.97)	0.002 (2.33)
Government R&D expenditure lagged 2 years	0.39 (4.51)	0.61 (3.73)	0.18 (1.24)	0.97 (3.04)
Summary statistics				
$\bar{R}^2$	0.991	0.964	0.993	0.958
Durbin-Watson	1.71	1.65	1.75	1.44

a. R&D expenditures, industry output, and GNP are measured in billions of 1982 dollars. The numbers in parentheses are *t*-statistics.

shipments were deflated by the GNP deflator and in the second by the recently developed computer price index, which reflects the very rapid decline in quality-adjusted computer prices based on the performance characteristics of computers.[12] In the first formulation the growth of the computer market was valued relative to the general price level (until 1984 this was the price index used to deflate computer shipments in government statistics). In the second formulation computer output was deflated by the new price index for computers. The results obtained with the computer output deflated by computer prices fit the statistical history better. The R&D investment equation is statistically more stable, its explanatory power is stronger, and it does not have the large negative constant compensated by possibly exaggerated coefficients for changes in GNP and government R&D spending that were obtained with the computer output deflated by the GNP deflator.

The equation estimated for the period 1964–85 based on the computer price deflator indicates that total private R&D expenditure in communication equipment and electronic components rises by $43 million for each billion-dollar increase in telecommunications industry output and by $27 million for every billion-dollar increase in constant dollar shipments of the computer industry (deflated by computer prices). The estimated effect of a rise in GNP is positive and significant at the 5 percent level, amounting to 0.1 percent of the two-year change in GNP; the estimated effect of past government R&D spending in this industry is highly significant, amounting to 39 percent of government R&D expenditure lagged two years.

The results for 1970–85 are about the same as for 1964–85 for the computer and telecommunications output variables. However, the effects of GNP change and government R&D are much weaker in the shorter period, when computer shipments are deflated by the computer price index. The effects are larger when computer shipments are deflated by the GNP deflator.

An attempt was made to relate U.S. R&D investment in communication equipment and electronic components to real Japanese R&D investment in these industries. The number of researchers rather than R&D expenditure was used to measure Japanese R&D so as to avoid problems raised by volatile exchange rates. The data series for Japanese R&D researchers

12. David W. Cartwright, "Improved Deflation of Purchases of Computers," *Survey of Current Business*, vol. 66 (March 1986), pp. 7–10; Rosanne Cole and others, "Quality-Adjusted Price Indexes for Computer Processors and Selected Peripheral Equipment," *Survey of Current Business*, vol. 66 (January 1986), pp. 41–50; and Jack E. Triplett, "The Economic Interpretation of Hedonic Methods," *Survey of Current Business*, ibid., pp. 36–40.

in the electrical equipment industry constructed from the Japanese sources is shown in the last column of table 12. No significant results were obtained by introducing that data series into the U.S. R&D investment function. The coefficients for Japanese R&D were not significant, and other results were unchanged. Similarly, negative results were obtained for the proportion of U.S. patents in telecommunications and electronics granted to European and Japanese applicants. Likewise, defense purchases of durable goods by the U.S. government did not have any separate effect on private U.S. R&D investment in telecommunications and electronic components.

R&D Capital in Communication Equipment and Electronic Components

The analytical model for embodiment of R&D specified in equation 9 relates lagged expenditures for R&D to the real value of annual gains in total factor productivity. Such a series is constructed by first calculating a total factor productivity index and then estimating the annual relative changes in that index and the real value of annual productivity gains in the telecommunications industry.

The NIPA data for employment, fixed capital stock, real output, and the labor and nonlabor shares of real output were used to calculate the total factor productivity index. Employment is measured by the number of persons engaged in production during the year; output represents the real gross product originating, that is, value added; and capital stock is the net stock of nonresidential equipment and structures. The fixed capital stock was averaged for the beginning and the end of the year. The productivity index was derived by the Tornqvist formula using the two factor shares, with the total nonlabor share applied to the fixed capital stock. Consistent data for other forms of capital were not available. Output, input, and productivity data for 1948 through 1985 are shown in table 12.

Table 12 also shows the annual relative changes in the productivity index for 1948–85 and the annual gains in real value added in the telecommunications industry resulting from productivity increases. This series was derived by multiplying the relative change in the productivity index by output averaged for the current and the preceding years.

Perhaps the most noteworthy changes in the long-term industry trends indicated by these data are the abrupt cessation of the rather rapid growth of the output of the telephone and telegraph industry and a concurrent drop in its total factor productivity after 1983. These changes, however, need to be viewed with caution pending confirmation or revision of these data by the statistical agencies concerned because of changes in industry organization and in data reporting for these years.

The productivity increments in the value of output are not adjusted for cyclical or structural changes in output or for other effects. Therefore, it is necessary to stabilize the estimate of the distributed lag effects of R&D on productivity by the inclusion of additional variables. Because the output of the telecommunications industry is a part of the dependent variable, the output itself cannot be used as an explanatory variable to stabilize the distributed lag equation. Therefore, instead of the output, the annual changes in the output of the industry are included, both for the current year and with a one-year lag. Also included in the equation is a variable representing general business conditions, the Conference Board index of help-wanted advertising, which is a more sensitive concurrent indicator of general business fluctuations than other indicators.

Estimates of the distributed lag structures obtained by three different methods, restricted and partially restricted quadratic Almon lags and a twelve-year unrestricted ordinary least squares (OLS) lag estimate, are shown in table 2. The lag estimates are qualitatively consistent with one another. The sum of the lag coefficients obtained by the different methods is similar. The restricted polynomial and the twelve-year OLS structure smoothed by moving averages indicate a relatively long lag period, with the R&D expenditures four to eight years earlier having the largest effect on productivity gain in the current year. The coefficients of the cyclical variables included in the equation are statistically significant, as is the sum of the distributed lag coefficients.

The length and the shape of the embodiment structure were explored further by estimating single-year lag coefficients for R&D expenditures. The results reported in table 3 also indicate the largest effects of R&D expenditures for the three- to seven-year lags, although the coefficients for earlier and for later years were not very different. The values of single-year coefficients are similar to the value of the sum of coefficients of the distributed lag structures shown in table 2, when normalized for units of measurement.

Five-year moving averages of the unrestricted twelve-year lag estimates are shown in table 2. The peak effect is in the year −6. These five-year moving averages—omitting the first two and the last one of the negative values and averaging the remaining last three moving averages to obtain a smoother pattern and to avoid a negative value—were then normalized to their sum. The following resulting proportions were used as weights to calculate the R&D capital stock:

Years lagged	*Moving average, smoothed*	*Weights*
−4	0.166	0.082
−5	0.423	0.210
−6	0.839	0.416
−7	0.197	0.098
−8	0.196	0.097
−9	0.196	0.097
Total	2.017	1.000

Table 2. *Estimates of a Lag Structure of R&D Effects on the Real Value of Annual Increments in Telecommunications Productivity*[a]

	Polynomial of 2d degree			
Variable	*Constrained at both ends*	*Constrained at the far end*	*Ordinary least squares*	*5-year moving averages*
Constant	0.75	0.47	1.21	. . .
	(1.14)	(0.69)	(1.10)	. . .
Change in the value of telecommunications output	0.89	0.96	0.75	. . .
	(9.61)	(9.33)	(4.46)	. . .
Change in the value of telecommunications output lagged 1 year	−0.39	−0.36	−0.26	. . .
	(3.76)	(3.55)	(2.21)	. . .
Help-wanted index	−1.87	−1.95	−2.93	. . .
	(2.56)	(2.73)	(2.74)	. . .
Private R&D expenditures, years lagged				
0	0.013	0.170	3.284	. . .
1	0.023	0.120	−4.348	. . .
2	0.032	0.088	1.885	−0.009
3	0.039	0.057	−2.446	−0.649
4	0.044	0.030	1.578	0.166
5	0.047	0.007	0.089	0.423
6	0.047	−0.010	−0.278	0.839
7	0.047	−0.023	3.173	0.168
8	0.044	−0.031	−0.369	−0.166
9	0.039	−0.034	−1.778	0.587
10	0.032	−0.033	−1.581	−0.532
11	0.023	−0.027	3.488	. . .
12	0.013	−0.016	−2.422	. . .
Summary statistics				
$\bar{R}^2$	0.819	0.828	0.89	. . .
Durbin-Watson	2.03	1.99	2.97	. . .
Sum of lag coefficients	0.44	0.30	0.28	. . .
	(2.37)	(1.44)	(n.a.)	. . .

a. Based on the data for 1964–85. Value of output is measured in billions of 1982 dollars. The numbers in parentheses are *t*-statistics.

Table 3. *Regression Results for Estimates of Single-Year Lag Effects of R&D in Communication Equipment and Electronic Components on the Real Value of Annual Increments in Telecommunications Productivity*

Number of years lagged	*Coefficient*	*t-statistic*	$\bar{R}^2$
0	0.318	2.99	0.84
1	0.339	2.64	0.83
2	0.362	2.42	0.82
3	0.441	2.62	0.83
4	0.490	2.74	0.83
5	0.490	2.50	0.82
6	0.465	2.25	0.81
7	0.432	2.06	0.81
8	0.354	1.70	0.79
9	0.338	1.68	0.79
10	0.397	2.05	0.81
11	0.415	2.18	0.81
12	0.391	2.05	0.81

A number of sensitivity analyses were performed to test the estimate of the lag structure for possible effects of two other variables, the private R&D expenditure by the office equipment industry and government spending for R&D in the communication equipment and electronics industries. The results were negative. No significant effects for these other variables were observed, and the lag coefficients estimates as shown in table 2 were not affected by their introduction.

Stock of R&D Capital in the Communication Equipment and Electronic Components Industries

The stock of R&D capital for communication equipment and electronic components was calculated by the following two expressions:

$$R(0) = R(-1) + \Delta R(0), \tag{13a}$$

$$\Delta R(0) = 0.082\,RD(-4) + 0.210\,RD(-5) + 0.416\,RD(-6) + 0.098\,RD(-7) + 0.097\,RD(-8) + 0.097\,RD(-9). \tag{13b}$$

The coefficients in the expression for ΔR (0) are the embodiment weights discussed above. The resulting series of R&D capital stock applicable to the telecommunications industry is shown in table 12. The company R&D

data series for the period 1950–85 was used to estimate the annual additions to the R&D capital stock beginning with 1959. Because R&D data were not available for earlier years, an initial value of the R&D capital stock of $5 billion (1982 dollars) was assumed for 1958 so as to approximate the effect of R&D performed in communication equipment and electronic components up to that year. This approximation was developed in relation to an estimate of R&D investment in all electrical machinery made in my previous study.[13] This estimate was introduced in order to avoid starting the capital stock series with a zero value, which is obviously incorrect.

Because the distributed lag estimate used in this series is subject to considerable uncertainty, an alternate series of R&D capital stock was calculated based on a single-year lag. The single-year lag estimates in table 3 indicate the strongest results for the year -4. Accordingly, the alternate R&D stock was estimated as above with the second expression changed to $\Delta R(0) = RD(-4)$. Both series are shown in table 12.

The Productivity Growth Function: Results

A simple, strong relationship exists between the total factor productivity described earlier and the stock of R&D capital. The stock of R&D capital statistically explains the productivity index of the telecommunications industry very well. The $\bar{R}^2$ is above 0.99 for 1964–85 (table 4).

The estimate was made in both the logarithmic and the Box-Cox form. In the latter formulation, an exponential parameter, λ, rather than a predetermined logarithmic or linear functional form was estimated from the data. The value of this parameter was zero for logarithmic and one for linear forms. The λ was estimated for 1964–85 to be -0.49, which indicates a curvature somewhat greater than in the logarithmic equation. Numerically the elasticity coefficient for the R&D stock in the Box-Cox estimate of 0.34 is similar to 0.31 estimated in the logarithmic equation, suggesting that the logarithmic fit can be accepted as a reasonably good approximation to reality.

Estimates of the two equations for 1970–85, which are also shown in table 4, have slightly higher elasticity coefficients, 0.36 and 0.33 respectively, and are generally similar to the estimates for 1964–85.

These results show that R&D performed in the communication equipment and electronic components industries is closely related to the tech-

13. Terleckyj, "Growth Model."

Table 4. *Estimates of the Effects of the R&D Capital Stock in Communication Equipment and Electronic Components on Total Factor Productivity in the Telecommunications Industry*[a]

Variable	*Natural log estimates*		*Box-Cox estimates*	
	1964–85	*1970–85*	*1964–85*	*1970–85*
Constant	−3.36	−3.55	−3.70	−3.83
	(55.85)	(29.22)	(69.73)	(32.63)
Ln(R&D stock in telecommunications equipment and electronic components)	0.31	0.33	0.34	0.36
	(52.84)	(28.44)	(65.99)	(31.72)
Summary statistics				
$\bar{R}^2$	0.99	0.98	1.00	0.99
Durbin-Watson	1.59	1.97	2.09	2.13
λ	...	...	−0.49	−0.83

a. Productivity index 1982 = 1.00. The numbers in parentheses are *t*-statistics.

nology determining productivity in the communications service industry as measured by the Tornqvist index, which assumes constant returns to scale. Apparently this technology is embodied in the fixed capital stock of the telecommunications industry.

The observed relationship between R&D stock and productivity, however, is based on the correlation of two smooth trends. It needs to be examined further. Not much additional light can be shed by an analysis of annual changes in capital stock, because the capital stock is derived from a distributed weighting of past R&D expenditures, and the annual changes in the resulting R&D capital stock have little variance. Changes in total factor productivity have somewhat more variation, but they may also be influenced by other factors.

The relationship between the stock of R&D capital and productivity was alternatively analyzed in terms of a production function. The output of the telecommunications industry was estimated as a function of three inputs: employment, fixed capital, and R&D capital. Two functional forms were estimated, a log-additive form

$$\ln Q = \ln A + a_1 t + a_2 \ln L + a_3 \ln K + a_4 \ln R; \quad (14)$$

and a variable coefficient form in which the R&D stock also appears in the exponents of labor and capital:

$$\ln Q = \ln A + a_1 t + a_2 \ln L + a_3 \ln K + a_4 \ln R + a_5 R \ln L + a_6 R \ln K. \quad (15)$$

Besides the three inputs, the production function was also estimated with and without an autonomous time trend.

The statistical results support the log-additive formulation without the time trend. The estimates of the variable exponent coefficients were not significant. And no significant estimates of an autonomous (that is, unrelated to R&D stock) time trend in total factor productivity were obtained for the telecommunications industry. Estimates of the log-additive production function are shown in table 5.

Telecommunications output is strongly related to labor input, with an elasticity coefficient of 0.49 estimated for 1964–85 in the equation, which also includes the fixed and R&D capital stocks. This estimate is similar to the share of compensation in real value added during that period. Fixed capital stock and the stock of R&D capital are also significantly related to output.

There is, however, a strong interaction between the estimates of the contributions of the two stocks of capital to output. When R&D stock is included with the other two inputs, the coefficient of the fixed stock is much smaller than when it is omitted, and vice versa. The results obtained with all three inputs appear to be most plausible, though the elasticity coefficient of 0.72 estimated for fixed capital seems high. This estimate may reflect some improvements in the productivity of equipment induced by R&D and some return to capital other than plant and equipment. Even so, an elasticity of output with respect to the R&D stock of about 0.2 is broadly consistent with the estimates of 0.3 obtained for total factor pro-

Table 5. *Ordinary Least Squares Estimates of a Logarithmic Production Function for the Telecommunications Industry*[a]

Variable	*1964–85*			*1970–85*		
Constant	−5.08	−4.98	−5.64	−5.60	−4.30	−8.02
	(14.86)	(13.07)	(10.11)	(5.27)	(5.23)	(8.96)
Ln(employment)	0.49	0.55	0.41	0.55	0.45	0.77
	(7.34)	(7.82)	(3.68)	(3.98)	(3.32)	(5.07)
Ln(fixed telecommunications stock)	0.72	1.00	. . .	0.64	1.01	. . .
	(6.04)	(44.64)	. . .	(3.03)	(36.17)	. . .
Ln(R&D stock)	0.19	. . .	0.66	0.24	. . .	0.65
	(2.44)	. . .	(29.40)	(1.76)	. . .	(30.46)
Summary statistics						
$\bar{R}^2$	0.999	0.999	0.997	0.996	0.996	0.994
Durbin-Watson	2.37	2.00	0.79	2.50	2.00	1.58

a. The numbers in parentheses are *t*-statistics.

ductivity, shown in table 4, in which the sum of labor and capital coefficients is restricted to unity.

In subsequent research very similar results were obtained with a number of different specifications of the production function for telecommunications, which were restricted to constant returns to scale.[14] The evidence seems to favor the interpretation that these returns represent the product of R&D capital rather than returns to scale, and that there is not sufficient evidence to reject the log-additive formulation. The coefficient for R&D capital stock estimated in this later research is virtually the same as that shown in table 4 for the Tornqvist index. These results strongly indicate the existence of a distinct and identifiable contribution of R&D capital to the output of telecommunications services, even if some uncertainty remains about the magnitude of this effect.

Price of Telecommunications Services

The relative prices of outputs of different industries have generally been strongly and negatively correlated with, and apparently influenced by, the relative changes in their total factor productivities.[15] The price of telecommunications output has also been highly correlated with the productivity of the industry, but only since about the mid-1960s. As already mentioned, before 1964 the effect of the relative unit cost index (the inverse of the productivity index) on price was much smaller, as shown in table 6. The table compares price equations fitted for the periods 1950–63 and 1964–85, as well as for 1970–85. There are great differences between the relationships estimated for 1950–63 and those for the later periods. Table 12 contains the data used in this analysis.

During 1964–85 relative productivity largely determined relative price in the telecommunications industry. But the estimates based only on relative productivity, or its inverse, relative unit cost, contain substantial autocorrelation in their residuals, which show a cyclical pattern. On closer examination, this cycle seemed to be related not to the business cycle, but perhaps to variations in the policy environment. The cyclical pattern of residuals was eliminated by introducing a dummy policy variable that had the value of 1 during Republican administrations and 0 during Dem-

14. David M. Levy and Nestor E. Terleckyj, "Problems in Identifying Returns to R&D in an Industry," *Managerial and Decision Economics* (forthcoming 1989).

15. Kendrick, assisted by Pech, *Productivity Trends in the United States*.

Table 6. *Price Equation for Telecommunications Output*[a]

Variable	*1950–63*		*1964–85*		*1970–85*	
Constant	1.19	1.19	−0.30	−0.42	−0.38	−0.32
	(6.71)	(6.45)	(5.24)	(8.46)	(3.48)	(4.82)
Relative unit cost index[b]	0.39	0.39	1.33	1.40	1.40	1.30
	(3.74)	(3.57)	(28.20)	(36.97)	(14.43)	(20.57)
Political party variable lagged 1 year[c]	. . .	−0.0057	. . .	0.06	. . .	0.08
	. . .	(0.31)	. . .	(4.38)	. . .	(4.88)
Summary statistics						
$\bar{R}^2$	0.500	0.459	0.974	0.986	0.932	0.974
Durbin-Watson	1.39	1.42	0.78	1.89	0.77	2.61

a. The numbers in parentheses are *t*-statistics.
b. See equation 11.
c. If Republican, 1; if Democratic, 0.

ocratic administrations, with a one-year lag. This variable did not change the estimated effect of the relative unit cost on price.

Essentially, the same relationship holds for the period 1970–85, with a somewhat larger coefficient for the policy variable. But the policy variable has no effect in the earlier period 1950–63. It is not possible to interpret this policy effect here.

The observed shift in the 1960s confirms the results obtained in my earlier study.[16] The hypothesis of no structural change in the regression coefficients for relative unit cost and the policy variables between the two subperiods is decisively rejected. Although the reasons for this shift are not clear, its practical significance is that the relationships determining the price of telecommunications services are not stable over the longer period. This result is important because many analyses of the industry have implicitly treated the period as homogeneous.

Demand for Telecommunications Output

The demand for the output of telecommunications services can be expected to depend on its relative price and on the real output of the economy. The demand equations incorporating these variables were estimated in a logarithmic form with constant elasticity. The statistical results are shown in table 7. The price elasticity estimated by the ordinary least squares method was −0.9 for 1964–85 and −0.8 for 1970–85, and the GNP elasticity was 1.5 and 1.4, respectively. However, these equations had large serial correlation in the residuals. They were then reestimated

16. Terleckyj, "Growth Model."

Table 7. *Demand Equations for Telecommunications Output Using Logarithmic Functions*[a]

Variable	*Ordinary least squares* 1964–85	*Ordinary least squares* 1970–85	*Generalized least squares* 1964–85	*Generalized least squares* 1970–85	*Two-stage least squares*[b] 1964–85	*Two-stage least squares*[b] 1970–85
Constant	−8.03	−7.25	−5.22	−5.24	−0.35	−1.21
	(3.90)	(3.45)	(2.89)	(2.73)	(0.27)	(0.99)
Price of telecommunications relative to the private economy	−0.93	−0.84	−1.20	−1.00	−1.80	−1.53
	(4.13)	(3.79)	(5.59)	(4.33)	(12.57)	(11.42)
GNP	1.53	1.43	1.18	1.18	0.59	0.69
	(6.03)	(5.53)	(5.32)	(4.99)	(3.69)	(4.57)
Summary statistics						
$\overline{R}^2$	0.987	0.972	0.977	0.989	0.997	0.995
Durbin-Watson	0.71	0.75	1.30	1.35	1.19	1.66
ρ (OLS serial correlation)	. . .	. . .	0.672	0.663	. . .	. . .

a. All amounts in 1982 dollars. The numbers in parentheses are t-statistics.

b. The price of telecommunications relative to the private business economy was estimated by the following equations:

$$\text{1964–85: } \mathrm{Ln}(P_T) = \underset{(0.98)}{1.65} + \underset{(4.42)}{1.02} \cdot \mathrm{Ln}(UC_T) - \underset{(0.97)}{0.20} \cdot \mathrm{Ln}(\mathrm{GNP})$$

$$\text{1970–85: } \mathrm{Ln}(P_T) = \underset{(0.44)}{0.97} + \underset{(3.76)}{1.18} \cdot \mathrm{Ln}(UC_T) - \underset{(0.43)}{0.12} \cdot \mathrm{Ln}(\mathrm{GNP}),$$

where UC_T is the relative unit cost index for telecommunications.

by the generalized least squares method, which produced substantially higher estimates of price elasticity, lower estimates of GNP elasticity, and reduced serial correlation in the residuals. But these estimates of the demand equation may be subject to simultaneity bias, in that the price variable may be correlated with the errors in the residual of the estimated equation. In an attempt to correct for this possible bias, two-stage least squares was used to estimate the demand equation with productivity and GNP as instruments for price. The results are shown in the third pair of columns of the equations in table 7. In these final estimates the price elasticity is much larger, and the effect of GNP is considerably smaller, than in the earlier equations. The two-stage least squares estimate of the equation is probably to be preferred.

The demand for telecommunications output may also be expected to depend on the stock of computers in use because computers generate substantial amounts of data communications that require the services of the telecommunications carriers. Accordingly, two versions of the demand equation for telecommunications output were estimated. In one, only the relative price of telecommunications services and GNP were introduced as explanatory variables. The other included also the real stock of office equipment, as estimated recently for the U.S. private economy by the Bureau of Economic Analysis, using the newly revised price deflator for computers. (The stock of computers in government, nonprofit institutions, and households has not been estimated.) But no significant result was obtained for a possible effect of the computer stock on demand for communications output in any of these formulations, and the business sector stock of computer equipment was omitted from further analysis. There do not appear to be strong linkages between computer and telecommunications markets.

A Model Simulation of the Growth of the Telecommunications Industry

In this section, I utilize the relationships just estimated to model the historical growth of the telecommunications industry from 1964 through 1985. The relationships chosen include the R&D investment equation from table 1, the two R&D embodiment relationships shown in equations 13(a) and 13(b), the total factor productivity equation from table 4, and the equations for price of and demand for telecommunications output used in the two-stage estimation of demand in table 7. The equations used in this model are as follows:

$$RD_T = -0.419 + 0.043 \cdot Q_T + 0.00067 \cdot [GNP - GNP(-2)]$$
$$+0.39 \cdot GRD_T(-2) + 0.027 \cdot Q_C$$
$$R_T = R_T(-1) + \Delta R_T,$$
$$\Delta R_T = \quad 0.082 \cdot RD_T(-4) + 0.210 \cdot RD_T(-5)$$
$$+ 0.416 \cdot RD_T(-6) + 0.098 \cdot RD_T(-7)$$
$$+0.097 \cdot RD_T(-8) + 0.097 \cdot RD_T(-9),$$
$$\ln(TFP_T) = -3.36 + 0.31 \cdot \ln(R_T),$$
$$\ln(P_T) = \quad 1.65 + 1.02 \cdot \ln(UC_T) - 0.20 \cdot \ln(GNP),$$
$$\ln(Q_T) = -0.35 - 1.80 \cdot \ln(P_T) + 0.59 \cdot \ln(GNP),$$

where

RD_T = private research and development in the telecommunications industry (billions of 1982 dollars)
R_T = stock of private research and development in the telecommunications industry (billions of 1982 dollars)
ΔR_T = annual change in R_T (billions of 1982 dollars)
TFP_T = total factor productivity in the telecommunications industry (1982 = 1.00)
P_T = relative price of telecommunications output
Q_T = output of the telecommunications industry (billions of 1982 dollars)
GNP = gross national product (billions of 1982 dollars)
GRD_T = federal research and development expenditures in telecommunications (billions of 1982 dollars)
UC_T = relative unit cost of telecommunications
Q_C = output of computers (billions of 1982 dollars).

The endogenous (dependent) variables are private R&D spending, the stock of private R&D capital, industry productivity, the relative price of telecommunications output, and demand for telecommunications output. The exogenous (independent) variables are GNP, productivity of the private sector, government funding of R&D in the telecommunications industry, and the output of the computer industry.

The model was tested by a dynamic historical simulation for its power to predict the actual behavior of the industry over the period for which it was estimated. The actual values of the endogenous variables were used only for the first year; after that, the values produced by the model were used. Actual values were used throughout for the exogenous variables.

If in such models the predicted values track the actual values of the endogenous variables closely and without systematic divergence over time, the model can be accepted as having good predictive power. A model incapable of tracking the historical experience has to be rejected. Ultimately, the predictive power of the model should be tested for periods outside that for which it was fitted. However, such tests are not practical here.

Some of the results of the historical simulation are shown in figures 1 and 2. Because of the lag relationships present in the model, the historical simulation was run from 1959 in order to have separate predicted values for 1964. These results show that the telecommunications industry model has good predictive power over the historical period in which it was fitted. The values of the endogenous variables predicted by the model did not diverge far from their actual values and in most cases varied on both sides

Figure 1. *Productivity in Telecommunications, 1964–85*

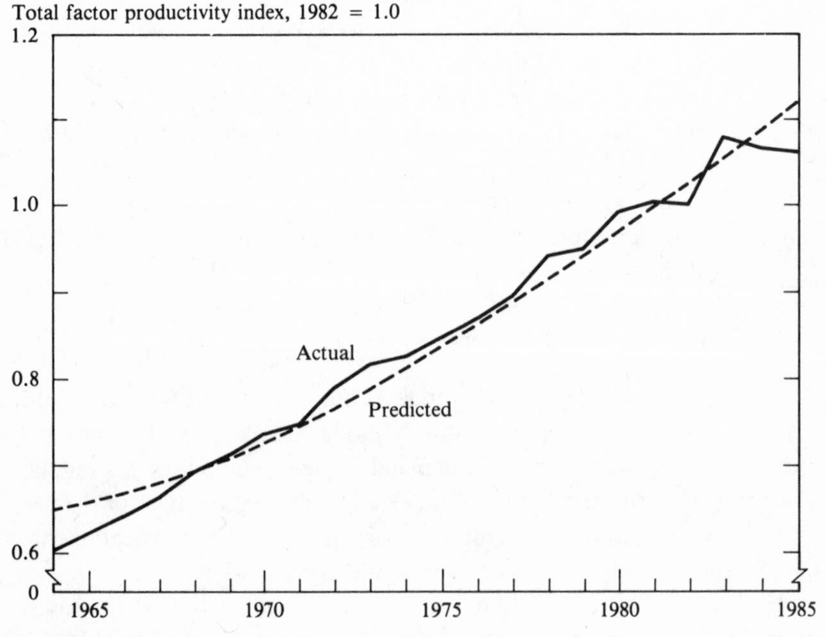

Source: Author's calculations.

Figure 2. *Relative Price of Telecommunications, 1964–85*

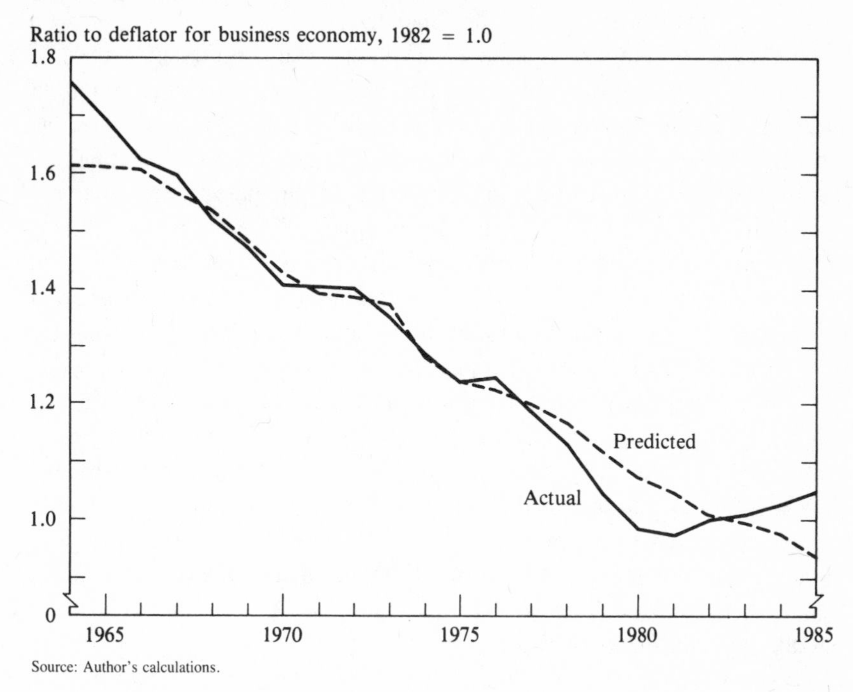

Source: Author's calculations.

of the actual path. The predicted end point values were within 5.3 percent for total factor productivity of the industry, 10.5 percent for the relative price of telecommunications, and 12.7 percent for output. The data for the last two years may be distorted by changes in the industry and in data reporting occasioned by the breakup of AT&T. In any event the overall track record of the model is good even without a correction.

Application to the U.S. Computer Industry

Even among technology-intensive industries the computer industry is unique in the relative size of its R&D and in the sustained and extraordinarily high rates of growth in its real output and decline in its product prices. Until recently these rates of change were not measured in government statistics, although several private research studies suggested that official statistics seriously understated them.

Thanks to recent measurement efforts and the new Bureau of Economic Analysis price index, it is now possible to study the growth of the computer industry quantitatively and to attempt to organize the relationships determining its growth into a consistent model.[17]

Aside from being a manufacturing rather than a service industry, the computer industry differs from telecommunications in two other important respects. The telephone industry originated entirely in the private sector, government involvement in its growth and operation being limited largely to regulation. Defense R&D and procurement purchases were relatively small and began rather late in the development of the industry.

Computers, in contrast, were originally developed for defense use. In its first decade the U.S. private computer industry was almost entirely dependent on government sales.[18] Throughout the history of the industry the government has consistently financed the development of and provided the initial market for new top lines of computers. Moreover, it is a true world industry. Foreign markets and international competition are very important, and the life cycles of products are short and getting shorter. Underlining the international nature of the industry is the fact that foreign sales exceeded 40 percent of the total sales of U.S. computer companies over the last ten years.[19] And as of 1985 more than half of all the patents granted in the United States for office machinery and equipment have gone to foreign investors.

Aside from the new price index, the data for the computer industry are not well developed or organized. Approximations must be made using data for office equipment, business machines, and "machinery except electrical" (for government-funded R&D). No separate capital stock data are available to permit analysis of the productivity of the computer industry directly. But the rate of price decline, which has been very rapid, can be used as an approximate indicator of increases in productivity. My results are tentative, given the limitations of these data.

Real output of computers, measured by the value of shipments deflated by the Commerce Department's new price index for computers, is shown in table 13. This measure of computer output grew very rapidly, averaging 37 percent a year over the twenty-year period 1965–85. The price of computers also declined rapidly in both nominal and real terms. The drop

17. Cole and others, "Quality-Adjusted Price Indexes"; Kenneth Flamm, *Targeting the Computer: Government Support and International Competition* (Brookings, 1987); and Cartwright, "Improved Deflation of Purchases of Computers."

18. Barbara Goody Katz and Almarin Phillips, "The Computer Industry," in Richard R. Nelson, *Government and Technical Progress: A Cross-Industry Analysis* (Pergamon Press, 1982), pp. 162–232; and Kenneth Flamm, *Creating the Computer: Government, Industry, and High Technology* (Brookings, 1988), pp. 29–79.

19. Flamm, *Creating the Computer*, p. 101.

in real price averaged 21 percent a year between 1965 and 1985, and 20 percent from 1972 to 1985.

Table 8 shows the reciprocal of the real computer price index (deflated by the GNP deflator). It can serve to indicate computer industry productivity given the assumption that the relative computer price reflects only its real cost. This amounts to assuming that improvements in productivity in the computer industry as embodied in the productive capacity of computers are passed along to computer buyers in the form of lower prices per unit of computer capability. Such an assumption seems plausible for long-term trends. The growth rate of productivity imputed from the computer price index averaged 26 percent a year between 1965 and 1985, and 25 percent from 1972 to 1985.

Private R&D expenditures by the U.S. office equipment industry are shown in table 13, together with the corresponding stock of private R&D capital in this industry and the expenditures and the stock of government R&D in the machinery industry.

Table 8. *Productivity in the U.S. Computer Industry, 1965–85*
1982 = 1.000

Year	*Implicit productivity indicator: 1/relative price index*	*Percent change in implicit productivity indicator*
1965	0.019	41.9
1966	0.030	54.9
1967	0.040	33.3
1968	0.054	35.4
1969	0.066	22.1
1970	0.077	16.4
1971	0.095	22.9
1972	0.114	20.8
1973	0.134	17.6
1974	0.186	38.6
1975	0.225	20.7
1976	0.273	21.6
1977	0.338	23.4
1978	0.427	26.5
1979	0.538	26.2
1980	0.730	35.8
1981	0.877	20.0
1982	1.000	13.9
1983	1.333	34.2
1984	1.572	17.2
1985	1.971	25.4

Sources: David W. Cartwright, "Improved Deflation of Purchases of Computers," *Survey of Current Business*, vol. 66 (March 1986), p. 8; and Kenneth Flamm, "Targeting Technology," draft (Brookings, 1985), table 2A-5.

Indicators of the international position of the U.S. computer industry show the increasing force of foreign competition (table 13). Imports of business machines increased from 22 percent of exports in 1967, to 34 percent in 1975, then to 83 percent in 1985. The proportion of U.S. patents granted to non-U.S. inventors in the office equipment category increased from 19 percent in 1965, to 36 percent in 1975, to 52 percent in 1985.

In the empirical model that follows, private investment in computer R&D, productivity growth, and demand are analyzed. Private investment in office equipment R&D may be analyzed in relation to the past growth of computer output, to government R&D expenditures in the machinery industry, and to the size of the foreign computer technology base measured by the proportion of U.S. office equipment patents granted to non-U.S. inventors. Productivity growth in the U.S. computer manufacturing industry, reflected in changes in computer prices, may be examined by relating it to the stock of private R&D expenditures in the U.S. office equipment industry, to the stock of government spending for machinery industry R&D, and to the stock of private and government R&D in communication equipment and electronic components. The possible effect of foreign technology on U.S. computer productivity is also considered. Demand for U.S. computer output is then analyzed in terms of such determinants as computer prices, GNP, defense procurement, and U.S. foreign trade in business machines.

Investment in Office Equipment R&D

Investment in office equipment R&D rose very rapidly between 1965 and 1985 (table 13). Real private R&D expenditure by U.S. office equipment companies grew at a rate of 9.0 percent a year during 1965–85, increasing more than 5½ times over the period. The growth rate averaged 9.9 percent in the years 1980–85. This sustained growth in office equipment R&D was supported by the rapid growth of industry's output. It evidently was stimulated also by foreign competition in office equipment in general, and in computers in particular. This competition has intensified greatly in recent years. Another factor that may have stimulated private investment in computer technology has been government-funded R&D in the computer industry.

Statistical estimates of the R&D investment function for the U.S. office equipment industry are shown in table 9. Both the level and the increases in computer output have significant estimated effects on R&D spending.

Table 9. *Estimates of the Investment Function for Private R&D Expenditures in the U.S. Office Equipment Industry*[a]

Variable	*1965–85*		*1972–85*	
Constant	−0.91 (4.39)	−1.27 (3.59)	−1.28 (1.45)	−1.79 (1.81)
Computer output lagged 1 year	0.016 (3.62)	0.013 (2.59)	0.014 (1.77)	0.010 (1.11)
Change in computer output	0.035 (2.54)	0.042 (2.85)	0.034 (1.98)	0.041 (2.26)
Foreign share in U.S. office equipment patents (percent)	0.11 (14.87)	0.11 (15.02)	0.12 (4.59)	0.12 (4.75)
Government R&D in machinery industry lagged 2 years	. . . . . .	0.39 (1.24)	. . . . . .	0.46 (1.09)
Summary statistics				
$\bar{R}^2$	0.986	0.986	0.969	0.970
Durbin-Watson	1.41	1.49	1.23	1.36

a. Output and R&D are measured in billions of 1982 dollars; the numbers in parentheses are *t*-statistics.

A very strong effect is estimated for the foreign share of the office equipment patents for the period 1965–85, indicating an expenditure of approximately $0.11 billion (in 1982 dollars) for each percentage point increase in the share of foreign patents. Unfortunately, government-funded R&D is not reported separately for the office equipment industry for most years. However, government-funded R&D in all machinery except electrical, which consists largely of R&D in the office equipment industry (between 76 and 96 percent in the years for which the data were reported), can be used as a proxy for government-funded R&D in the computer industry. The coefficient for government R&D in all machinery industries was not significant, but it was large and positive. The results for 1972–85 were similar to those for 1965–85.

R&D Stocks and the Price of Computer Output

Because labor and capital data are not available for the office equipment industry, it is not practical to construct a productivity index for that industry. The effect of R&D on the price of computer output was therefore examined directly. Since the industry has usually been competitive and the price of its output has been dropping at an extraordinarily rapid pace, it appears reasonable to view the rate of the price decline as an indicator of productivity increase.

In linking the growth of computer productivity to office equipment R&D, it was not feasible to use a distributed lag of past R&D expenditures, because a stable structure could not be estimated from the data. Therefore, a four-year embodiment lag was assumed for R&D, and the R&D capital stock was calculated with this lag from annual R&D expenditures (in 1982 dollars) and an assumed initial-year value of $2.5 billion in 1961. Table 10 shows estimates of the computer productivity equation, with the stock of private R&D, exports and imports of business machines, and an autonomous time trend as explanatory variables. The only stable effect across time periods is that of the stock of private office equipment R&D. The time trend was insignificant when foreign trade variables were included. The latter were not significant in the period 1972–85.

Computer industry productivity was also related to other stocks of R&D capital: the stock of government-funded R&D in machinery, except electrical, and stocks of private and government-funded R&D in the communication equipment and electronic components industries. It was also related to the stock of foreign-origin U.S. patents in office equipment

Table 10. *Estimates of the Determinants of Productivity in the Computer Industry*[a]

Variable	*1967–85*			*1972–85*		
Constant	−6.97	−2.72	−5.34	−7.51	−4.47	−6.35
	(67.17)	(4.51)	(20.37)	(80.75)	(3.01)	(14.01)
Ln (private office equipment R&D stock)	1.93	0.38	1.26	2.09	0.85	1.65
	(56.17)	(1.17)	(12.55)	(73.53)	(1.40)	(9.02)
Time	...	0.26	...	...	0.13	...
	...	(7.09)	...	...	(2.05)	...
Ln (exports of business machines)	...	...	0.30	...	...	0.05
	...	...	(2.47)	...	...	(0.70)
Ln (imports of business machines)	...	...	0.04	...	...	0.12
	...	...	(0.43)	...	...	(0.91)
Summary statistics						
$\bar{R}^2$	0.995	0.999	0.998	0.998	0.998	0.998
Durbin-Watson	0.48	1.63	2.00	1.97	2.45	1.75

a. The productivity variable is the natural logarithm of the inverse of real computer prices. The numbers in parentheses are *t*-statistics.

lagged four years. However if the stock of private R&D in the office equipment industry is included as an independent variable, none of these other variables adds to the explanatory power of the equation. The U.S. stock of private R&D investment in the office equipment industry alone provided the best explanation for the trend in computer industry productivity. There was no discernible spillover from investments in communications technology.

Demand for Computers

Demand for computer output is related to computer price and to variables defining the principal markets for computers: specifically, business gross product, federal defense procurement, and net exports of business machines. Data for net exports are available beginning in 1967. The ratio of foreign-origin to U.S.-origin office equipment patents was also included in the equation as a possible indicator of the relative competitiveness of foreign computer products in the absence of more direct price or quality information.

The results are displayed in table 11 for the periods 1967–85 and 1972–85. They show a strong effect for computer price, with a large negative price elasticity of about −1.4. The estimated effects of business gross product were insignificant in both periods. This result could be interpreted

Table 11. *Estimates of the Demand Equation for U.S. Computer Output*[a]

Variable	*1967–85*		*1972–85*	
Constant	−3.50	0.95	−2.72	−2.96
	(0.66)	(1.02)	(0.49)	(1.85)
Computer price relative to business economy	−1.37	−1.45	−1.42	−1.41
	(9.55)	(13.91)	(8.62)	(14.49)
Business gross product	0.54	. . .	−0.03	. . .
	(0.86)	. . .	(0.05)	. . .
Defense procurement	0.34	0.31	0.76	0.76
	(2.64)	(2.52)	(3.71)	(3.98)
Ratio of foreign to U.S. office equipment patents	−0.57	−0.62	−1.44	−1.42
	(1.67)	(1.86)	(2.46)	(3.18)
Net exports of business machines	0.09	0.08	0.22	0.22
	(1.18)	(1.09)	(2.57)	(2.73)
Summary statistics				
$\bar{R}^2$	0.998	0.998	0.998	0.998
Durbin-Watson	1.45	1.51	2.54	2.53

a. All variables are in logarithms; monetary amounts are in constant dollars. The number in parentheses are *t*-statistics.

to mean that the impact of changes in the size of the U.S. business economy was overwhelmed by the large declines in computer prices.

In contrast, the size indicators for the more specialized computer markets, defense procurement and net exports of business machines, were highly significant. The effect of the ratio of foreign-origin to domestic-origin U.S. office equipment patents was also significant, but negative. One possible interpretation of the negative effect is that the patent ratio represents the implicit price or effectiveness of foreign computers relative to U.S. computers. But more analysis is needed, including the effects of foreign prices and exchange rates.

A Model Simulation of the Growth of the Computer Industry

The relationships describing the growth of the computer industry were assembled into an industry growth model. Only four 1972–85 period equations were included: one each for R&D expenditure, R&D capital stock, the inverse of computer price, and demand for computers. These equations are as follows:

$$\begin{aligned} RD_C &= -1.79 + 0.01 \cdot Q_C(-1) + 0.041 \cdot [Q_C - Q_C(-1)] \\ &\quad + 0.12 \cdot FPSHARE + 0.46 \cdot RD_{GN}(-2), \\ R_C &= R_C(-1) + RD_C(-4), \\ \ln(1/P_C) &= -7.51 + 2.09 \cdot \ln(R_C), \\ \ln(Q_C) &= -2.96 - 1.41 \cdot \ln(P_C) + 0.76 \cdot \ln(GDEF) \\ &\quad - 1.42 \cdot \ln(FPRATIO) + 0.22 \cdot \ln(X_C), \end{aligned}$$

where

$FPRATIO$ = Ratio of foreign-origin to U.S.-origin office equipment patents granted in the United States

$FPSHARE$ = U.S. office equipment patents of foreign origin as percentage of total

$GDEF$ = U.S. defense procurement (billions of 1982 dollars)

RD_{GN} = federally funded R&D expenditure in nonelectrical machinery industry (billions of 1982 dollars)

P_C = U.S. computer price index divided by deflator for gross domestic product of private business (1982 = 1.00)

Q_C = output of computer industry (billions of 1982 dollars)

RD_C = private research and development expenditures in office equipment industry (billions of 1982 dollars)

R_C = stock of private computer R&D capital (billions of 1982 dollars, 1961 = 2.5)

X_C = U.S. net exports of business machines (billions of 1982 dollars).

The exogenous variables are the foreign share in U.S. office equipment patents, government R&D in the machinery industry lagged two years, defense procurement, net exports of business machines, and the ratio of foreign to U.S. office equipment patents. The simulation period is 1972 to 1985, corresponding to the years for which the new price index series is available.

The results of the model simulation, some of which are shown in figures 3 and 4, appear to track the experience of the computer industry even more closely than that for telecommunications. The end points of the simulated values are within 1.5 percent of the actual R&D expenditure, 2.6 percent for R&D capital, 9.3 percent for computer price, and 5.8 percent for computer output. From this one may tentatively conclude that the behavior of the computer industry is consistent with the growth model for the high-technology industries as outlined here. Little relationship between the computer and communications industries is evident.

Figure 3. *Relative Price of Computers, 1972–85*

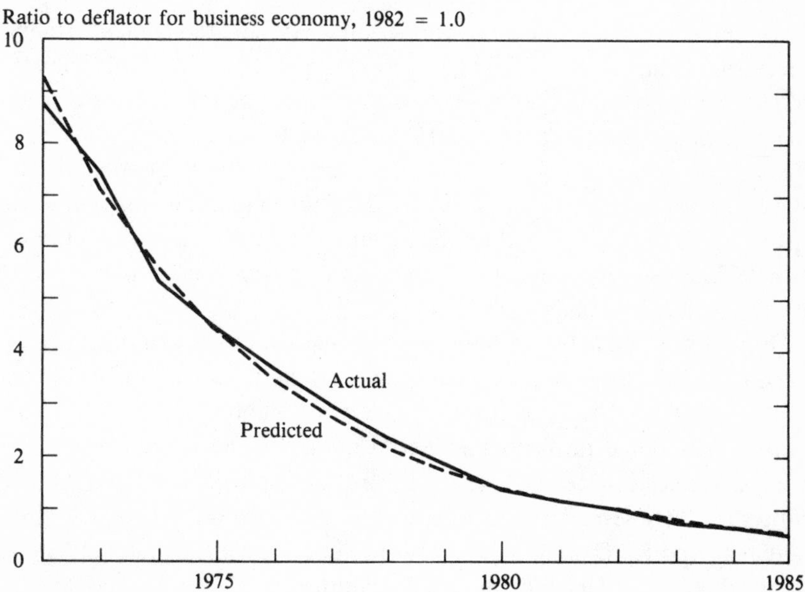

Source: Author's calculations.

Figure 4. *Demand for Computer Output, 1972–85*

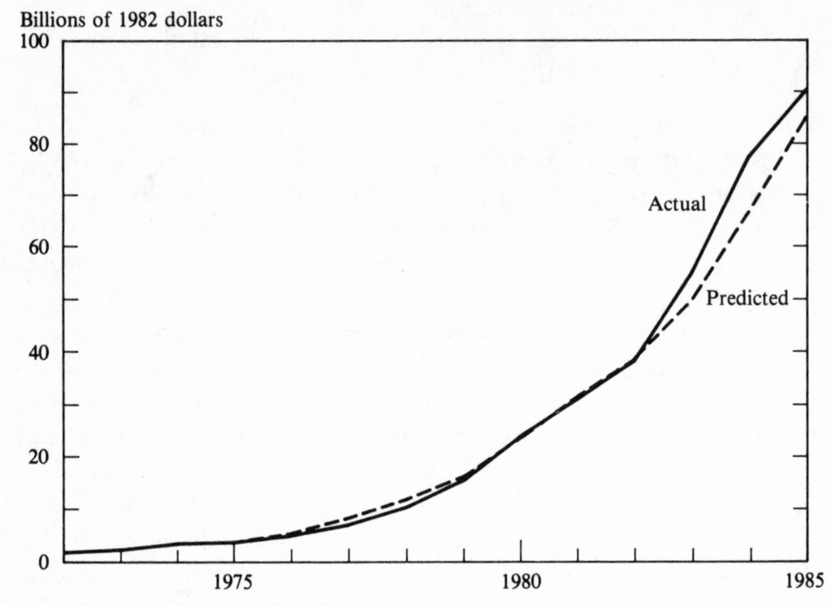

Source: Author's calculations.

Concluding Observations

Both the telecommunications services industry and the computer manufacturing industry have followed a growth pattern typical of technology-intensive industries. In both cases, growth-reinforcing influences run from the growth of output to private R&D investments, which accumulate into the R&D capital stock, which in turn induces productivity improvements. Productivity increases, equivalent to reductions in real unit costs, lead to reductions in real prices. Price reductions stimulate more than proportionate increases in demand, which in turn induce further increases in R&D investments, and so on.

Both R&D intensity and the rates of change in prices and output have been much greater in the computer industry than in telecommunications. There are other differences between these two industries as well. In my model, only domestic factors appear to have influenced growth in U.S. telecommunication services, whereas growth in the computer industry was influenced by international factors as well. Historically, government procurement and R&D were significant only in computers, while regulation affected only the telecommunications industry.

R&D Investments

R&D investments in both communication equipment and electronic components and in the office equipment industry are clearly linked to the levels of output from sectors that make use of the results of this R&D. Specifically, real private R&D expenditure in the communication equipment and electronic components industries rises by an estimated 3 to 4 percent of the increase in the real value of telecommunications output and by 3 percent of the increase in the real value of computer shipments. Private R&D expenditures in the U.S. office equipment industry (mainly computer R&D) rise by about 1.5 percent of the increase in computer output lagged by one year and by an additional 3.5 percent of the increase in computer output from the preceding year.

Private spending on R&D in communication equipment and electronic components also appears to depend on government R&D expenditures. Private R&D in telecommunications and electronic components is also positively related to past increases in GNP. No international influence on R&D investment in communication equipment and electronic components could be found.

Investment in private R&D in the office equipment industry is apparently not clearly related to government-funded computer R&D. However, private computer R&D appears to be affected by the share of patents of foreign origin in total U.S. patents issued in the office equipment field, indicating that U.S. R&D expenditures are sensitive to foreign competitive pressure.

Accumulation of R&D Capital

Annual productivity increases in the telecommunications service industry responded to private R&D investments in previous years. The largest effect was estimated for a six-year lag, but precise estimates were sensitive to specification. The estimated timing of lag effects suggests that the concentration of industrial R&D expenditures is in the development stage. The lead time for relatively less expensive but more fundamental and conceptual research may be much longer; however, it could not be identified statistically.

R&D and Productivity

R&D capital appeared to have a substantial impact on output and productivity in telecommunications services. Using the inverse of the newly developed computer price index of the U.S. Department of Commerce as an implicit productivity index, one may conclude R&D capital stock also had a large effect on productivity in the computer industry.

Productivity and Prices

Since 1964 there appears to have been a very clear relationship between the rate of productivity change and the rate of change in the real price of telecommunications services. Because productivity has grown more rapidly in the telecommunications industry than in the general economy, the relative price of telecommunications services has also been declining. Before 1964, though, only a weak relation existed between changes in productivity and price. The real price of telecommunications showed little change. The reasons for this break are unclear and need to be investigated in some detail. There were changes in the regulatory behavior and in the regulatory climate at the state and federal level in the 1960s that may have promoted a lowering of prices in line with productivity change. Changes in technology may also have influenced the behavior of the industry and its regulators.

Price Changes and Growth of Demand

Demand functions were estimated for both telecommunications services and computer equipment. In both cases very high price elasticities were estimated: −1.4 for the computer industry, and −1.6 to −1.8 for telecommunications services. The estimate for the computer industry is very similar to other estimates.[20]

The high price elasticity of computers helps explain how very large price declines for computers—averaging over 20 percent annually—contributed to the 35 percent annual growth rates in output sustained over the past two decades. Similarly, the high price elasticity of demand for telecommunications services may explain why, in spite of the substantial slowdown in the growth of the economy after the mid-1960s, the annual growth rate of telecommunications output after 1964 continued unchanged (6.7 percent) in 1964–85, or even higher (7.4 percent) from 1964 to 1983.

In addition to price declines, other factors also influenced the demand for the output of these industries. One was the size of the market. The

20. Flamm, *Targeting the Computer*, p. 31.

income elasticity of demand as finally estimated for telecommunications services was low at 0.6 to 0.7, but statistically highly significant. The demand for computers was apparently unrelated to private business product as a whole. However, significant output elasticities were estimated for two more narrowly defined markets: defense procurement (0.76) and net U.S. exports of business machines (0.22).

Convergence of Telecommunications and Computer Industries

The potential convergence of the telecommunications and computer industries is a matter of tremendous importance to business executives, regulators, and international economic policymakers. However, no statistically significant linkages were found between the growth processes in these two industries, despite the obvious role of electronics technology innovation in both industries.

The Role of Government

Government has shaped growth in the telecommunications industry by supporting research in communication equipment and electronic components. Each dollar of government R&D appears to have induced 70 cents to one dollar of private R&D. By contrast, government R&D expenditures have had little effect on private R&D expenditures in the office equipment industry. However, computer output and (indirectly) private computer R&D have been positively influenced by defense procurement.

We know from detailed analyses of the computer industry, by Flamm and by Katz and Phillips, that the U.S. government has funded development of successive generations of the top lines of new computer technologies and also provided a ready market for its products. Only later were these products adapted to commercial markets. The development of computer technology in the United States has thus been supported by the defense needs of the U.S. government. Recently, firms in other countries, particularly in Japan, have succeeded in absorbing U.S. computer technology and in developing frontier technologies of their own. However, the substantial market for new products utilizing defense-sponsored technology remains a unique feature of the growth of the U.S. computer industry. It is doubtful that without government support for research and its purchases of successive new state-of-the-art products, the growth of the computer industry could have proceeded at anything near the rate observed. By contrast, the development of telecommunications technology, at least in its formative years, was almost entirely the effort of the private sector.

International Linkages

U.S. computer output appears to be significantly and negatively related to the ratio of foreign to domestic office equipment patents granted in the United States. This linkage apparently represents a foreign competitive effect and might be interpreted as an indicator of the relative quality or price of foreign computers. However, any hypothesis along these lines needs to be tested using actual data on prices for foreign computers.

Private R&D investment in computers appears to be related to the proportion of U.S. office equipment patents granted to foreign inventors. Foreign competition has thus spurred U.S. investment in computer R&D and contributed to productivity increases and declines in computer prices. No such international linkages could be identified in the telecommunications industry. U.S. telecommunications seems to lack both a clear link to the computer industry and the international orientation that so pervades the U.S. computer industry.

Table 12. *Data for the Telecommunications Service Industry, 1948–85*
Billions of 1982 dollars unless otherwise specified

Year	Output	Employment (thousands)	Net stock of fixed capital	Total factor productivity index (1982 = 1.000)	Productivity changes: Annual change (percent)	Productivity changes: Real value of increment	Real price (1982 = 1.000)	Real unit costs (1982 = 1.000)
1948	7.9	694	29.0	0.387	7.6	0.558	n.a.	n.a.
1949	8.2	688	31.5	0.390	0.9	0.073	1.819	n.a.
1950	8.6	666	33.9	0.403	3.2	0.270	1.903	1.769
1951	9.4	691	36.2	0.419	3.9	0.355	1.833	1.775
1952	9.9	715	38.5	0.421	0.4	0.037	1.885	1.785
1953	10.8	739	41.2	0.437	3.8	0.396	1.932	1.748
1954	10.9	731	43.7	0.431	−1.4	−0.148	1.916	1.760
1955	11.8	741	46.5	0.449	4.3	0.489	1.866	1.754
1956	12.6	784	50.2	0.449	0.0	−0.004	1.841	1.764
1957	13.5	800	54.0	0.460	2.4	0.307	1.811	1.723
1958	14.0	748	57.2	0.479	4.1	0.567	1.826	1.667
1959	15.2	719	60.2	0.513	7.3	1.061	1.824	1.608
1960	15.9	724	64.0	0.518	0.9	0.139	1.823	1.609
1961	16.7	711	67.9	0.530	2.4	0.386	1.837	1.603
1962	18.1	706	72.3	0.556	4.8	0.843	1.789	1.574
1963	19.8	703	77.1	0.586	5.4	1.027	1.774	1.528
1964	21.3	717	82.0	0.603	2.9	0.590	1.759	1.519
1965	23.3	748	87.6	0.623	3.4	0.751	1.693	1.503
1966	25.6	789	93.7	0.644	3.2	0.794	1.623	1.489
1967	27.8	817	99.8	0.664	3.2	0.851	1.597	1.440
1968	30.2	825	105.9	0.693	4.4	1.277	1.522	1.391
1969	33.2	886	113.0	0.712	2.7	0.869	1.473	1.345
1970	36.8	949	121.2	0.736	3.3	1.170	1.407	1.284
1971	38.6	944	129.2	0.748	1.6	0.591	1.404	1.272
1972	42.3	954	136.8	0.790	5.6	2.283	1.401	1.244
1973	45.7	975	145.7	0.817	3.4	1.492	1.353	1.239
1974	48.0	985	155.2	0.826	1.1	0.537	1.288	1.175
1975	49.8	959	162.9	0.846	2.4	1.185	1.240	1.141
1976	51.8	941	169.9	0.868	2.6	1.342	1.245	1.145
1977	55.0	950	178.0	0.895	3.1	1.645	1.188	1.131
1978	60.7	983	188.8	0.941	5.2	2.993	1.131	1.091
1979	65.1	1,036	202.7	0.949	0.8	0.526	1.046	1.071
1980	71.1	1,062	216.2	0.991	4.4	3.002	0.985	1.013
1981	75.1	1,091	229.3	1.003	1.2	0.894	0.975	1.011
1982	77.3	1,109	240.8	1.000	−0.3	−0.244	1.000	1.000
1983	83.2	1,051	251.0	1.079	7.9	6.355	1.010	0.945
1984	83.5	1,035	261.4	1.067	−1.2	−0.965	1.028	0.979
1985	83.8	1,000	272.3	1.062	−0.4	−0.358	1.050	0.979

Sources: R&D expenditures are from National Science Foundation. R&D price deflator is from Edwin Mansfield, "Price Indexes for R and D Inputs, 1969–1983," *Management Science*, vol. 33 (January 1987), p. 125. The number of researchers in Japan is from Office of the Prime Minister, Bureau of Statistics, *Report on the Survey of Research and Development, 1985* (Tokyo, 1985), and earlier years. Other data are from U.S. Department of Commerce, Bureau of Economic Analysis. Stocks of R&D capital are author's calculations. Productivity changes are author's calculations based on productivity and output data.

Continued on next page

Table 12—(*continued*)

	R&D in communication equipment and electronic component industries[a]						
	Expenditures		*Stocks of R&D capital*[b]				
Year	*Federally funded*	*Company-funded*	*Company-funded distributed lag structure*	*Company-funded 4-year lag*	*Govern-ment-funded, 4-year lag*	*R&D price deflator electrical equipment industry (1982 = 1.000)*	*Number of researchers in Japanese electrical equipment industry*
1948	n.a.	n.a.	n.a.	n.a.	. . .	n.a.	n.a.
1949	n.a.	n.a.	n.a.	n.a.	. . .	n.a.	n.a.
1950	n.a.	0.443	n.a.	n.a.	. . .	n.a.	n.a.
1951	n.a.	0.479	n.a.	n.a.	. . .	n.a.	n.a.
1952	n.a.	0.519	n.a.	n.a.	. . .	n.a.	n.a.
1953	n.a.	0.561	n.a.	n.a.	. . .	n.a.	n.a.
1954	n.a.	0.619	n.a.	n.a.	. . .	n.a.	n.a.
1955	n.a.	0.676	n.a.	n.a.	. . .	n.a.	n.a.
1956	n.a.	0.732	n.a.	n.a.	. . .	n.a.	n.a.
1957	1.773	0.787	n.a.	n.a.	. . .	0.292	n.a.
1958	2.062	0.848	5.000	5.000	6.100	0.298	n.a.
1959	2.801	1.009	5.559	5.676	7.267	0.305	n.a.
1960	3.042	1.225	6.170	6.408	8.037	0.310	6,398
1961	2.981	1.500	6.835	7.195	9.810	0.313	5,636
1962	3.352	1.614	7.555	8.043	11.872	0.320	5,843
1963	3.716	1.733	8.340	9.052	14.672	0.325	7,266
1964	3.732	1.828	9.219	10.276	17.714	0.330	10,065
1965	3.656	1.994	10.249	11.776	20.695	0.339	8,629
1966	4.063	2.339	11.464	13.390	24.048	0.351	10,411
1967	4.147	2.580	12.869	15.123	27.763	0.361	11,352
1968	4.031	2.623	14.418	16.952	31.495	0.379	12,925
1969	3.895	2.785	16.109	18.946	35.151	0.400	14,362
1970	3.379	2.815	17.946	21.222	39.214	0.420	14,970
1971	3.356	2.841	19.969	23.802	43.360	0.441	18,937
1972	3.371	2.995	22.214	26.424	47.391	0.457	18,753
1973	3.120	3.127	24.660	29.209	51.286	0.483	19,374
1974	2.505	3.021	27.233	32.024	54.666	0.527	20,613
1975	2.157	2.910	29.944	34.865	58.021	0.579	23,822
1976	2.162	3.079	32.735	37.860	61.392	0.611	24,511
1977	2.173	3.232	35.602	40.986	64.513	0.647	28,550
1978	2.147	3.445	38.574	44.007	67.018	0.698	27,210
1979	2.503	3.819	41.590	46.917	69.175	0.760	28,897
1980	2.395	4.154	44.582	49.996	71.337	0.850	30,997
1981	2.286	4.465	47.594	53.228	73.510	0.938	38,830
1982	2.573	4.837	50.718	56.673	75.657	1.000	41,819
1983	2.779	5.864	53.976	60.492	78.160	1.050	45,542
1984	3.279	6.529	57.449	64.646	80.558	1.091	53,641
1985	3.692	6.880	61.226	69.111	82.845	1.127	54,558

n.a. Not available.

a. The R&D price deflators for the periods of 1957–68 and 1984–85 were derived by linking the Mansfield deflator for electrical equipment (1982 = 1.00) to GNP price deflators.

b. As discussed in the text, an initial value of $5 billion for company-funded stocks was estimated.

Table 13. *Data for the Computer Industry, 1965–85*

Billions of 1982 dollars unless otherwise specified

				R&D expenditures		*Stocks of R&D capital*[a]	
Year	*Value of ship-ments*	*Computer price index (1982 = 1.00)*	*Real computer price (1982 = 1.00)*	*Company-funded (office equipment industry)*	*Federally funded (non-electrical machinery industry)*	*Company-funded (office equipment industry)*	*Federally funded (non-electrical machinery industry)*
1965	0.174	18.18	51.65	1.141	0.790	4.992	13.175
1966	0.379	12.10	33.33	1.191	0.817	5.744	14.022
1967	0.443	9.31	25.01	1.275	0.897	6.601	14.793
1968	0.687	7.19	18.47	1.365	0.902	7.553	15.593
1969	0.973	6.17	15.13	1.695	0.653	8.694	16.383
1970	1.048	5.55	13.00	1.740	0.624	9.884	17.200
1971	1.315	4.74	10.58	1.909	0.709	11.159	18.097
1972	1.742	4.08	8.76	2.052	0.862	12.524	18.999
1973	2.227	3.69	7.45	2.292	0.867	14.219	19.652
1974	3.440	2.91	5.37	2.656	0.946	15.960	20.276
1975	3.551	2.65	4.45	2.924	0.858	17.868	20.985
1976	4.911	2.31	3.66	3.000	0.843	19.920	21.848
1977	6.896	2.00	2.96	3.299	0.709	22.212	22.714
1978	10.408	1.69	2.34	3.540	0.529	24.868	23.661
1979	15.574	1.46	1.86	3.763	0.426	27.792	24.519
1980	23.925	1.18	1.37	4.009	0.755	30.792	25.362
1981	31.114	1.07	1.14	4.093	0.738	34.091	26.071
1982	38.191	1.00	1.00	4.722	0.857	37.631	26.600
1983	55.140	0.77	0.75	4.987	1.101	41.395	27.026
1984	77.539	0.68	0.64	5.724	1.126	45.404	27.781
1985	90.505	0.56	0.51	6.415	1.387	49.496	28.519

Sources: Data for the first seven columns from National Science Foundation, from Kenneth Flamm, "Targeting Technology," draft (Brookings, 1985), table 2A-5. and from David W. Cartwright, "Improved Deflation of Purchases of Computers," *Survey of Current Business*, vol. 66 (March 1986), p. 8. International data (last seven columns) are from the U.S. Department of Commerce, Bureau of Economic Analysis, and Patents and Trademark Office.

Continued on next page

Table 13—(*continued*)

	Number of U.S. patents granted for office equipment inventions				*U.S. trade in business machines*		
Year	*To U.S. inventors*	*To foreign inventors*	*Total*	*Percent foreign*	*Exports*	*Imports*	*Net exports*
1965	1,276	306	1,582	19.34	n.a.	n.a.	n.a.
1966	1,487	335	1,822	18.39	n.a.	n.a.	n.a.
1967	1,383	345	1,728	19.97	0.579	0.126	0.453
1968	1,078	303	1,381	21.94	0.606	0.151	0.455
1969	1,237	365	1,602	22.78	0.754	0.215	0.539
1970	1,168	394	1,562	25.22	0.950	0.267	0.683
1971	1,751	618	2,369	26.09	0.897	0.283	0.614
1972	1,538	664	2,202	30.15	0.986	0.359	0.627
1973	1,424	616	2,040	30.20	1.512	0.564	0.948
1974	1,269	643	1,912	33.63	2.190	0.699	1.491
1975	1,112	625	1,737	35.98	2.048	0.701	1.347
1976	1,170	645	1,815	35.54	2.355	0.927	1.428
1977	1,185	627	1,812	34.60	3.060	1.131	1.929
1978	1,080	594	1,674	35.48	4.066	1.662	2.404
1979	835	539	1,374	39.23	5.514	1.952	3.562
1980	999	668	1,667	40.07	8.334	4.076	4.258
1981	1,023	723	1,746	41.41	10.148	5.000	5.148
1982	1,060	869	1,929	45.05	11.008	6.166	4.842
1983	1,067	910	1,977	46.03	15.503	11.743	3.760
1984	1,290	1,060	2,350	45.11	22.006	19.299	2.707
1985	1,283	1,400	2,683	52.18	26.061	21.562	4.499

n.a. Not available.

a. Estimates assume no depreciation and a single four-year embodiment lag.

Appendixes to Kenneth Flamm's Paper

Appendix A: Methodology

Decomposition of Computer Price Change

The starting point is a set of estimates of price per unit bandwidth developed for computer components.[1] For disk and tape storage units, high and low estimates of the cost of bandwidth were developed, differing in the way in which storage capacity is treated as a determinant of effective bandwidth.[2]

From knowledge of the production technology that transforms inputs of component bandwidth into computer system bandwidth, a corresponding cost function may be derived, giving the cost of computing bandwidth as a function of the prices of component inputs. This cost function will be approximated with a homogeneous translog cost function (a flexible functional form that provides a second-order approximation to an arbitrary, twice continuously differentiable linear homogeneous cost function). This implies constant returns to scale in the provision of computer bandwidth, that is, that doubling all inputs of component bandwidth doubles system bandwidth.[3] It can then be shown that the translog cost function is exactly indexed by the Tornqvist price index,[4] defined by

$$\text{(A-1)} \quad \ln c(p^1) - \ln c(p^0) = \sum_i ((s_i^1 + s_i^0)/2)(\ln p_i^1 - \ln p_i^0),$$

1. These may be found in Kenneth Flamm, *Targeting the Computer: Government Support and International Competition* (Brookings, 1987), app. A.

2. Ibid.

3. Recent analysis of data used to support increasing returns in computing power ("Grosch's Law") show that conclusion to be an artifact of classification bias. A corrected analysis of these same data does not reject the hypothesis of constant returns to scale in computing capacity. See Haim Mendelson, "Economies of Scale in Computing: Grosch's Law Revisited," *Communications of the ACM*, vol. 30 (December 1987), pp. 1066–72.

4. Since the translog function is a second-order approximation to an arbitrary cost function, and the translog function is exactly indexed by the so-called Tornqvist index, the Tornqvist index is a so-called *superlative* price index. See W. E. Diewert, "Exact and Superlative Index Numbers," *Journal of Econometrics*, vol. 4 (May 1976), pp. 115–45.

where p^1 is a vector of input prices in period 1, p_i^1 is element i of that vector, s_i^1 is the share of input i in total expenditure on inputs used in producing computer systems in period 1, and superscript 0 is used to denote the same variables in period 0.

The Tornqvist index has a convenient property: changes in the natural logarithms of the index are just the weighted sum of the changes in the natural logarithms of the input prices, weighted by the average share (over two adjacent periods) of that input in total expenditure. First differences in natural logs are approximately equal to the percentage rate of change of a variable over time. Thus the approximate rate of decline in overall system cost is easily broken down into the respective contributions of decline in each component price.[5]

Two additional points should be made. First, the construction of the component price indexes and the interpretation of the cost of bandwidth as the price of bandwidth assume that the price paid for bandwidth is equal to its average cost, that is, that computer systems are produced in competitive (or contestable) markets. Second, allowing for the possibility of a significant degree of substitution among components in the provision of system bandwidth is firmly rooted in the history of computer design. As particular components have become cheaper, they have in fact been substituted for relatively more expensive components in order to provide higher performance at lower cost. Buffers made from cheap memory can speed up relatively slow input-output and storage devices, and inexpensive slave processors, put to work supervising the operations of peripheral devices, can free a central processor to carry out its primary functions.

Modeling Incremental Cost in Transmission and Switching

It is helpful to write out the total cost of a transmission system as

$$C(r, x) = C_T(x) + C_D(r, x), \tag{A-2}$$

where C_T is the cost of the terminal equipment required at the ends of a transmission route, which for a particular type of system generally does not vary much with the distance covered by the route, and C_D is the portion of transmission system cost that varies with the distance covered (this may include the cost of *repeaters*, amplifiers required to maintain signal quality that must be installed at fixed intervals).

5. The Tornqvist index for system bandwidth price plotted here is virtually identical to the Fisher "ideal" index developed in *Targeting the Computer*. Since both are flexible second-order approximations to arbitrary functions, this is not particularly surprising.

The portion of cost that varies with distance is a function of r, the number of physical route miles spanned by the system, and x, the width or cross section of the communications link, expressed in number of circuits. In general, for a particular transmission technology, empirical data suggest C_D can be expressed as

$$C_D(r, x) = f(x)xr \tag{A-3}$$

with $f(x)$ a function giving the cost per circuit mile of a system with cross section x. Both functions C_T and f will vary with the precise technology used. Choice of a least-cost transmission technology will generally depend on the physical distance to be covered (r) and the capacity of the communications link (x).

Figure A-1 compares annual first costs (that is, initial installation and material, excluding maintenance and repair) per circuit mile for a variety of technologies available in 1950, over the range of system capacities with which they could be utilized. Figure A-2 is a similar curve constructed in 1975.

Both diagrams (known as Dixon-Clapp curves), separated by twenty-five years of technological innovation, show costs per circuit mile de-

Figure A-1. *Relative Line Costs versus System Capacity, Dixon-Clapp Curve, 1950s*

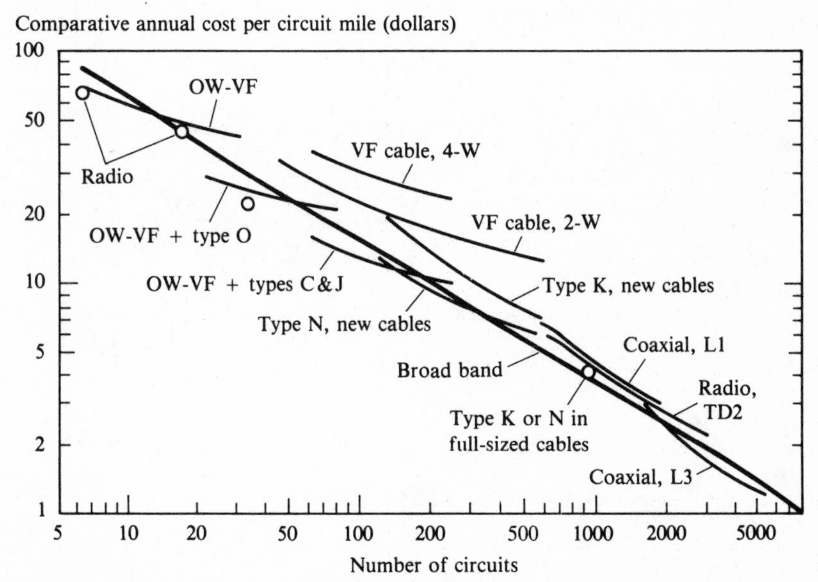

Source: E. F. O'Neill, ed., *A History of Engineering and Science in the Bell System*, vol. 7: *Transmission Technology, 1925–75* (Murray Hill, N.J.: AT&T Bell Laboratories, 1985), pp. 778–79.

Figure A-2. *Economies of Cross Section for Transmission, Dixon-Clapp Curve, Facility Line Costs, 1975*

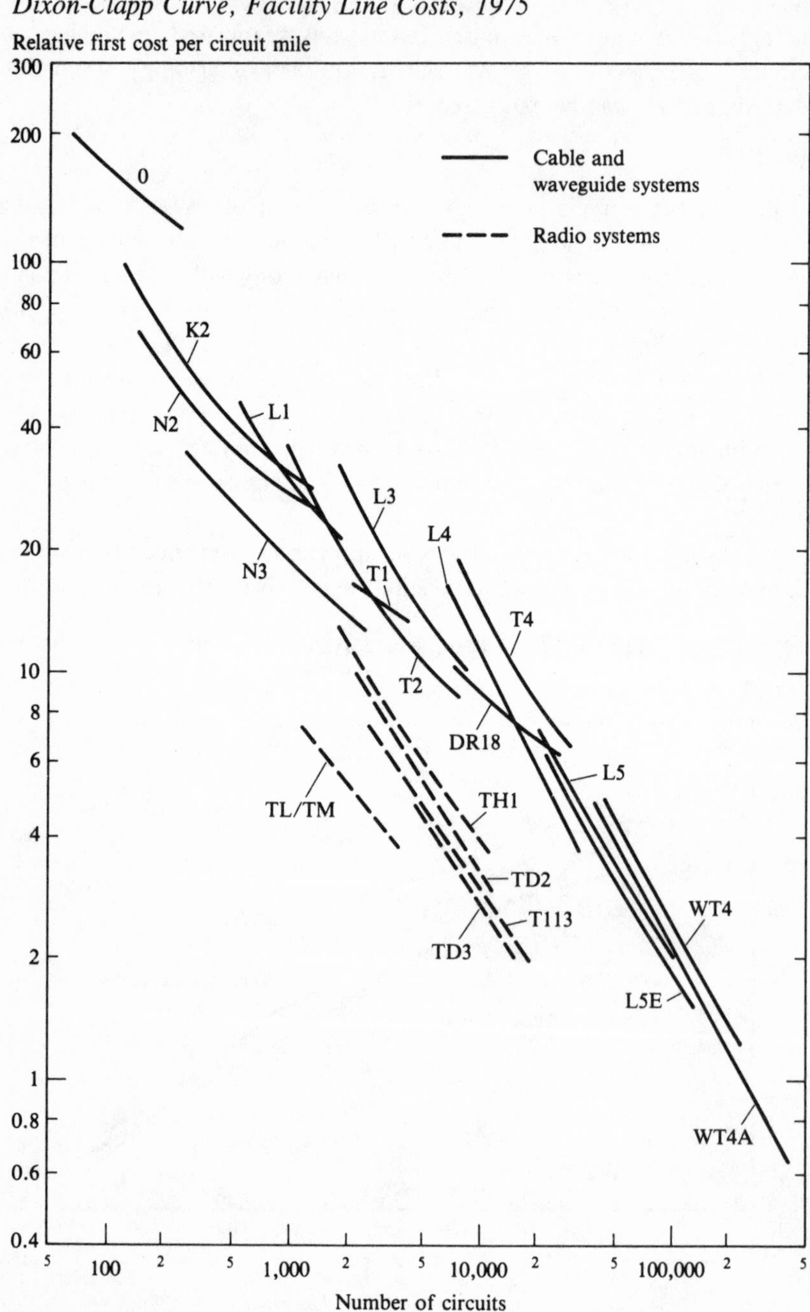

Source: Same as figure A-1.

clining by about a factor of 100 with a thousandfold increase in the number of circuits on a line. The Dixon-Clapp curves essentially depict an "average" function f as in equation A-3, for a number of discrete technologies, over the relevant ranges of circuit cross sections. Note that because the Dixon-Clapp curve is traced out by movement between distinct technologies, it can be considerably steeper than the relation f that holds true for any individual technology.

Measuring Incremental-Capital Cost in Transmission

It is not clear that one should necessarily hold cross section constant when assessing the impact of technological change. Much of the technological advance in transmission systems has been focused on pushing along the Dixon-Clapp curve, down and to the right, and the increase in cross section—along with the lower cost it brings—is an outcome of technological progress. One can, however, assume that the slope of the Dixon-Clapp curve has remained relatively constant over time (as seems to have been the case in figures A-1 and A-2), control for changes in cross section, and attribute the residual decline in cost to vertical shifts in the Dixon-Clapp curve.

To be more precise, one can argue that change in transmission cost is given approximately by

$$dC_D \simeq d(f(x)xr), \quad \text{(A-4)}$$

based on equation A-3 and the assumption that termination costs are a small part of new investment in transmission capacity. I shall further assume that the total cost per route mile of a link with cross section x is given by

$$f(x)x = \alpha\, x^{\epsilon+1}, \quad \text{(A-5)}$$

with parameter ϵ an estimate of the (assumed constant) elasticity of transmission cost per circuit mile with respect to cross section (the slope of the Dixon-Clapp curve), and α a constant cross-section measure of vertical shifts in the Dixon-Clapp curve, reflecting the level of cost at some moment in time for a fixed cross section. After plugging equation A-5 into A-4 and some simple calculus, one has

$$\alpha \simeq dC_D/[x^{\epsilon}(\epsilon r\, dx + d(xr))]. \quad \text{(A-6)}$$

Note that if there were no economies of cross section (that is, $\epsilon = 0$), the expression would reduce to incremental cost divided by incremental circuit miles $[d(xr)]$.[6] The differentials in this relation will be approximated by annual changes in the variables.

However, the Dixon-Clapp curve is best interpreted as a relation that obtains "averaging" across transmission technologies, at any moment in time. There are then two potential causes for the vertical shifts measured by α. One is technological change, reducing the cost of a route mile for every given cross section and shifting this "average" Dixon-Clapp curve in.

The other cause for changes in α is variation in the mix of route lengths for new transmission capacity. That is, particular transmission technologies corresponding to different route lengths (long versus short hauls) may yield points lying somewhat off the Dixon-Clapp relation, which is "averaged" across all transmission technologies. However, since expansion in medium- and long-haul capacity, including carrier systems of all kinds, has dominated the growth of the national network over the postwar period, one can probably overlook variation in route length as a source of change in costs.

Appendix table A-1 reports an attempt to measure incremental switching and transmission cost using historical AT&T data. "Switching" activity was assumed proportional to total local calls plus ten times long-distance messages.[7] "Transmission" capacity was taken as proportional to total circuit miles of telephone grade carrier (on open wire, cable, and point-to-point radio) installed in the Bell System. Incremental costs were calculated as follows: for switching capacity, annual change in the net book value of central office switching equipment, and for transmission costs, annual change in the net book value of central office circuit and radio equipment, plus change in the net book value of toll and exchange lines.[8]

6. Or if one were to attribute all reductions in cost per circuit mile to the effects of technological change.

7. The factor of 10 is applied to long-distance messages because the duration of long-distance calls is about 2.5 times that of a local call, and a long-distance call goes through perhaps 4 times as many switches. Therefore the total erlangs of installed switching capacity required for a long-distance message would be about 10 times that of a local call.

The 2.5 figure derives from noting that the average duration of a local call was about 3 minutes in the Bell System in the 1970s, compared with about 8 paid minutes for a long-distance call. See G. E. Schindler, Jr., ed., *A History of Engineering and Science in the Bell System*, vol. 3: *Switching Technology, 1925–1975* (Murray Hill, N.J.: AT&T Bell Laboratories, 1982), p. 295; and AT&T, *Bell System Statistical Manual, 1950–1981* (June 1982), p. 805. The assumption that 4 times as many switches are involved boils down to 2 local switches (at each end) and 2 toll switches for a long-distance message, versus a single local switch for a local call. Roberts takes 3 to 4 interswitch "hops" as a reasonable average estimate for messages routed through a nationwide communications network. Lawrence G. Roberts, "Data by the Packet," *IEEE Spectrum*, vol. 11 (February 1974), pp. 46–51.

8. "Average cross section" was estimated as total circuit miles of telephone carrier plus total route miles of wire (coaxial, pairs, and other), divided by total route miles of wire and radio. All are from the *Bell System Statistical Manual* except the last item (microwave radio

Table A-1. *Cost Measures of Incremental Capacity, 1951–81*
1970 = 1

Year	*Transmission* $\epsilon = 0$	$\epsilon = -0.4$	$\epsilon = -0.5$	*Switching*
1951	5.70	4.98	4.83	0.73
1952	2.61	2.59	2.62	1.00
1953	1.52	1.58	1.63	1.36
1954	1.86	1.83	1.84	0.80
1955	4.36	3.28	3.04	0.41
1956	4.08	3.28	3.10	0.68
1957	3.18	2.67	2.55	0.93
1958	1.41	1.44	1.46	0.79
1959	1.77	1.60	1.57	0.43
1960	1.55	1.44	1.42	0.59
1961	1.41	1.39	1.40	1.07
1962	1.08	1.12	1.15	0.55
1963	1.57	1.50	1.49	0.85
1964	0.96	0.99	1.01	0.70
1965	1.15	1.13	1.14	0.57
1966	0.60	0.66	0.68	0.59
1967	0.93	0.98	1.00	0.85
1968	0.75	0.80	0.81	0.70
1969	0.54	0.60	0.62	0.56
1970	1.00	1.00	1.00	1.00
1971	1.13	1.12	1.12	1.18
1972	1.02	1.03	1.04	0.96
1973	1.40	1.28	1.25	0.96
1974	1.84	1.64	1.59	1.05
1975	1.64	1.65	1.65	1.59
1976	2.38	1.97	1.87	0.93
1977	1.72	1.70	1.69	0.87
1978	2.26	1.97	1.90	0.57
1979	0.96	1.09	1.13	0.91
1980	0.91	1.09	1.15	1.06
1981	1.07	1.27	1.33	1.34

Source: Author's calculations based on historical AT&T data. See text for description of methodology.

route miles), which is from the Federal Communications Commission, *Statistics of Communications Common Carriers* (Washington, D.C.: GPO, various years).

Net book value of central office equipment and of toll and exchange lines is reported in the *Bell System Statistical Manual, 1950–1981*, p. 602. Central office equipment was disaggregated into switches and into circuit and radio equipment by ''reverse engineering'' the weights used in the calculation of the TPIs, which are reported in *Bell System Telephone Plant Indexes*, Report SL/83-05-11 (May 31, 1983), group B, sec. 2, p. 1.

The reverse engineering was accomplished as follows. For each category and subgroup in the index, constant dollar expenditure weights were reported as representative of spending for a particular period. These weights were assigned to the median year for that period (or two years, if an even number of years), and intermediate-year weights calculated by simple linear interpolation. The left endpoints (first years) were held constant through the year when a weight was first assigned; the right endpoints (last years) were interpolated forward using the growth

The elasticity of cost per circuit mile with respect to cross section (ϵ) was taken to be from about -0.4 to -0.5, based on the data depicted in figures A-1 and A-2. (A value for $\epsilon = 0$ is also shown for comparison.) For transmission, α is calculated; for switching, cost per increment in switching capacity.

By noting that an index that ignores change in cross section ($\epsilon = 0$) exceeds the calculated series that controls for change in cross section ($\epsilon = -0.4$ or -0.5) only when cross section falls, it can be concluded that reductions in average cross section generally seem to have coincided with the spikes in incremental cost (α), particularly in the early 1950s and mid-1970s.[9] Since installation of lower capacity links probably accompanied a shift to shorter route lengths, this suggests that the spikes may in part reflect bursts of investment in shorter haul transmission facilities.

An Econometric Model of Switch Prices, 1970–82

Selected information on switch prices and characteristics[10] collected by the Rural Electrification Administration to monitor loans was used in this

---

rate calculated before the year of the last set of weights. Because the reported weights were constant dollar expenditure shares, that is, $c_i = P_i^0 Q_i / \Sigma(P_i^0 Q_i)$, with Q_i the fixed quantity weight for item i over the weighting period, and P_i^0 the price for item i in year 0, by multiplying this expression by $(P_i^t/P_i^0)/(\Sigma_i(P_i^t/P_i^0)c_i))$ an estimate of current dollar expenditure weights was produced. These were used to disaggregate current dollar central-office equipment spending into switch and transmission equipment components.

9. It is readily shown that if using an $\epsilon < 0$ reduces estimated α, then average cross section must have declined. From equation A-6, using $\alpha_{\epsilon=k}$ to denote the estimate of α produced when ϵ is assumed to take value $k \leq 0$, we have

$$\frac{\alpha_{\epsilon = -c}}{\alpha_{\epsilon = 0}} = x^c \, [d(xr)/(d(xr) - crdx)].$$

Since $x \geq 1$ and r, $c > 0$, then $x^c \geq 1$. Also, $d(xr)$ is always positive in our data. So to have the ratio fall short of 1, $crdx$ must be negative, which implies $dx < 0$, that is, cross section has declined.

10. From the viewpoint of performance, the crucial characteristics of a switching system are the number of incoming calls that can be processed (a limitation dependent on the power of the processor or controller in the switch), the traffic load (in erlangs) that can be handled by the switching matrix, and the number of lines or trunks that can be physically interfaced to the switching network. In practice, switching systems are engineered so that the number of lines or trunks that can be connected is rarely a binding constraint on the system; for given processor power, maximum call attempts and traffic load are linked by a relation involving the average duration of a call, which varies with customer mix. Since the advent of stored program control, an additional characteristic of a switch has been the range of so-called features (speed calling, call waiting, call forwarding) that it supports. Since these are written into the control software, processor power may be a good proxy for the potential range of features that can be added to a switch of given size.

For constant-offered traffic load per customer line or trunk, and constant customer mix, cost per line is a measure of the cost of switch functionality. This, perhaps, is why cost per line is commonly used as a summary measure of the cost of switching functionality.

analysis.[11] Among the data included were total price for the equipment covered by a contract; whether the bid was competitive, negotiated, or reflected certified construction; the manufacturer of the equipment; the number of lines and trunks attached to the system; the number of digital trunks, if any, after 1981; the number of local loop adapters (actually, an REA estimate derived by dividing loop adapter expenditure by a fixed adapter price estimate in any given year); the number of push-button lines; whether the bid price included installation cost; whether the system had common control or processor control; and whether automatic number identification (ANI), local automatic message accounting (LAMA), or common mode operation (CMO) were included as system features in the bid. Processor-controlled systems normally include ANI as a standard feature and do not require CMO, and since systems installed after 1981 normally included processor control, ANI and CMO features are no longer reported after that year. Also, a separate feature cost for ANI is sometimes (but not always) reported in years before 1981.

Data on switch prices were assembled from the REA equipment loan files for the years 1970, 1976, and 1982. Switch prices were treated as being generated by a multi-output cost function, where the outputs produced were switch characteristics—lines, trunks, and options. The cost function can be thought of as a unit cost function, if switch markets are competitive or contestable, and there are constant returns to scale in switch production. Alternatively, the cost function may be interpreted as a hedonic function, a parametric relation between switch costs and characteristics, perceived by users and producers. If markets are competitive or contestable so price always equals average cost, the hedonic function will coincide with the producer's cost function. The multi-output cost function for any year was assumed to take the form

$$\text{(A-7)} \qquad C(L, T, i, j, k, b) = c(L, T, i; p', t)\, d(j, k, b).$$

That is, the cost of a switch unit configured with L lines, T trunks, vector i of optional features, installed by manufacturer j in state k, and using bid method b (competitive, negotiated, or certified) is the product of two functions. One (c) varies over time, reflecting changes in input prices (p')

See Bell Communications Research, "Traffic Environment," *LSSGR Issue 1*, 2d rev. (June 1985), sec. 17; and Groupe des Ingénieurs du Secteur Commutation du CNET (GRINSEC), *North-Holland Studies in Telecommunications*, vol. 2: *Electronic Switching* (Amsterdam: Elsevier Science Publishers, 1983), pp. 287–90.

11. The loan operations of the REA are described in Barry P. Bosworth, Andrew S. Carron, and Elisabeth S. Rhyne, *The Economics of Federal Credit Programs* (Brookings, 1987), pp. 121–26.

and technology (indexed by t) that alter the contours of this function. The other (d) does not, reflecting the assumption that interstate cost differentials, based on transportation, labor, and other installation costs, and the bid-specific and firm-specific cost factors remain relatively constant.

Since additional information on the cost of the ANI feature was often available, this information was incorporated into the estimation procedure for 1970 and 1976 by specifying another equation:

$$\text{(A-8)} \quad \text{Cost of ANI} = [c(L, T, i) - c(L, T, i')]\, d(j, k, b)$$
$$= g(L, T, i)\, d(j, k, b) \text{ if ANI cost available,}$$

where i' is the vector of switch characteristics with ANI zeroed out.[12] As before, function g will vary from year to year with changes in input prices and technology. The actual system of equations estimated was

$$\text{(A-9)} \quad C(L, T, i, j, k) = [c(L, T, i') + g(L, T, i)]\, d(j, k, b).$$

$$\text{(A-10)} \quad \text{Cost of ANI} = g(L, T, i)\, d(j, k, b).$$

A first-order approximation (linear in natural logarithms) of the form

$$\text{(A-11)} \quad c = \exp\,(b_0 + b_1 \ln L + b_2 \ln T + b_3 i')$$

was used, along with similar loglinear specifications for functions g and d.[13] The estimation was accomplished after transforming equation A-10 to logs, but leaving equation A-9 in its original untransformed state. The coefficients of all variables other than state and manufacturer were allowed to differ from year to year, in order to reflect changes in input prices and technology.

Effects of firm, state, and bid method were estimated as displacements from constants reflecting values for negotiated contracts between Stromberg-Carlson (53 percent of the 371 contracts in my sample)[14] in Minnesota (9.4 percent of all contracts). Of the 371 contracts in the sample, 107 were for 1970, 88 were for 1976, and 176 for 1982.

12. Since not all switches can be upgraded to ANI, a dummy variable reflecting the use of ANI, whether or not a separate cost for ANI was available, was included in the specification.

13. T includes an analog equivalent for digital T-1 trunks in 1982, computed as 24 times the number of such trunks (which include the equivalent of 24 analog circuits), then added to the sum of other analog trunks. The share of such digital trunks in all trunks was included as an additional variable in vector i', to control for cost differences between digital and analog interfaces to trunks. The ratio of push-button lines to total lines, and local loop adapters to lines, were also included as variables in i'.

14. Followed by Northern Telecom with 21 percent. Northern Telecom accounted for 44 percent of 1982 transactions, none in other years.

This nonlinear system was estimated by applying Zellner's "seemingly unrelated regression" technique, that is, by imposing cross-equation constraints on coefficients and accounting for possible covariances among equations in the residual error terms to produce more efficient estimates. The estimated coefficents, along with approximate standard errors, are reported in appendix table A-2.[15] The estimated coefficients are of reasonable magnitude and signs, and they are for all major variables (excluding some of the state and other dummy variables) statistically significant (I reject the null hypothesis that they are zero). The results seem to indicate that there were economies of scale in the provision of switching capacity (the inverse of the sum of the cost elasticities with respect to outputs of trunk and line interface capacity is greater than one).[16]

The Transition through Divestiture, 1982–85

Procurement data covering the 1982–85 period were obtained on magnetic tape from the REA. Because these data were made available in a different and more detailed form than the data found in the published reports of previous years, a somewhat different model was used to estimate quality-adjusted price change.

To begin, all options added to the basic switch were generally priced separately. The cost of a basic switch, stripped of all options (for example, LAMA, push-button calling, firebars, loop extenders, custom calling) could be calculated by deducting from the switch price the costs of all such features included in the package. A "stripped" switch cost, including only the costs of basic switching services, and the actual configuration of lines and trunks (further broken down as incoming, outgoing, two-way, and digital) was thus calculated. Note as well that by 1982, virtually all

15. All reported standard errors use White's heteroskedasticity-consistent estimator. See Halbert White, "A Heteroskedasticity-Consistent Covariance Matrix Estimator and a Direct Test for Heteroskedasticity," *Econometrica*, vol. 48 (May 1980), pp. 817–38.

16. In the functional form used here to approximate the cost function, optional switch outputs other than qualitative features and ordinary analog lines (L) and trunks (T) are expressed in terms of multiples of either L or T; for example, z replaces optional output Z as an argument in the cost function where $z = Z/T$. Thus $C(L, T, Z) = C(L, T, zT) \equiv C'(L, T, z)$, with $C'_T = C_T + C_z z$. Therefore $C'_T T/C = C_T T/C + C_z Z/C$. The apparent elasticity of the cost function C' with respect to output T equals the true elasticity of cost C with respect to T, summed with the elasticities of cost C with respect to all other outputs that enter the cost function scaled as multiples of T.

A consistent estimate of the inverse of the sum of cost elasticities with respect to all outputs (and approximate standard errors) based on the parameter estimates for L and T in appendix table A-2 is 1970, 1.21 (.26); 1976, 1.13 (.099); and 1982, 1.31 (.043). Economies of scale may have increased over these dozen years.

Table A-2. *Econometric Models of Switch Prices, 1970–82*[a]

Parameter	*Coefficient*	*Standard error*[b]
Common to both equations		
States		
Alabama	−0.6512799E-01	0.9913553E-01
Alaska	0.3162304	0.1213432**
Arizona	−0.5561068E-01	0.9345634E-01
Arkansas	−0.5017854E-02	0.1119683
California	0.7667642E-01	0.1129680
Colorado	−0.1117967	0.8640576E-01
Florida	0.5513092E-01	0.9458544E-01
Georgia	0.1813385E-01	0.7976628E-01
Idaho	−0.4519519E-01	0.1420877
Illinois	−0.5102670E-01	0.7314635E-01
Indiana	0.9671030E-01	0.1131260
Iowa	−0.1155599	0.1071100
Kansas	0.1870256	0.8041490E-01**
Kentucky	−0.9532073E-01	0.8634992E-01
Louisiana	−0.6472291E-01	0.1106729
Michigan	−0.3737001E-01	0.7121398E-01
Mississippi	0.2956435	0.9202766E-01**
Missouri	−0.2296303	0.1185343
Montana	0.8213803E-01	0.1074528
Nebraska	0.7084450E-01	0.7658158E-01
New Hampshire	0.6405738E-01	0.1069673
New Mexico	0.1676177	0.1647315
New York	−0.1279013	0.9761091E-01
North Carolina	−0.8515452E-01	0.1937924
North Dakota	−0.4334668E-01	0.6791860E-01
Ohio	0.1207530	0.1039651
Oklahoma	0.2046420E-01	0.9701480E-01
Oregon	0.8104940E-01	0.8248904E-01
Pennsylvania	0.1163264	0.1511226
South Carolina	0.2695538	0.9703356E-01**
South Dakota	−0.1156382	0.8641446E-01
Tennessee	0.2756565E-03	0.1016002
Texas	−0.2112023E-01	0.1516713
Utah	0.2078100	0.9889376E-01**
Washington	0.7092643E-01	0.1953229
West Virginia	0.3177754	0.1579324**
Wisconsin	0.1314566	0.9996238E-01
Wyoming	−0.5985539E-01	0.7715028E-01

Table A-2—*continued*

Parameter	*Coefficient*	*Standard error*[b]
Type of Contract		
Competitive	−0.4459455E-01	0.4886993E-01
Certified construction	0.3465657E-01	0.1000320
Manufacturers		
Automatic Electric	0.4815281E-03	0.1071285
GTE	−0.1520768	0.1807621
Harris	−0.6796321E-01	0.1943700
ITT	−0.1383000	0.7838669E-01*
NEC	−0.1851636E-01	0.6070628E-01
North Electric	−0.1503870	0.1248631
Northern Telecom	0.1197236E-01	0.5920334E-01
Redcom Labs	−0.9117526	0.1961454**
Siemens	−0.7142867E-01	0.1288388
ANI equation[c]		
Lines, 1970	0.4116586	0.6095531E-01**
Lines, 1976	0.1632980	0.1326891
Trunks, 1970	−0.6170634E-01	0.5512606E-01
Trunks, 1976	−0.1253833	0.1119803
Constant for ANI[d]	8.605459	0.6193512**
Shift terms:		
Year 1970	−2.101648	0.7179903**
Common control, 1976	0.3459872	0.1401700**
Common control, 1970	−0.1477041	0.1139288
Loop adapters/lines, 1970	−0.9574637E-01	0.4609525
Loop adapters/lines, 1976	−0.6510737	0.5809805
Not installed	−0.1647994	0.1737737
Common mode, 1970	−0.1127976	0.8330439E-01
Common mode, 1976	−0.1750930	0.1864209
Switch equation[e]		
Lines, 1970	0.7486358	0.6177233E-01**
Lines, 1976	0.6306868	0.9079706E-01**
Lines, 1982	0.5031599	0.6189131E-01**
Trunks, 1970	0.7817041E-01	0.4976197E-01
Trunks, 1976	0.2519865	0.9003361E-01**
Trunks, 1982	0.2591921	0.6700024E-01**
Constant for switch[f]	8.271253	0.2633970**
Shift terms:		
Year 1970	−2.057439	0.4594291**
Year 1976	−1.477304	0.5111385**
No processor, ANI, 1970	−0.1002826E-01	0.8627517E-01
No processor, ANI, 1976	0.9766483E-01	0.8209887E-01

Table A-2—*continued*

Parameter	*Coefficient*	*Standard error*[b]
Common control, 1970	0.5750057	0.1754056**
Common control, 1976	0.3035446	0.1042269**
Processor control, 1976	0.2860849	0.2006224
LAMA, 1976	0.6871641E-01	0.1135140
LAMA, 1982	0.1117554	0.1217040
Loop adapters/lines, 1970	0.4746573	0.3020654
Loop adapters/lines, 1976	−0.5772041	0.2340914**
Loop adapters/lines, 1982	−0.4384886	0.7744208
Not installed	−0.2647485	0.8487914E-01**
Share push-button lines	−0.6306263E-01	0.4712698E-01
Share digital trunks	−0.1305034	0.7411482E-01*
Common mode, 1970	−0.1252635	0.8394802E-01
Common mode, 1976	0.1476776	0.7210306E-01**

Source: Author's calculations.

Abbreviations: ANI is automatic number identification; CMO is common mode operation; LAMA is local automatic message accounting.

* Significant in a two-tailed test at 10 percent significance level.

** Significant in a two-tailed test at 5 percent significance level.

a. Numbers designated E+ or E− must be multiplied by the appropriate exponent of 10; for example, E+13 by 10^{13}, E−01 by 10^{-1}. Log of likelihood function is −4338.61. Number of observations is 371.

b. Asymptotic heteroskedasticity-consistent estimates.

c. For ANI equation, the sum of squared residuals is 3.50038; standard error of the regression is 0.971339E-01.

d. Minnesota, negotiated, Stromberg-Carlson, common control, 1976.

e. For switch equation, the sum of squared residuals is 0.193815E+13; standard error of the regression is 72278.1.

f. Minnesota, negotiated, Stromberg-Carlson, processor control, 1982.

switches sold were processor-controlled. Thus "unbundled," fairly homogeneous switch costs for a variety of manufacturers were constructed over this period.[17]

A total of 567 such observations on stripped-switch cost were available, over a period of four years. The measured switch "outputs" were a relatively small number of items, including incoming lines, incoming trunks, outgoing trunks, two-way trunks, and digital trunks. As before, the underlying model was taken to be

$$C(L, T, j, k, b) = g(L, T)\, d(j, k, b), \qquad \text{(A-12)}$$

17. For a small number of contracts (15), it was not possible to unbundle the cost of LAMA from the total price reported. These contracts were dropped from the sample used. In another case, an obvious error in reporting total contract cost was spotted. A total of 16 observations from an initial sample of 635 contracts were therefore dropped for these reasons. An additional 52 observations on switches manufactured by two manufacturers—Harris and Redcom—were dropped for two reasons. First, unlike the other manufacturers, they did not sell any RSTs to go with their switches, suggesting that these machines were not strictly comparable to those capable of being configured with RSTs. And second, preliminary analysis of the data suggested that these manufacturers' switches were priced a great deal lower than similarly configured switches from the other manufacturers, again suggesting that these were not really comparable equipment. A grand total of 567 contracts were thus used in the estimation.

that is, total cost of a switch with L lines and a vector T of different types of trunks equals some function g of these characteristics multiplied by another function d reflecting other cost factors specific to manufacturer j, state k, and type of bidding procedure b. It was again assumed that g varies over time, with changes in input prices and technology, while function d is constant over time.

Because the measured switch characteristics consisted of a much smaller number of items, after stripping out the costs of various extras, it was feasible to use a more flexible, second-order approximation to model function g. The production technology for switches, with joint outputs of lines and various types of trunks, was assumed to change over time, with technical innovations. Cost function g_t is therefore indexed with t to indicate that it changes from period to period.

It was also assumed that the impact of technological change can be approximated with a finite number of time-varying parameters embedded in the production function, which change from year to year as the production technology changes. (For example, Hicks-Neutral, labor-augmenting, or capital-augmenting technical change can each be represented with only a single embedded parameter whose value changes over time.) Along with input prices, these form a finite set of parameters that take on fixed values that change from one year to the next but do not vary across the sample for any given year.

Two distinct flexible functional forms capable of yielding a second-order approximation were used. One was a simple quadratic function. With this functional form, to produce outputs of L lines and T trunks (and collapsing the time-varying set of technology and price parameters into a single time-varying parameter, V, for expositional convenience only)

$$\text{(A-13)} \quad g_t(L, T) = g(L, T, V_t) = a + b_V V_t + b_L L + b_T T + e_{LT} LT + e_{LV} V_t L + e_{TV} T V_t + f_L L^2 + f_T T^2 + f_V V_t^2.$$

Since V_t has a constant value that varies only from year to year, however, this can be rewritten as

$$g_t(L, T) = (a + b_V V_t + f_V V_t^2) + (b_L + e_{LV} V_t) L + (b_T + e_{TV} V_t) T + e_{LT} LT + f_L L^2 + f_T T^2,$$

or,

$$\text{(A-14)} \quad g_t(L, T) = a_t + b_{Lt} L + b_{Tt} T + e_{LT} LT + f_L L^2 + f_T T^2,$$

that is, we have a flexible second-order approximation, with time-varying coefficients on the linear terms.

In estimating this functional form, all variables have been transformed into deviations from their sample mean over the 1982–85 period.[18] The constant a_t in equation A-14 is then an estimate of price for a switch with a configuration identical to the sample mean, and the coefficients of the linear terms can be interpreted as the marginal costs of those outputs or characteristics in a specific year at the sample mean configuration. Function d was taken to be a simple linear function, and manufacturer, state, and bidding procedure effects were all expressed as deviations from a Northern Telecom switch, installed in Minnesota, with a negotiated price.

An alternative to the quadratic functional form is the translog approximation, which is essentially identical to equation A-14 but with natural logarithms of both cost and outputs used in place of the original variables. One possible complication is that some of the outputs (for example, digital, two-way, incoming, and outgoing trunks) are sometimes zero, and therefore cannot be arguments in a simple translog cost function. The solution adopted here is a generalization of that used in the simpler loglinear approximations estimated over 1970–82. The four original trunk outputs were transformed into four new output measures: total channels (that is, digital times 24 plus incoming plus outgoing plus two-way), and fraction of total channels accounted for by digital, outgoing, and two-way trunks (which obviously range from zero to one). Function g_t was written with arguments of the form e^S for these "share" output measures, that is, for a single such share measure, S,

$$\text{(A-15)} \quad \ln g_t(L, T, e^S) = a_t + b_{Lt} \ln L + b_{Tt} \ln T + b_{St} S + e_{LT} \ln L \ln T + e_{LS} \ln L\, S + e_{TS} \ln T\, S + f_L (\ln L)^2 + f_T (\ln T)^2 + f_S S^2.$$

When a translog version of equation A-14 is written out, all input and output measures except these share measures are transformed to natural logs, while the share measures show up in untransformed form. Note that if all the second-order terms in equation A-15 are restricted to equal zero, the functional form—a first-order loglinear approximation—is equivalent to the function used over the 1970–82 period. Thus results for four functional forms are reported, in an attempt to gauge how robust these results are to specification: a quadratic form, a translog form, and linear and loglinear forms (which can be obtained by imposing restrictions on the coefficients of the quadratic and translog).

18. There was no obvious trend in switch configuration over the four years examined.

The estimates reported earlier in text table 5 were generally nonlinear functions of parameters estimated in equations A-14 and A-15;[19] full and complete regression results may be found in appendix tables A-3 and A-4. As before, all approximate standard errors reported use White's heteroskedasticity-consistent estimator. Formal statistical tests lead to a rejection of the first-order approximations in favor of the more complex second-order approximating functions.[20]

Some of the other estimated parameters deserve further comment. In all cases, competitive and certified bidding practices seem to result in lower costs, compared with negotiated contracts. And certain states—Idaho, Kansas, Oregon, and Puerto Rico—consistently show much higher costs than Minnesota, the standard of comparison, regardless of the functional form chosen, while South Dakota is cheaper. Other state-specific cost differentials, while sometimes statistically significant, vary with functional form.

Prices for Remote-Switching Terminals

Since RSTs have only one measured characteristic—number of lines under their control—approximating a cost function is much simpler than is the case for standard switches. The same four functional forms were used to estimate variations of equations A-14 and A-15. As before, formal statistical tests lead us to reject use of the simpler, first-order approximations.[21] The results are reported in appendix tables A-5 and A-6. Coefficients for variables other than price were quite similar to those for standard switches. Competitive and certified bidding lowered cost relative to negotiated bids.

19. For the specifications using logarithms, these are calculated at the log of the mean value, not the mean of the logs of the values.

20. The likelihood statistics reported in appendix tables A-3 and A-4 allow us to construct likelihood ratio tests for the restrictions on the second-order approximations implied by the first-order approximating functions. For the restrictions on the translog function giving the loglinear, -2 times the difference in log likelihoods (which is distributed asymptotically as a chi-squared statistic, with degrees of freedom equal to the number of restrictions, 15) is 99.3; for the restrictions on the quadratic giving the linear form, the test statistic is 122. At the 1 percent significance level, the critical point for this distribution is about 30.6. Thus in both cases we reject the simpler approximations in favor of the more complex second-order forms.

21. The asymptotic *t*-statistic for the quadratic term is -8.2 in the quadratic function, 15.4 in the translog function.

Table A-3. *Econometric Models of Stripped-Switch Cost, 1982–85: Quadratic and Linear Approximations*[a]

Parameter	*Quadratic approximation*[b]		*Linear approximation*[c]	
	Coefficient	*Standard error*[d]	*Coefficient*	*Standard error*[d]
Year constants				
1982	458386.3	18216.25**	476284.6	24395.69**
1983	451179.2	16180.34**	440452.7	17134.83**
1984	381860.6	18952.58**	383706.2	20829.60**
1985	347302.3	19920.11**	346857.2	18451.74**
First-order terms				
Lines, 1982	196.2815	13.88192**	187.6514	17.19395**
Lines, 1983	213.8873	12.95531**	229.3711	15.60039**
Lines, 1984	156.2824	11.93569**	214.4096	18.00832**
Lines, 1985	138.6054	14.39514**	169.8059	15.22756**
Incoming trunks, 1982	457.2005	556.7673	1281.542	348.5004**
Outgoing trunks, 1982	2156.197	524.6959**	−287.7119	427.4658
Two-way trunks, 1982	290.6270	413.8723	580.4454	322.3231*
Digital trunks, 1982	5044.471	1842.472**	8620.101	3118.310**
Incoming trunks, 1983	388.2527	1751.028	158.1633	1108.022
Outgoing trunks, 1983	1357.857	1662.833	−203.3664	840.6961
Two-way trunks, 1983	508.6604	444.2455	700.8033	339.2443**
Digital trunks, 1983	8957.889	1674.796**	6118.758	1372.245**
Incoming trunks, 1984	−1420.226	1650.520	−2763.926	1424.646*
Outgoing trunks, 1984	2971.182	1411.362**	3052.059	1136.801**
Two-way trunks, 1984	−509.5368	433.3413	−776.2457	247.6976**
Digital trunks, 1984	5992.940	1491.819**	1208.117	695.6497*

Incoming trunks, 1985	−1464.029	1021.383	−708.2610	733.3106
Outgoing trunks, 1985	2660.813	1622.816	1180.100	1212.862
Two-way trunks, 1985	496.2247	393.7119	482.5701	296.5121
Digital trunks, 1985	8170.754	1645.088**	4660.008	790.6651**
Second-order terms				
Lines × incoming	−0.1776690	0.2771012	. . .	. . .
Lines × outgoing	−0.3035395	0.2319385	. . .	. . .
Lines × two-way	0.5456212E-01	0.1759097	. . .	. . .
Lines × digital	0.3330215	0.5796733	. . .	. . .
Incoming × outgoing	7.607799	6.432702	. . .	. . .
Incoming × two-way	−8.835877	13.07268	. . .	. . .
Incoming × digital	65.19439	34.44601*	. . .	. . .
Outgoing × two-way	13.45632	10.82449	. . .	. . .
Outgoing × digital	51.75142	23.75574**	. . .	. . .
Two-way × digital	−36.56126	18.33846**	. . .	. . .
Lines^2	0.6528297E-02	0.3944081E-02*	. . .	. . .
Incoming^2	2.286136	2.040591	. . .	. . .
Outgoing^2	−17.00381	6.458876**	. . .	. . .
Two-way^2	−0.7756399	2.089896	. . .	. . .
Digital^2	−23.56117	8.449665**	. . .	. . .
States				
Alabama	0.5769292E-03	0.4951345E-01	−0.4535750E-01	0.4741299E-01
Alaska	0.4611753	0.1265845**	0.4140932	0.1142983**
Arizona	−0.3933631E-01	0.6813887E-01	0.4701084E-01	0.6339067E-01
Arkansas	0.8045222E-01	0.9154439E-01	0.8407555E-01	0.8855755E-01
California	0.2596256E-01	0.9187893E-01	−0.9904657E-01	0.8796501E-01
Colorado	−0.8853281E-01	0.7624791E-01	−0.6991591E-01	0.8623646E-01
Florida	−0.3063404E-01	0.7028689E-01	−0.2650240E-01	0.6837097E-01

Table A-3—*continued*

Parameter	*Quadratic approximation*[b]		*Linear approximation*[c]	
	Coefficient	*Standard error*[d]	*Coefficient*	*Standard error*[d]
Georgia	0.9396924E-02	0.4914299E-01	0.3230403E-01	0.5928328E-01
Idaho	0.3598597	0.7925720E-01**	0.2095901	0.9582591E-01**
Illinois	−0.5751363E-01	0.5859561E-01	0.3324505E-01	0.6709252E-01
Indiana	0.1297430	0.1056331	0.2149227	0.1194006*
Iowa	−0.1082028	0.5759129E-01*	−0.9754431E-01	0.6046820E-01
Kansas	0.1320726	0.6145427E-01**	0.1431043	0.7700111E-01*
Kentucky	0.5303803E-01	0.6219452E-01	0.3558703E-01	0.6998241E-01
Louisiana	0.8336280E-01	0.7131704E-01	0.1077058	0.7676885E-01
Maine	0.2106547	0.1140891*	0.1344104	0.9397786E-01
Massachusetts	−0.1314564	0.4310291E-01**	−0.1629903	0.4785759E-01**
Michigan	0.4248393E-01	0.6407314E-01	0.5628026E-01	0.6830575E-01
Mississippi	−0.7148170E-01	0.4589801E-01	−0.1001048	0.4772347E-01**
Missouri	−0.8601763E-01	0.7600780E-01	−0.6763138E-01	0.7095676E-01
Montana	0.3716119E-01	0.8621488E-01	0.4312083E-01	0.9250147E-01
Nebraska	−0.4717007E-01	0.5454596E-01	−0.3587168E-01	0.5590607E-01
New Hampshire	0.1092162	0.8600894E-01	0.1616576	0.8022348E-01**
New Mexico	−0.1265276E-01	0.6090743E-01	−0.9759215E-01	0.9131017E-01
New York	−0.2129546	0.6931620E-01**	−0.2272190	0.6728417E-01**
North Carolina	0.3327634E-01	0.5680364E-01	0.1792393E-02	0.5900815E-01
North Dakota	−0.1090536	0.5130731E-01**	−0.2755511E-01	0.6681385E-01
Ohio	−0.9082777E-01	0.9213137E-01	−0.9575241E-01	0.8630713E-01
Oklahoma	−0.4900275E-01	0.4292784E-01	−0.1436143E-01	0.5331205E-01
Oregon	0.1739387	0.8126887E-01**	0.1620280	0.8126642E-01**

Pennsylvania	−0.1075116E-01	0.9949048E-01	0.9692607E-01	0.1051225
Puerto Rico	0.9638208E-01	0.1336980	0.2729790	0.1348163**
South Carolina	−0.6250840E-01	0.6405232E-01	−0.4285301E-01	0.5608382E-01
South Dakota	−0.2822464	0.3425830E-01**	−0.2851667	0.4278892E-01**
Tennessee	−0.5885233E-01	0.6361070E-01	−0.8424533E-01	0.7321927E-01
Texas	0.3540424E-01	0.3858015E-01	0.4155600E-01	0.5170890E-01
Utah	0.6345087E-01	0.1505949	0.1983641E-01	0.1588940
Vermont	0.1917093	0.7124943E-01**	0.1754147	0.8738344E-01**
Washington	0.7144288E-01	0.1458823	0.1489010	0.1242785
West Virginia	0.4810895E-01	0.4770049E-01	0.1262569	0.6009714E-01**
Wisconsin	−0.9325134E-01	0.4365620E-01	0.8921010E-02	0.5051494E-01
Wyoming	−0.3596674E-01	0.7770156E-01	0.5002112E-01	0.8824291E-01
Manufacturers				
CIT-Alcatel	−0.2726297	0.1167460**	−0.3641733	0.1002717**
GTE	−0.4297474E-01	0.1114876	−0.1569921	0.1085914
ITT	−0.9298431E-01	0.6484258E-01	−0.9233715E-01	0.7568295E-01
NEC	−0.8029492E-01	0.4396417E-01*	−0.1008963	0.5425183E-01**
Stromberg-Carlson	0.2826784E-01	0.2513770E-01	−0.6650621E-03	0.2679958E-01
Type of Bid				
Competitive	−0.1035177	0.2376164E-01**	−0.1084289	0.2727389E-01**
Certified	−0.5398173E-01	0.4072291E-01	−0.6908077E-01	0.4350190E-01

Source: Author's calculations.

* Significant in a two-tailed test at 10 percent significance level.

** Significant in a two-tailed test at 5 percent significance level.

a. Numbers designated E+ or E− should be multiplied by the appropriate exponents of 10; for example, E+13 by 10^{13}, E−1 by 10^{-1}.

b. For quadratic approximation, sum of squared residuals is 0.313205E+13; standard error of the regression is 80862.3; R-squared is 0.940815; number of observations is 567; F-statistic (87, 479) is 87.5198; and log of likelihood function is −7164.11.

c. For linear approximation, sum of squared residuals is 0.388808E+13; standard error of the regression is 88716; R-squared is 0.926528; number of observations is 567; F-statistic (72, 494) is 86.5231; and log of likelihood function is −7225.41.

d. Asymptotic heteroskedasticity-consistent estimates.

Table A-4. *Econometric Models of Stripped-Switch Cost, 1982–85: Translog and Loglinear Approximations*[a]

Parameter	*Translog approximation*[b]		*Loglinear approximation*[c]	
	Coefficient	*Standard error*[d]	*Coefficient*	*Standard error*[d]
Year constants				
1982	11.05767	1.259154**	8.537424	0.2680827**
1983	11.22964	1.318852**	9.031813	0.5344001**
1984	10.16295	1.460806**	7.779910	0.6516277**
1985	11.05300	1.099336**	9.225390	0.3784265**
First-order terms				
Lines, 1982	0.1379179	0.2911451	0.5383391	0.4944130E-01**
Lines, 1983	0.1037690	0.2931521	0.4606778	0.3789842E-01**
Lines, 1984	0.6515308E-01	0.2911352	0.4232971	0.3402615E-01**
Lines, 1985	0.1293297	0.2602810	0.3589963	0.6969114E-01**
Total trunks, 1982	−0.2259290	0.2567720	0.1444325	0.5141451E-01**
Share outgoing, 1982	−1.225593	1.706726	0.4922723E-01	0.2344013
Share two-way, 1982	−1.895436	1.026827*	0.1223457E-01	0.1491321
Share digital, 1982	1.167654	1.268935	−0.1731104	0.1542917
Total trunks, 1983	−0.2544985	0.2637353	0.1671383	0.3576620E-01**
Share outgoing, 1983	−1.014856	1.939764	−0.1647936	0.8734737
Share two-way, 1983	−1.741423	1.167164	−0.1276360	0.4704419
Share digital, 1983	1.384498	1.424499	−0.2339823	0.4632806
Total trunks, 1984	−0.2431971	0.2557313	0.1637454	0.3582118E-01**
Share outgoing, 1984	1.014604	2.131504	2.332501	1.045399**
Share two-way, 1984	−0.6327155	1.276869	1.124888	0.6181253*
Share digital, 1984	2.544455	1.534461*	1.182626	0.5899431**

Total trunks, 1985	−0.3300654	0.2410161	0.1523718	0.7991466E-01*
Share outgoing, 1985	−1.120858	1.362881	0.1217307	0.3535821
Share two-way, 1985	−1.606497	1.013963	0.1118260E-01	0.2276808
Share digital, 1985	1.502703	1.151117	0.7871114E-01	0.1956957
Second-order terms				
Total trunks^2	0.5244190E-01	0.1834511E-01**	...	...
Lines^2	0.6015700E-01	0.9597804E-02**	...	...
Share outgoing^2	−0.5626819	0.7172036	...	...
Share two-way^2	1.568145	0.7254447**	...	...
Share digital^2	−2.006297	0.7473893**	...	...
Lines × total trunks	−0.6480191E-01	0.1817136E-01**	...	...
Lines × outgoing	−0.1243683	0.3544773	...	...
Lines × two-way	−0.2483350	0.2067193	...	...
Lines × digital	−0.6518389E-01	0.1941312	...	...
Total × outgoing	0.5309325	0.4120631	...	...
Total × two-way	0.4246265	0.2091953**	...	...
Total × digital	0.2507488	0.2086368	...	...
Outgoing × two-way	2.848894	1.164791**	...	...
Outgoing × digital	−3.791254	1.444034**	...	...
Two-way × digital	−0.6939359	0.9156715	...	...
States				
Alabama	−0.1011243	0.8434864E-01	0.4997493E-01	0.7529739E-01
Alaska	0.1616723	0.1403633	0.2597666	0.1447984*
Arizona	0.2064634	0.1415369	0.1861306	0.1255281
Arkansas	0.8897717E-01	0.9344142E-01	0.9676477E-01	0.1037762
California	−0.1832561E-01	0.1241344	−0.5909641E-01	0.1238254
Colorado	−0.3146274E-01	0.9590991E-01	0.6185770E-01	0.1020664
Florida	0.1199812	0.8851592E-01	0.2212629	0.1262920*

Table A-4—*continued*

Parameter	*Translog approximation*[b]		*Loglinear approximation*[c]	
	Coefficient	*Standard error*[d]	*Coefficient*	*Standard error*[d]
Georgia	−0.1996473E-01	0.5378264E-01	0.3589735E-02	0.5322838E-01
Idaho	0.2029591	0.8668155E-01**	0.1744558	0.1026299*
Illinois	−0.4164224E-01	0.7137878E-01	0.5755768E-01	0.7360074E-01
Indiana	0.1094610	0.9165778E-01	0.6825474E-01	0.9258897E-01
Iowa	−0.1549301	0.5264200E-01**	−0.1628663	0.5323502E-01**
Kansas	0.1409695	0.5959287E-01**	0.1427601	0.5797573E-01**
Kentucky	−0.5214781E-01	0.9086834E-01	−0.6146354E-01	0.9057033E-01
Louisiana	0.8716005E-01	0.9747809E-01	0.1340954	0.9162730E-01
Maine	0.6460110E-01	0.1630322	0.6405002E-01	0.1957399
Massachusetts	−0.8634269E-01	0.5987551E-01	0.9018935E-02	0.6110298E-01
Michigan	0.4414790E-01	0.5925689E-01	0.6114569E-01	0.6080945E-01
Mississippi	−0.1432515	0.7413446E-01*	0.7641591E-01	0.7863221E-01
Missouri	0.7874730E-01	0.6500409E-01	0.8618680E-01	0.6191959E-01
Montana	0.1394263E-01	0.9494450E-01	0.3123859E-01	0.9627036E-01
Nebraska	0.1060771	0.5916480E-01*	0.1153745	0.5310026E-01**
New Hampshire	0.1375987	0.1323895	0.7798241	0.3917077**
New Mexico	0.5197226E-01	0.7300091E-01	0.9037500E-01	0.9256298E-01
New York	−0.9590326E-01	0.1006415	−0.1061249	0.9206913E-01
North Carolina	0.1532351	0.6117337E-01**	0.1744960	0.6488814E-01**
North Dakota	−0.2841616E-01	0.7728661E-01	−0.6864987E-01	0.7164592E-01
Ohio	0.3577063E-01	0.7932198E-01	0.2106737E-01	0.6730445E-01
Oklahoma	−0.6107843E-01	0.5879966E-01	−0.5954281E-01	0.5495282E-01
Oregon	0.2470897	0.8033529E-01**	0.2302419	0.6942100E-01**

Pennsylvania	0.1260447	0.9164749E-01	0.2260541	0.1227829*
Puerto Rico	0.4028018	0.1240806**	0.4604653	0.1245492**
South Carolina	0.2263269	0.8735675E-01**	0.3067941	0.9923869E-01**
South Dakota	−0.2654157	0.5045378E-01**	−0.2572911	0.5096176E-01**
Tennessee	0.1480214E-02	0.8595034E-01	0.1075205	0.1033159
Texas	0.1028976	0.5405329E-01*	0.1078470	0.5771444E-01*
Utah	0.8777502E-01	0.1062274	0.1051710	0.9348572E-01
Vermont	0.8374049E-01	0.7132304E-01	0.1571580	0.7894750E-01**
Washington	0.2540243	0.1732113	0.1824214	0.1530778
West Virginia	0.6800483E-01	0.6504285E-01	0.2262349E-01	0.6867453E-01
Wisconsin	−0.7427131E-01	0.8020560E-01	−0.7780826E-01	0.8166925E-01
Wyoming	−0.3964661E-0	0.7001949E-01	−0.2760609E-01	0.6632765E-01
Manufacturers				
CIT-Alcatel	−0.2332978	0.9342314E-01**	−0.3599375	0.1516489**
GTE	−0.3019573E-01	0.8993001E-01	−0.6393207E-02	0.1009127
ITT	0.8001594E-02	0.9631611E-01	0.6314873E-01	0.1075526
NEC	−0.1962010E-01	0.5806433E-01	−0.7977238E-01	0.7057822E-01
Stromberg-Carlson	0.2491555E-01	0.3168926E-01	0.3533711E-01	0.3268941E-01
Type of Bid				
Competitive	−0.8974510E-01	0.2443725E-01**	−0.6977442E-01	0.2701991E-01**
Certified	−0.5481642E-01	0.4398454E-01	−0.4291996E-01	0.4516097E-01

Source: Author's calculations.

* Significant in a two-tailed test at 10 percent significance level.

** Significant in a two-tailed test at 5 percent significance level.

a. Numbers designated E+ or E− should be multiplied by the appropriate exponents of 10; for example, E−01 by 10^{-1}. Dependent variable is natural log (cost).

b. For translog approximation, sum of squared residuals is 30.6181; standard error of the regression is 0.252826; R-squared is 0.878908; number of observations is 567; F-statistic (87, 479) is 39.9616; and log of likelihood function is 22.9322.

c. For loglinear approximation, sum of squared residuals is 36.479; standard error of the regression is 0.27; R-squared is 0.855726; number of observations is 567; F-statistic (72, 494) is 40.6949; and log of likelihood function is −26.7268.

d. Asymptotic heteroskedasticity-consistent estimates.

Table A-5. *Econometric Models of Remote-Switching Terminal (RST) Cost, 1982–85: Quadratic and Linear Approximations*[a]

Parameter	*Quadratic approximation*[b]		*Linear approximation*[c]	
	Coefficient	*Standard error*[d]	*Coefficient*	*Standard error*[d]
Year Constants				
1982	137204.9	7560.502**	127781.9	7558.752**
1983	142716.2	7324.064**	133357.1	7422.353**
1984	129992.7	5230.430**	122311.0	5410.404**
1985	131878.6	6343.445**	125438.7	6520.191**
First-order terms				
Lines, 1982	250.7490	16.01788**	222.0123	15.93793**
Lines, 1983	263.7009	15.19795**	227.5207	14.86730**
Lines, 1984	190.5968	10.96694**	171.6914	12.37856**
Lines, 1985	189.9108	11.56374**	160.0422	12.07365**
Second-order terms				
$Lines^2$	−0.1459333E-01	0.1777383E-02**	. . .	. . .
Type of Bid				
Competitive	−0.1711646	0.2759877E-01**	−0.1465385	0.2985942E-01**
Certified	−0.2373450	0.6299879E-01**	−0.2008662	0.6451248E-01**
Manufacturer				
CIT-Alcatel	−0.3007329	0.5119693E-01**	−0.2417760	0.5472485E-01**
GTE	−0.2189178	0.3515727E-01**	−0.1689438	0.3855713E-01**
ITT	−0.1426125	0.4827894E-01**	−0.7806655E-01	0.5421669E-01
NEC	−0.6639027E-01	0.8846800E-01	0.4396514E-01	0.9639314E-01
Stromberg-Carlson	−0.6320835E-01	0.3154909E-01**	−0.1446553E-02	0.3543826E-01
Vidar	0.1561128	0.1216123	0.2720532	0.1268628**

States				
Alabama	0.2673914	0.1234535**	0.2675334	0.1254446**
Alaska	0.4685268	0.9730679E-01**	0.5292018	0.1010200**
Arizona	0.3001876E-01	0.7216957E-01	0.3139854E-01	0.7639598E-01
Arkansas	0.1368550	0.1513681	0.1653124	0.1445120
California	0.7217776E-01	0.1061881	0.7166019E-01	0.1157544
Colorado	0.2770836	0.1607982*	0.3168474	0.1463703**
Florida	0.3472846	0.8470040E-01**	0.3149218	0.8734740E-01**
Georgia	0.1672150	0.7110771E-01**	0.1074368	0.7415242E-01
Illinois	0.2448460	0.7753597E-01**	0.2548503	0.8032830E-01**
Indiana	0.1761190	0.7152009E-01**	0.1635828	0.8835505E-01*
Iowa	0.2077667	0.1372267	0.2273123	0.1415305
Kansas	0.6089581E-01	0.1309698	0.7953683E-01	0.1368438
Kentucky	0.1985776	0.8389955E-01**	0.1926430	0.9840791E-01*
Louisiana	0.4255215	0.9005042E-01**	0.4190682	0.9740213E-01**
Maine	0.4654728	0.1273049**	0.4521532	0.1449335**
Michigan	0.1882926	0.8864623E-01**	0.2229895	0.1060717**
Mississippi	0.8055159E-01	0.5261932E-01	0.4930341E-01	0.5935815E-01
Missouri	0.3101013	0.1641906*	0.3218665	0.1806774*
Montana	0.1084760	0.1022367	0.9857415E-01	0.9311983E-01
Nebraska	0.2209679	0.7290353E-01**	0.2020784	0.7842204E-01**
New Hampshire	0.1502173	0.6797151E-01**	0.1218240	0.7253523E-01*
New Jersey	0.1500747	0.5912442E-01**	0.7377451E-01	0.7846596E-01
New Mexico	0.1188118	0.2720084	0.1113046	0.2675943
New York	0.2246094	0.7669805E-01**	0.2174897	0.8320415E-01**
North Carolina	0.2641335	0.8090215E-01**	0.2449870	0.8958802E-01**

Table A-5—*continued*

Parameter	*Quadratic approximation*[b]		*Linear approximation*[c]	
	Coefficient	*Standard error*[d]	*Coefficient*	*Standard error*[d]
North Dakota	0.2407889E-01	0.7690552E-01	−0.1432307E-01	0.6958068E-01
Ohio	−0.3073639E-01	0.5871127E-01	−0.2812917E-01	0.5770722E-01
Oklahoma	0.2094386	0.5212280E-01**	0.2315644	0.5542416E-01**
Oregon	0.2580434	0.7452540E-01**	0.1299587	0.6734734E-01**
Pennsylvania	0.8449948E-01	0.5922242E-01	0.2543842E-01	0.6170860E-01
Puerto Rico	0.5982560	0.9414397E-01**	0.4386954	0.9383850E-01*
South Carolina	0.7765533E-01	0.6820567E-01	0.6131988E-01	0.7512255E-01
Tennessee	0.8059518E-01	0.6500703E-01	0.4794589E-01	0.8083130E-01
Texas	−0.2288865E-01	0.8424467E-01	−0.6164722E-01	0.8762216E-01
Utah	0.3524277	0.7949538E-01**	0.3389619	0.8736105E-01**
Virginia	0.2275776	0.1567415	0.2252389	0.1729342
Vermont	0.2438455	0.1431026*	0.2151051	0.1424224
Washington	0.4330708	0.5602104E-01**	0.4035396	0.6174997E-01**
West Virginia	0.1822872	0.1434228	0.1815116	0.1562492
Wisconsin	0.7802290E-01	0.5181424E-01	0.8315911E-01	0.5923911E-01
Wyoming	0.1308803	0.5565254E-01**	0.8947223E-01	0.6064702E-01

Source: Author's calculations.

* Significant in a two-tailed test at 10 percent significance level.

** Significant in a two-tailed test at 5 percent significance level.

a. Numbers designated E+ or E− should be multiplied by the appropriate exponents of 10; for example E+12 by 10^{12}, E−01 by 10^{-1}.

b. For quadratic approximation, dependent variable is cost; sum of squared residuals is 0.413307E+12; standard error of the regression is 30106.1; R-squared is 0.946327; number of observations is 514; F-statistic (57, 456) is 141.050; and log of likelihood function is −5999.18.

c. For linear approximation, sum of squared residuals is 0.464287E+12; standard error of the regression is 31873; R-squared is 0.939706; number of observations is 514; F-statistic (56, 457) is 127.189; and log of likelihood function is −6029.07.

d. Asymptotic heteroskedasticity-consistent estimate.

Table A-6. *Econometric Models of Remote-Switching Terminal Cost, 1982–85: Translog and Loglinear Approximations*[a]

Parameter	Translog approximation[b]		Loglinear approximation[c]	
	Coefficient	Standard error[d]	Coefficient	Standard error[d]
Year constants				
1982	10.54071	0.2900607**	7.349270	0.2207994**
1983	9.834907	0.2778428**	7.347941	0.2645195**
1984	10.91533	0.3025100**	8.335501	0.3016584**
1985	11.16473	0.1187903**	9.882060	0.3800063**
First-order terms				
Lines, 1982	−0.3467862	0.7903006E-01**	0.7476207	0.3596178E-01**
Lines, 1983	−0.2230665	0.7608480E-01**	0.7336201	0.4693154E-01**
Lines, 1984	−0.4154708	0.7705197E-01**	0.5594418	0.5022510E-01**
Lines, 1985	−0.4661774	0.4867263E-01**	0.2962206	0.6429048E-01**
Second-order terms				
Lines^2	0.9026649E-01	0.5873562E-02**	. . .	. . .
Type of bid				
Competitive	−0.1154291E-01	0.2972947E-02**	−0.1924326E-01	0.3701831E-02**
Certified	−0.3305902E-01	0.6410876E-02**	−0.4669457E-01	0.7796857E-02**
Manufacturer				
CIT-Alcatel	−0.2450669E-01	0.5474336E-02**	−0.2179232E-01	0.8270159E-02**
GTE	−0.5399005E-03	0.5736584E-02	0.4870859E-02	0.5718305E-02
ITT	−0.6819376E-03	0.4318554E-02	0.6349493E-02	0.5199118E-02
NEC	0.2613268E-02	0.9931743E-02	0.9775444E-02	0.5944735E-02
Stromberg-Carlson	0.4863944E-02	0.2558102E-02*	0.5247597E-02	0.3472766E-02
Vidar	0.8698316E-02	0.8726229E-02	−0.9498419E-02	0.1192455E-01

Table A-6—*continued*

Parameter	*Translog approximation*[b]		*Loglinear approximation*[c]	
	Coefficient	*Standard error*[d]	*Coefficient*	*Standard error*[d]
States				
Alabama	0.2302221E-01	0.1033736E-01**	0.2316709E-01	0.1178839E-01**
Alaska	0.2657149E-01	0.7634092E-02**	0.2281725E-01	0.7807975E-02**
Arizona	−0.2545525E-01	0.1770527E-01	−0.2401149E-01	0.1820312E-01
Arkansas	0.3866542E-02	0.1365629E-01	0.1562980E-01	0.2043299E-01
California	0.1269872E-01	0.6597216E-02*	0.2048214E-01	0.8351131E-02**
Colorado	0.2406727E-01	0.1147496E-01**	0.1177602E-01	0.1018621E-01
Florida	0.2241855E-01	0.5269123E-02**	0.2417433E-01	0.6501791E-02**
Georgia	0.1847200E-01	0.6132783E-02**	0.3672710E-01	0.7820241E-02**
Illinois	0.2020536E-01	0.6413759E-02**	0.2466401E-01	0.7228814E-02**
Indiana	−0.6152978E-02	0.6419161E-02	−0.7014060E-03	0.7617616E-02
Iowa	0.2072707E-01	0.1035175E-01**	0.1632969E-01	0.1147225E-01
Kansas	0.1824958E-01	0.1340196E-01	0.1383758E-01	0.7840992E-02**
Kentucky	0.8408557E-02	0.6576537E-02	0.2675897E-01	0.9184299E-02**
Louisiana	0.3841027E-01	0.9481674E-02**	0.3683875E-01	0.1068228E-01**
Maine	0.1091501E-01	0.8901176E-02	0.2038828E-01	0.1016854E-01**
Michigan	0.1597007E-01	0.7017574E-02**	0.1618682E-01	0.6931925E-02**
Mississippi	0.1112176E-02	0.5571290E-02	0.1401158E-01	0.7344746E-02*
Missouri	0.1985847E-01	0.1193232E-01*	0.2556572E-01	0.1529603E-01*
Montanta	0.4484823E-02	0.1040198E-01	0.1024209E-01	0.1088876E-01
Nebraska	0.1130103E-01	0.6224673E-02*	0.1172243E-01	0.9278187E-02

New Hampshire	0.2093754E-01	0.6807724E-02**	0.3004556E-01	0.7525402E-02**
New Jersey	−0.1819745E-02	0.8737481E-02	0.2467450E-01	0.1163779E-01**
New Mexico	−0.1450651E-02	0.2023934E-01	−0.3703140E-02	0.2067774E-01
New York	0.8944252E-02	0.5817900E-02	0.7764195E-02	0.7305341E-02
North Carolina	−0.7824576E-03	0.7147913E-02	−0.3959901E-02	0.8885500E-02
North Dakota	0.6313265E-02	0.7523244E-02	0.6274297E-02	0.8309142E-02
Ohio	−0.7366398E-02	0.5467207E-02	−0.5770715E-02	0.6716059E-02
Oklahoma	0.1415834E-01	0.4678329E-02**	0.1777052E-01	0.5523644E-02**
Oregon	0.1860389E-01	0.6435202E-02**	0.3137269E-02	0.7357519E-02
Pennsylvania	−0.3172505E-02	0.5921079E-02	0.2093519E-02	0.6852945E-02
Puerto Rico	0.2061831E-01	0.9621325E-02**	0.3403705E-01	0.8863140E-02**
South Carolina	−0.4533778E-02	0.6268488E-02	0.2082569E-02	0.6998467E-02
Tennessee	0.7228102E-02	0.6061951E-02	0.1982309E-01	0.9781035E-02**
Texas	0.7964420E-02	0.7347396E-02	0.2572194E-02	0.1037748E-01
Utah	0.9250464E-02	0.8009206E-02	0.1172212E-01	0.8717423E-02
Vermont	0.3090277E-01	0.1734009E-01*	0.3423720E-01	0.1208950E-01**
Virginia	0.1424526E-01	0.1109751E-01	0.3886530E-02	0.1403062E-01
Washington	0.2532016E-01	0.4869472E-02**	0.2886782E-01	0.5984755E-02**
West Virginia	0.3577751E-02	0.8167920E-02	0.8615959E-02	0.1058515E-01
Wisconsin	−0.6608485E-02	0.5379397E-02	−0.6103981E-02	0.6006423E-02
Wyoming	−0.4669029E-02	0.6361785E-02	−0.2985793E-03	0.7663530E-02

Source: Author's calculations.

* Significant in a two-tailed test at 10 percent significance level.

** Significant in a two-tailed test at 5 percent significance level.

a. Numbers designated by E+ or E− should be multiplied by the appropriate exponents of 10; for example, E−01 by 10^{-1}, E−02 by 10^{-2}.

b. For translog approximation, dependent variable is cost; sum of squared residuals is 34.6991; standard error of the regression is 0.27585; R-squared is 0.871444; number of observations is 514; F-statistic (57, 456) is 54.2297; and log of likelihood function is −36.5888.

c. For loglinear approximation, sum of squared residuals is 53.3673; standard error of the regression is 0.34172; R-squared is 0.802281; number of observations is 514; F-statistic (56, 457) is 33.1136; and log of likelihood function is −147.223.

d. Asymptotic heteroskedasticity-consistent estimate.

Appendix B: Data

THIS appendix contains estimates by Bell Communications Research (Bellcore) of switch prices for local offices (table B-1), and toll and tandem switching centers for selected years (table B-2).

Table B-1. *Estimates by Bellcore of Local Office Switch Prices, Selected Years, 1972–84*

	Small offices[a] *(1–5,000 lines)*		*Medium-sized offices*[b] *(5,000–20,000 lines)*		*Large offices*[c] *(more than 20,000 lines)*	
Year	*Switch Type*	*Dollars per line*	*Switch type*	*Dollars per line*	*Switch type*	*Dollars per line*
1972	5-XB	257	2-ESS	149	1-ESS	140
1973	...	...	2-ESS	115	1-ESS	148
1975	5-XB	291	2-ESS	165	1-ESS	155
1976	...	...	2-ESS	168	...	...
1978	3-ESS	223	2B-ESS	169	1A-ESS	151
1979	10A-ESS	227	2B-ESS	166	1A-ESS	150
1980	...	...	1-ESS	158	...	...
1981	DMS-10	285	...	...	1A-ESS	165
1982	DMS-10	304	DMS-100	206	1A-ESS	189
1984	5-ARSM	195	5-ESS	168	5-ESS	141
1984[d]	...	...	5-ESS	183	5-ESS	153

a. Dollars-per-line figure assumes an office of 2,500 lines; engineered, furnished, and installed (EF&I) price for an all POTS (plain old telephone service) Office; 2.8 originating and terminating (O + T) hundred call seconds (CCS) per line; 1.3 originating and incoming (O + I) calls per line; 12:1 line-to-trunk ratio; all analog lines; 50 percent intraoffice (IAO) calling; and 35 percent digital trunks.

b. Calculations assume a 12,500-line office; costs for POTS-only offices are EF&I; 3.2 O + T CCS/line; 1.7 O + I calls per line; 9:1 line-to-trunk ratio; all analog lines; and 33 percent of traffic is IAO.

Before 1978 all trunking is analog; trunking goes from 30 percent to 100 percent digital between 1978 and 1983. The 1A-ESS figures give a credit for not having to use a D bank. The cost per line for the 5-ESS has two numbers, one assuming a $15 per line credit for D bank equipment, and the other assuming no credit.

c. Cost-per-line figures based on 25,000-line office; costs are EF&I for POTS-only office; 3.3 O + T CCS/line; 1.8 O + I calls per line; 10:1 line-to-trunk ratio; and 30 percent IAO traffic. All trunking before 1978 is analog; trunking grows from 20 percent to 80 percent digital in 1984. Cost-per-line figures for the 5-ESS reflect a $15 per line credit for D banks that are not needed in the 5-ESS.

d. Alternative calculation for 1984 assuming no $15 per line credit for D bank equipment.

Table B-2. *Estimates by Bellcore of Toll and Tandem Switching Center Prices, Selected Years, 1965–82*

	15,000-trunk center		*30,000-trunk center*		*60,000-trunk center*	
Year	*Switch type*	*Dollars per trunk*	*Switch type*	*Dollars per trunk*	*Switch type*	*Dollars per trunk*
1965	4A-XB	840	4A-XB	840	4A-XB	840
1972	4A-XB	1,123	4A-XB	1,123	4A-XB	1,123
1974	4A-XB	1,223	4A-XB	1,223	4A-XB	1,223
1976	4-ESS	738	4-ESS	589	4-ESS	515
1980	4-ESS	578	4-ESS	437	4-ESS	367
1982	4-ESS	525	4-ESS	409	4-ESS	352

Appendix C: Depreciation and Innovation

ASSUME R is the regulated rate of return applied to book value to determine total allowed profit, and r is the utility's cost of capital. In period 0, the utility has the choice of scrapping a machine with book value B_0, and replacing it with a new machine costing E_0. Prices remain constant over time. We shall assume that the replacement cost of the older machine is also E_0, so that there is no incentive, aside from depreciation factors, to switch technologies in order to increase absolute profit on given outputs. Both machines require an annual depreciation and maintenance expenditure of dE_0 to be kept in operation, while the utility company is granted a depreciation rate $s < d$ by regulators, to determine the book value of its investment for rate-of-return setting procedures.

Assume that in period 0 the machine chosen has book value B_0, market or economic value (or replacement cost) E_0, with $B_0 = aE_0$, $a \geq 1$. If the utility installs a new machine, $a = 1$. If the utility sticks with the old machine, and book depreciation rates fall short of economic depreciation rates, book value will exceed economic value and $a > 1$.

In period 1, when the new (or old) setup actually goes into operation, book value will be given by

$$B_1 = B_0 + dE_0 - sB_0 = B_0 (1 - s) + dE_0, \tag{C-1}$$

that is, last period's book value plus required replacement investment less book depreciation. Similarly,

$$B_2 = B_1 (1 - s) + dE_0, \tag{C-2}$$

and so forth, and solving, we have

$$B_t = \left[a (1 - s)^t + d \sum_{i=1}^{t} (1 - s)^{i-1} \right] E_0. \tag{C-3}$$

In the limit, as t gets very large, B_t approaches $(d/s)\, E_0$; note that $d/s > 1$ whenever economic depreciation is greater than book depreciation.

If the utility sticks with the old equipment, the present discounted value of its future stream of profits, assuming it earns the fixed rate of return, R, on the book value of its capital, is just

$$V^{O} = \sum_{t=1}^{\infty} (1 + r)^{-t} [RB_t^{O} - dE_0], \qquad \text{(C-4)}$$

the discounted sum of allowed returns on capital less the depreciation costs associated with that capital. Because book value exceeds economic value, $B_0^O = aE_0$, with $a > 1$.

If the firm goes with the new methods,

$$V^{N} = \left(\sum_{t=1}^{\infty} (1 + r)^{-t} [RB_t^{N} - dE_0] \right) - E_0 + zB_0^O. \qquad \text{(C-5)}$$

Now $B_0^N = E_0$, because initial book value equals purchase price. E_0 is subtracted to reflect purchase cost of the new machine at time 0, and zB_0 is revenue from scrapping of the old machine, where $z \geq 0$. If proceeds from selling the old machine are counted against profit on the rate base, or the scrapped item is not resold, then $z = 0$. If the scrapped machine is sold for its economic value, $z = 1/a$. In the unlikely event the utility can talk the ratepayer into granting it the book value as additional profit in compensation for scrapping the machine, z might even rise as high as 1. The realistic case is probably $z = 0$.

One may then calculate $V^0 - V^N$, the gains from sticking with the old technology. More algebra gives us

$$V^{O} - V^{N} = [R(a - 1)((1 - s)/(r + s)) + 1 - za]E_0. \qquad \text{(C-6)}$$

This will be positive as long as

$$z < (1/a) + R(1 - (1/a))((1 - s)/(r + s)). \qquad \text{(C-7)}$$

Since both terms in this last expression are positive, and with any reasonable story about what happens when the old machine is scrapped $z \leq 1/a$, profits are always maximized by sticking with the old over the new. Buying a new machine means writing down the existing excess of book value over economic value (as long as book depreciation falls short of economic depreciation), and the present discounted value of the returns that extra book value would bring in years to come would be lost.

Appendix D: Trade-offs in Network Design

THIS appendix briefly shows how the number of users, their geographic dispersion, and the costs of switching and trunking interact in determining the least-cost communications network design. In particular, within this idealized framework, one can show that economies of cross section in transmission are a necessary but not sufficient condition for a decentralized, geodesic "ring" configuration to be the most economic network architecture. Otherwise, a centralized "star" network is always less costly than a geodesic "ring" network.

I assume there are $2N$ users to be connected on the network and they are uniformly distributed on a circle of radius r. The size of r is thus a measure of their geographic dispersion. I shall also assume that the network is engineered so that all users can communicate simultaneously without any blocking.

A choice will be made between two idealized network structures, a centralized star and a geodesic ring. (See figure D-1.) A switch handling N pairs of connections is assumed to have cost function $g(N)$ with $g' > 0$. A transmission link with cross section N (that is, a "thickness" of N channels) is assumed to have a cost per unit of distance of $h(N)$, with $h' > 0$.

In the star network, $2N$ transmission links of cross section 1 are linked to a single central switch capable of handling N pairs of connections. On the ring network, in the worst case, all $2N$ nodes must have a switch installed that is capable of handling N pairs of connections, and transmission links to its neighboring nodes must also have cross section N. Otherwise, if all nodes on the ring wished to communicate with their most distant opposite node, some pair of users might be blocked.

Thus, on the ring, switching costs are $2N\, g(N)$; on the star they are $g(N)$. I shall assume a large number of users, so that the transmission plant needed on the ring is approximated well by a circle of radius r and cross section N, and total trunking cost is $2\pi r\, h(N)$. On the star, $2N$ trunks

Figure D-1. *Idealized Network Architectures*

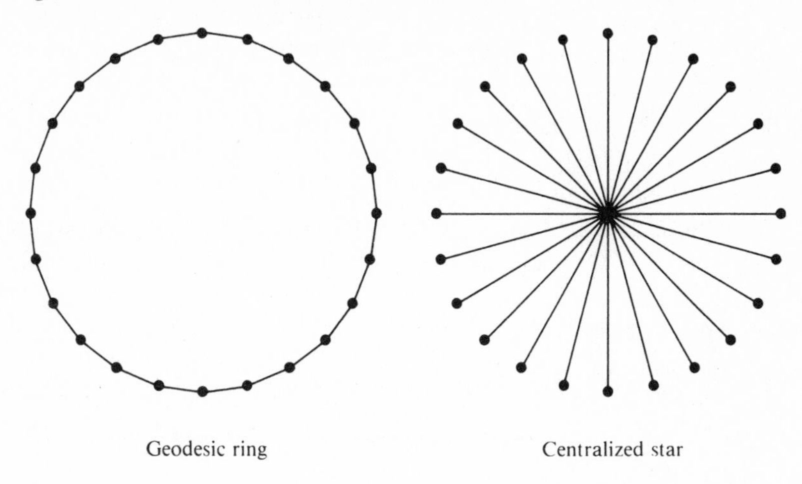

Geodesic ring Centralized star

of length r and cross section 1 are needed, so trunking cost is $2Nr\ h(1)$.

Total network cost with the geodesic ring architecture is given by

$$2N\ g(N) + 2\pi r\ h(N).$$

With the star network, total cost is $g(N) + 2Nr\ h(1)$.

Deducting the star's total cost from that of the ring, one has

$$g(N)\ (2N - 1) + 2r(\pi h(N) - Nh(1)).$$

The first part of this expression is always positive. The second part may be written

$$2r((\pi - 1)h(N) + h(N) - Nh(1)),$$

which will necessarily be positive unless $h(N) - Nh(1) < 0$; that is, unless there are economies of cross section in transmission. In that case the second expression can be negative, and thus the difference in cost can be negative. Intuitively, n rings of circumference $2\pi r$ are lengthier than n spokes of length r, and are therefore more costly, unless there are economies of cross section.

I conclude that unless there are economies of cross section in transmission, a centralized network will always be cheaper than a geodesic network. If there are economies of cross section, the geodesic network *may* be cheaper, particularly for a large r (where a thick ring is being substituted for a hub with many thin and very long spokes). In particular, if N and h are such that $\pi h(N) - Nh(1)$ is negative, with a sufficiently large r, a ring will become cheaper than a star. The geodesic network is thus likely to be economic with a geographically dispersed population of users.

Real networks, of course, are layered combinations of these two idealized topologies. Also, a ring has the advantage of limited redundancy, so if one link is cut, all users may still communicate.

However, if two links are cut, large subsets of users may be cut off from one another. On a star network, a cut link will definitely isolate one user, but each additional cut link will still isolate only one more user, not large groups of users.

A star is also easier to expand than a ring, since only the central switch capacity must be upgraded, and one additional link added. In contrast, expansion of a ring network requires adding capacity to all the switches and transmission links in the network.

Conference Panelists and Discussants

with their affiliations at the time of the conference

Stanley M. Besen *Rand Corporation*
Richard K. Blake *Siemens Communications Inc.*
Ashton B. Carter *Harvard University*
Alan G. Chynoweth *Bell Communications Research*
Rosanne Cole *International Business Machines Corporation*
Peter F. Cowhey *University of California at San Diego*
Robert W. Crandall *Brookings Institution*
Blaine Davis *AT&T*
Peter Evans *University of California at San Diego*
Kenneth Flamm *Brookings Institution*
Bailey M. Geeslin *Nynex Service Corporation*
Tetsushi Honda *Tokyo University*
B. R. Inman *Westmark Systems Inc.*
Daniel Kelley *MCI Telecommunications*
Ashoka Mody *AT&T Bell Laboratories*
William Moyer *C-TEC Corporation*
Jurgen Muller *Deutsches Institut für Wirtschaftsforschung*
Tsuruhiko Nambu *Gakushuin University*
Eli M. Noam *Columbia University*
Garth Saloner *Massachusetts Institute of Technology*
Frederic M. Scherer *Swarthmore College*
Lee L. Selwyn *Economics & Technology Inc.*
Luc Soete *University of Limburg, The Netherlands*
Kazuyuki Suzuki *Japan Development Bank*
Nestor E. Terleckyj *National Planning Association*
Nick von Tunzelman *University of Sussex*
Leonard Waverman *University of Toronto*

Index